D1122018

THE WHOLE SHEBANG

Books by Timothy Ferris

The Red Limit

Galaxies

SpaceShots

The Practice of Journalism

(with Bruce Porter)

Coming of Age in the Milky Way

World Treasury of Physics, Astronomy, & Mathematics

(Editor)

The Mind's Sky

The Universe and Eye

THE WHOLE SHEBANG

A State-of-the-Universe(s) Report

Timothy Ferris

Weidenfeld & Nicolson
LONDON

First published in Great Britain in 1997
by Weidenfeld & Nicolson

The author and publisher gratefully acknowledge permission to reprint material from the following works: F. D. Reeve, *Coasting* © 1995 by F. D. Reeve. David Weinberg, *The Dark Matter Rap: A Cosmological History for the MTV Generation* © 1992 by David Weinberg.

A catalogue reference is available from the British Library

Printed in Great Britain by Clays Ltd, St Ives plc

Weidenfeld & Nicolson
The Orion Publishing Group Ltd
Orion House
5 Upper Saint Martin's Lane
London, WC2H 9EA

FOR J.B.F.

The world is so full of a number of things,
I'm sure we should all be as happy as kings.

—R. L. STEVENSON

Contents

Preface

We . . . who are going to discourse of the nature of the universe, how created or how existing without creation, if we be not altogether out of our wits, must invoke the aid of gods and goddesses and pray that our words may be above all acceptable to them and in consequence to ourselves.

—PLATO[1]

Mistress of high achievement, O lady Truth,
do not let my understanding stumble
across some jagged falsehood.

—PINDAR[2]

WE LIVE in a changing universe, and few things are changing faster than our conception of it. The cosmos of our not-so-distant ancestors was small, static, and Earth-centered. By the middle of the twentieth century we had discovered that we are adrift in an expanding universe so large that light from its outer reaches takes more than twice the age of the earth to reach our telescopes. Looking ahead, we can see an emerging cosmology in which our universe turns out to be a great deal larger still, and to be but one among many sovereign universes.

Our conceptions of how the universe works have changed as well. For more than a thousand years it was thought that the heavens obeyed a different physics than pertains here on Earth. With the scientific renaissance that culminated in Isaac Newton's work it became clear that, on the contrary, the same natural laws rule the earth and the sky. The cosmos came to be viewed as a clockwork marvel, events following from causes like the tickings of brass cogs. The realm of the inexplicable—where dwell the gods of those dazzled by the unexplained—was thereafter relegated to the first moment of time, when the universe somehow blossomed into being. Then quantum chance reared its indeterminate face, as a creative agency that authored the first phenomena of cosmic time. So we are obliged to consider that even the largest systems are ruled by quantum precepts that govern nature on the smallest scales, and that the origin of the universe may itself have been a quantum flux. The constants of nature may have resulted from chance events that transpired when the infant universe fell like a broken angel from an unblemished originality. Old riddles of chaos and necessity, being and nonbeing, and perfection and imperfection are written with fresh urgency across the universal sky.

This book aims to summarize the picture of the universe that science has adduced as the second millennium A.D. draws to a close, and to forecast an exciting if unsettling new picture that may emerge in the near future. Therefore it contains both findings and speculations. In such a situation it is customary for the conscientious author to promise to carefully distinguish facts from imaginings—and certainly I have endeavored to do so. But in cosmology, the science concerned with the structure and history of the universe as a whole, the line dividing fact from fancy is not always as clear as in other fields. The expansion of the universe, for instance, which today is a well-established fact, began as a speculation. It emerged unbidden from Einstein's general theory of relativity, and originally seemed so improbable that Einstein himself did not believe it. Resistance to cosmic expansion survived for decades after Edwin Hubble in 1929 discovered evidence of it, and was not fully laid to rest until the *cosmic microwave background*, noise coming from the big bang itself, was detected, in 1965—an event that may be said to have established cosmology as a science.

When it comes to a frontier science like cosmology, where the scope of inquiry stretches out to the distant galaxies and down to the subatomic jitterbug from which they emerged, conservatism is not decidedly a virtue nor imagination a vice. The universe is cleverer than we are, and to investigate it we need to be creative as well as critical. It may seem crazy to imagine that most of the matter in our universe is composed of exotic subatomic particles of varieties never yet observed, or that there are billions of universes each subject to a different set of laws, or that the mind may be said to bring the universe into existence. But quite possibly such hypotheses are, as Niels Bohr often used to say in exploring the atom, "not crazy enough."[3]

The empirical spirit on which the Western democratic societies were founded is currently under attack, and not just by such traditional adversaries as religious fundamentalists and devotees of the occult. Serious scholars claim that there is no such thing as progress and assert that science is but a collection of opinions, as socially conditioned as the weathervane world of Paris couture. Far too many students accept the easy belief that they need not bother learning much science, since a revolution will soon disprove all that is currently accepted anyway. In such a climate it may be worth affirming that science really is progressive and cumulative, and that well-established theories, though they may turn out to be subsets of larger and farther-reaching ones—as happened when Newtonian mechanics was incorporated by Einstein into general relativity—are seldom proved wrong. As the physicist Steven Weinberg writes, "One can imagine a category of experiments that *refute* well-accepted theories, theories that have become part of the standard consensus of physics. *Under this category I can find no examples whatever in the past one hundred years.*"[4] Science is not perfect, but neither is it just one more sounding board for human folly.

Cosmology today is mostly conducted within the broad framework of the standard cosmological model, known as the "big bang" theory. For reasons specified in the opening chapters of this book, I expect the standard model to endure. This position may seem curious to readers of the many newspaper and magazine articles that have appeared during the past decade proclaiming that this or that observational finding has put the big bang theory in jeopardy. Such accounts seem to me to result from a misunderstanding of science generally and of the big bang theory in particular.

Science is not a static body of dogma, to stray from which is to risk having one's epaulets stripped off in a ceremony of banishment from the scientific community.[5] It is a self-correcting system of inquiry, in which errors—of which there are, of course, plenty—are sooner or later detected by experiment or by more careful analysis. Science is also a "bottom-up" system, in which grand pronouncements are arrived at not in an overarching, *sui generis* fashion but by building up inferences from many small cases. As a result science, while it can be exasperatingly detailed, is also pliant. Scientific findings, even the most imposing ones, customarily stumble into the world fraught with blunders that have to be worked out before they really begin to fly. They lack the satisfying, thunderclap certitude of religious and pseudoscientific dicta that admit to no error. But they are alive, and the withering of one branch of a theory does not necessarily mean that the theory as a whole is doomed.

The standard (or "big bang") cosmological model is broad and pliant. It comprises an arena within which many narrower theories and experimental programs thrive and compete. It is incomplete: Scientists don't yet know exactly how old the universe is, how big it is, how rapidly it expands, or how much matter is in it. (As the English astronomer royal, Martin Rees, remarks, "It's embarrassing that ninety per cent of the universe is unaccounted for.")[6] Nor is it clear how the matter we *do* see organized itself into stars and galaxies. There are a great many things we do not know. But it is quite possible that all these issues will be resolved, one way or another, without leaving the basic precepts of the standard model behind.

So we begin with the basic precepts of the standard model. They include these:

Physical laws adduced on Earth pertain throughout the observable universe. And a good thing, too. It would be a lot harder to do physics if, say, each galaxy had its own physics. Fortunately, scientists find that stars millions of light-years away from Earth are made of atoms identical with those here at home—although, of course, one encounters, in plasma jets, black holes, and other such exotic objects, much more extreme ramifications of physics than can be reproduced here at home.

The universe is expanding. Einstein's general relativity theory predicted that cosmic space should be either expanding or contracting. Evidence that it is expanding was then found, in the *red-shift* of light coming from galaxies. The only available and wholly consistent explanation for this phenomenon is that it is a Doppler shift—that is, one due to the recession of galaxies from our galaxy and from one another. The rate of expansion is not yet known exactly, but the correct value probably will turn out to be somewhere within 20 percent of contemporary estimates. What the expansion rate means in terms of the age of the universe depends on the geometrical model of the universe one adopts, but to a first approximation the expansion rate suggests an age for the universe of roughly 15 billion years. This fits well with what the astrophysicists estimate to be the ages of the oldest existing stars, about 14 billion years. There are, however, some recent observations that yield a smaller age for the universe. If these data hold up, cosmologists will face a chronological crisis.

The universe is isotropic and homogeneous. Isotropic means that it looks much the same in every direction. If you are swimming far out to sea, your view is isotropic: The sea looks the same in every direction, so you can't tell from the view which way you're looking. If you're swimming near the coast, so that you see land on one side and sea on the other, your view is *anisotropic*. Observationally the universe appears to be isotropic, with galaxies and clusters and superclusters of galaxies found in equal numbers in all parts of the sky except where clouds of dust and gas in our own galaxy obscure our view of space beyond. By *homogeneous*, astronomers mean that, while matter is collected locally into planets, stars, and galaxies, and while the galaxies in turn are clustered, on very large scales their distribution is smooth. If you scooped up a random piece of space the size of a star or a galaxy you'd get inhomogeneous results, sometimes netting stars or planets or nebulae, more often coming up with only space. But if you used a big enough scoop—one measuring, say, a billion light-years on a side—you'd get the same mix of galaxies and space no matter where you took your sample. In the 1980s it was discovered that clusters of galaxies are organized into giant bubbles measuring some 300 million light-years in diameter. That's nearly 3 percent of the radius of the observable universe. The existence of another, even larger level of hierarchy might call into question the assumption that the universe is homogeneous. But preliminary research indicates that the bubbles do indeed represent the top level of structure, so that on a scale of billions of light-years, matter is homogeneously distributed.

General relativity accurately describes the behavior of gravitation in the universe today. Einstein's theory describes gravitation as a warping of space in the presence of matter. Using the theory, it is therefore possible to model the overall shape of cosmic space if one knows the cosmic matter density: the more matter, the more acutely space is curved. For convenience, cosmologists describe the matter density by a single quantity, omega. If omega is greater than one, meaning that the universe is relatively dense, the cumulative gravitational force of all the galaxies will eventually halt cosmic expansion and the universe is destined to collapse. A universe with this kind of curvature is called *closed* and is analogous to a sphere. If omega is less than one, the universe is *open,* and will continue expanding forever. If omega equals one—a state known as *critical density*—the universe will continue expanding, but at an ever-slowing rate that will forever approach but never quite achieve stasis. Such a universe is called *flat.* Relativistic models of the universe are based on a four-dimensional *spacetime continuum.* The terms *closed, open,* and *flat* refer by analogy to spherical, hyperbolic, and plane geometric shapes in three dimensions.

One can attempt to measure the cosmic matter density in two ways—directly, by taking a census of all the seen and unseen stuff in the sky, or by observing the extent to which cosmic expansion is slowing as a result of the gravity exerted by the galaxies on one another. Scientists are working on both approaches, with results that are not yet conclusive but which indicate that the density is close to critical, meaning that the universe, insofar as we can observe it, is either flat or close to flat. This is puzzling: If the cosmic matter density was selected by chance, there is no more reason to expect it to come out at omega = 1 than to expect a pole vaulter's pole to remain standing, poised on its tip, for centuries following his vault. A possible explanation is provided by the inflationary hypothesis (about which more later).

Another premise of the standard model is that *the early universe was in a state of high density and high energy.* This is the "hot" big bang model. Describing it requires that we define a few terms.

The big bang theory holds that the universe began in a *singularity*—a state of infinite curvature of spacetime. In a singularity, all places and times are the same. Hence the big bang did not take place in a preexisting space; all space was embroiled *in* the big bang. Nor did the big bang happen in a remote location: It happened right where you are, and everywhere else. All places that exist today were originally the same place. Nor was it an explosion, as we usually think of explosions, since things did not fly out into space but remained where they were, while the surrounding space expanded. Some cosmologists use the term "big bang" to refer to the initial singularity, and "early universe" for the "hot," high-energy physics fest that ensued. Others use the term "big bang" more broadly, to refer as well to the hot universe as it evolved through the first seconds and minutes of time. (In this book we adopt the broader usage.) Since physicists understand thermonuclear reactions rather well, they can with some confidence work out what happened in the early universe. Their calculations predict, among other things, that photons released as the primordial material thinned and cleared should be detectable today as the cosmic microwave background. This prediction has been confirmed observationally.

The universe is evolving. Another important prediction yielded by nuclear physics is that the light elements hydrogen, helium, and lithium were made in the early universe, whereas the heavier elements were made later, in stars. This means, as one observer put it, that the periodic table is a *phylogeny*—a record of evolutionary development. If, as many theorists suspect, the constants of nature were decided by random "phase transitions" that took place during the first moment of time, then the laws of nature, too, are evidence of historical events. Evolution is creative: In an evolving universe, all events could not be predicted even if we knew the precise state of the early universe. Cosmology is an ongoing *story.*

Less well established, but of intense interest in cosmology these days, is the hypothesis that the very early universe may have undergone a period of extremely rapid expansion—an event called *inflation.* Although the putative "inflationary epoch" lasted only a fraction of a second, it would have made the universe very much larger than the *observable* universe, the part of it we can see at the present time. As a result, local space would appear to be flat regardless of its curvature, just as the earth looks flat from its surface.

Cosmology, once the province of mythology, is now a science. It differs from other sciences, however, in one important way. Other sciences thrive by comparing things—quarks with leptons, gas planets with rocky planets, chalk with granite. But cosmologists have only one universe to study. So it must be compared, not with anything concrete, but with theories and computer models of what other universes might be like, or how the real universe might have turned out differently. Therefore it may be very difficult to determine which aspects of our universe could not have been otherwise and which resulted from chance. The task poses a marvelous challenge to the imagination of the cosmologists, whose charter to envision alternative universes is enlivened by the possibility that other universes may actually exist.

Like the universe, cosmology is expanding. In 1977, the year that my first book, *The Red Limit*, was published, there were fewer than a hundred professional cosmologists in the world. Today there are more than a thousand. Behind this explosion in numbers stand revolutionary changes in technology. In 1977 astronomers studied deep space primarily by means of a few big, ground-based optical and radio telescopes, augmented by a few primitive satellites. Their data were collected slowly and painstakingly and were reduced by hand, sometimes with the help of expensive and cumbersome computers. Today, in addition to more and better radio and optical telescopes, scientists use a vast array of satellites, space probes, balloon-borne sensing devices, and desktop computers that extend their vision deep into space and time and far beyond the range of human sensory apparatus. The result has been a flood of information. By 1970 astronomers knew the approximate distances (having measured their redshifts) for about two thousand galaxies; by 1994 that figure had climbed to one hundred thousand galaxies and was going up fast. So many galaxies have now been imaged that there are not enough astronomers to classify them all; increasingly, that job is being done by neural networks and other "artificial intelligence" computer programs.

Thanks to the information explosion, cosmological theories

can be more strictly "constrained," as they say, by the data of observation. Twenty years ago, astrophysicists theorizing about how galaxies formed had only a handful of facts to work with: They knew, for instance, the approximate masses and sizes of normal galaxies, and something of the composition and ages of the stars that inhabit them. That was about it. Today, theorists working on galaxy formation can draw upon (and must contend with) computerized catalogues listing the types and locations of millions of galaxies and their observed profiles in visible light, infrared light, radio energy, X rays, gamma rays, and more. If cosmology does not yet have the exactitude and stature of, say, quantum electrodynamics, neither is it any longer a field so loosely speculative that its critics could get away with the dismissive gibe, "Cosmologists are often in error but seldom in doubt."

One can, of course, ask what difference cosmology makes to our everyday lives. The answer to this question, oddly enough, is that it seems to matter a lot. For some reason—and nobody seems to know just why—virtually every human society, from ancient Egyptians to Native Americans to the residents of just about every towering city and tiny village today, has developed models of the universe and explanations of how it came into being. And these models influence our thinking in ways that are not always readily apparent.

One product of the interaction between cosmology and daily life is the Declaration of Independence. Impressed by the elegant, clockwork precision of planetary motions revealed in Newton's laws, Western thinkers of a liberal bent dismissed God from his old role of personal intervention—since his services were no longer required to move the planets around—while retaining the concept of God as Creator of the universe. Having done so, they increasingly turned their attention to the study of nature as a way of appreciating God's marvelous design. From this concentration on natural law it was a short step to John Locke's assertion that there are also natural laws that address human beings and their governance. The thrust of these laws is, as Locke put it, that "the natural liberty of man is to be free from any superior power on earth, and not to be under the will or legislative authority of man, but to have only the law of Nature for his rule."[7] By the late eighteenth century Locke's ideas were so much in the air as to be echoed by Thomas Jefferson in the first sentence of the Declaration of Independence: When Jefferson wrote of "the separate and equal station to which the Laws of Nature and of Nature's God entitle" a people, he meant that human equality and the right to life, liberty, and the pursuit of happiness are *natural* laws, based in nature as are Newton's laws of gravity.[8] In this and many other instances, it is clear that the findings and speculations of cosmologists, though they deal in the main with events remote in time and space, can weave their way through society and influence everything from the innovations of literature and the arts to deliberations in legislatures and courts of law.

Religion, too, has long been entangled with cosmology, with results ranging from the belief that science is good because it reveals the Creator's beneficence ("The heavens declare the glory of God; and the firmament sheweth his handywork"—Psalms 19:1) to the tragicomedy of the Roman Catholic Church's persecution of the aging Galileo. So enduring is the habit of conflating religious with cosmological ideas that many physicists have of late succumbed to the temptation to invoke the name of God in describing purely scientific work. When the Berkeley astronomer George Smoot found, in the cosmic microwave background, evidence that the superclusters of galaxies began as quantum flux events in the infancy of the universe, he volunteered to reporters that the discovery was "like looking at the face of God."[9] James Trefil titled his book on the concept of laws of nature *Reading the Mind of God*. Leon Lederman titled his 1993 book *The God Particle*. (The particle to which he referred is more commonly known as the Higgs boson.) Stephen Hawking ended his *Brief History of Time* by speculating that the discovery of a unified theory of physics could lead to understanding "why it is that we and the universe exist," which in turn, he writes, would mean that "then we would know the mind of God."[10]

The psychological connections between religion and cosmology probably go too deep to be uprooted, but it may be worth keeping in mind that much of this God-mongering arises from the assumption that God is embodied in a set of equations. That may be the case but it is by no means self-evidently so. For a scientist to make such an assumption risks introducing religious controversies into cosmology, a science that has more than enough to do trying to figure out *how* the universe works without also flattering itself that it is going to tell us *why*. Religious systems are inherently conservative, science inherently progressive. So, though it goes without saying that there need be no hostility between these two systems of thought, neither does it seem likely or even desirable to imagine that they are headed for some sort of rapprochement. This may be an instance where good walls make good neighbors.

The strongest bond between cosmology and everyday life resides, in any event, not in the high carrels of religion and philosophy, but in the ability of science to pursue questions of universal simplicity about how things came to be as they are. The book you are holding in your hand is made of paper from trees. The ability of trees and people to live on Earth results from energy provided by the sun, from water (most likely bequeathed us by comets that struck our planet in its early days), and from geological processes inside the earth that drive volcanoes and churn fresh material up from the depths. Nuclear physics teaches us how the sun releases energy from its core and how radioactive heat powers geological activity at the center of the earth. Both lines of inquiry lead back to the phase of the big bang when the first atomic nuclei were built. Astrophysics explores how the sun and its planets formed in the collapse of gas clouds lining a spiral arm of the Milky Way galaxy, and this in turn opens questions of how galaxies themselves coalesced when the universe was young.

From the perspective of cosmic history, all scientific questions turn ultimately into narratives. The room in which I am writing these words looks out on a steep hillside in Sonoma County, California. The hill is steep because it is fairly young; erosion by wind and rain has not yet had time to wear it smooth. (Looking carefully, one can see the remnants of extinct volcanoes among the peaks that rim Sonoma's Valley of the Moon.) It's young because it stands near the San Andreas fault, in one of the most geologically active regions in the world. Why is the region active? Because it is here that the Pacific Plate, inching its way east, collides with the plate that supports the Americas. The Pacific Plate dives down, the Americas wrinkle upward, and fresh mountains are built. Plate motion carries the mountains north. Some of the rocks in these hills resided, millions of years ago, on the floor of the South Pacific from whence they migrated to Chile before turning left; now they are passing through, on their way to Alaska.

What drives the movement of the continental plates? Heat from deep within the earth, generated by the decay of radioactive atoms. Where did those atoms come from? From the interstellar cloud out of which the sun and its planets congealed, long ago. Prior to that, the atoms were adrift in space. But if these atoms decay, how come there were lots of them still around when the solar system formed, some 10 billion years after the big bang? Because they were freshly minted, in one or more stellar explosions that occurred shortly before the solar system came into existence. Where did *those* stars come from? From similar processes that occurred earlier in the history of the galaxy. Spiral galaxies like the Milky Way are star-making machines. Where did the *galaxy* come from? It,, too, condensed from a cloud of gas. And where did *that* cloud come from? From the cooling, darkening matter that emerged from the big bang, a billion years after the beginning of time.

At no point in this long chain of questions and answers does nature suggest that we can with just cause insert a knifeblade and declare that everything prior to *this* time, or larger than *this* scale in space, belongs to the universe, while everything earlier or smaller belongs to the realm of purely human affairs. Nothing that is human is purely human, and nothing that we see in the sky is purely cosmological. We are embroiled in the cosmos. All roads lead to cosmology, and the higher we climb, the farther we can see.

1
The Shores of Light

D'où venons-nous? Qui sommes-nous? Où allons-nous?
["Where do we come from? Who are we? Where are we going?"]

—*Title of a painting by Gauguin*

Little by little, time brings out each several thing into view, and reason raises it up into the shores of light.

—LUCRETIUS[1]

THIS BOOK will summarize what we know about the cosmos and how we know it, and will speculate about the directions cosmology may take in the future. We need first to outline how science arrived at its present understanding of the age, scale, and evolution of the universe.

Here's the story so far:

The ancient Greeks thought that the earth (which they understood to be a sphere) sat immobile at the center of the universe, orbited by concentric crystalline spheres to which were attached the sun, moon, planets, and stars. This model answered well to common sense: The stars do appear to circle the earth daily, while to advocate the alternative proposal—that this effect is produced by a rotation of the earth rather than of the starry sphere—was to encounter objections that were insurmountable at the time. (If the earth is spinning, why does a man who jumps straight up land in his footprints, rather than hundreds of yards to the west?) The geocentric cosmos was also aesthetically pleasing: It portrayed our world as a sphere set at the center of a nested set of spheres, a conception that resonated with Plato's conviction that the sphere is the most perfect of all geometric shapes, since it confines the largest possible volume within a given surface area.

This model was put together by two of the keenest minds of the fourth century B.C., the philosopher Aristotle and the astronomer Eudoxus, and it won widespread acceptance. But the Greeks were not content with simply admiring its splendors. They also expected the theory to account for the data of observation—to explain motions seen in the sky in the past and to predict those coming up in the future, especially such spectacular events as eclipses of the sun and moon and conjunctions of the planets. It is for this reason more than any other that we celebrate the Greeks as the precursors of modern science. Their skepticism set in motion the questioning, subversive, and perpetually dissatisfied spirit that is characteristic of science. The ultimate failure of their model proffers a cautionary lesson as well—that in cosmology a theory can be sensible and beautiful and also quite wrong. The geocentric cosmology of Aristotle and Eudoxus did not, in the long run, generate accurate predictions of the motions of the planets.

Better results were obtained by the more complicated model composed in the second century A.D. by Ptolemy at Alexandria. In the Ptolemaic universe each planet orbited in an epicycle—a small circle—centered on a point in its orbit around the earth, or even on a point in another epicycle. This was clever but highly abstract; Ptolemy himself viewed his model as merely a mathematical expedient. And it was so complicated that Ptolemy's name became a lasting epithet for theories regarded as unduly elaborate or insufficiently physical. Nevertheless the Ptolemaic universe reigned in the West for fourteen hundred years, until it was challenged by Copernicus.

Schoolchildren are still being taught that the sun-centered Copernican universe brought simplicity and light to cosmology in a single stroke. But the Copernican model in its original form was neither less complicated than Ptolemy's nor more accurate.

Copernicus assumed that the orbits of the planets are circular; consequently he too had to resort to epicycles. Copernicanism was favored by some astronomers, particularly younger scholars of a radical bent, not because it solved all their problems but because, by demonstrating that a heliocentric cosmology could compete with Ptolemy's geocentric one, it opened up fresh prospects for original thought. And the prospects were enormous—literally so. The Ptolemaic universe was inherently small: The sphere of stars that enclosed it had to spin once around the earth every day, and if the starry sphere was very big it would have to rotate at such tremendous speed that it might fly apart. But if Copernicus was right, then the very fact that the stars remain in the same place in the sky while the earth moves through its orbit, thereby altering its perspective on the stars, means that the stars must be far away. In this way the Copernican proposal threw back the walls surrounding the solar system, opening up a vast universe beyond.

But the Copernican model was afflicted by two major problems. Since it portrayed the planets as orbiting the sun in perfect circles, it was driven toward complexity and error. Planetary orbits are not circular but elliptical. Trying to predicate planetary motions on circular orbits is like trying to learn how a football bounces by bouncing a basketball. And since physics had not advanced much since the time of the Greeks, advocates of any geocentric model were still stumped by the old objections: If the earth rotates, why *don't* jumpers land to the west of their starting point and howling easterly winds constantly rake the surface of the planet, especially at the equator, where everything is moving east at a velocity of 1,000 miles an hour?

It fell to two of the leading scholars of the Renaissance to address these problems by correcting flaws in the Copernican cosmology and marrying it to terrestrial physics. Johannes Kepler and Galileo Galilei were both talented writers whose books carried their ideas into the mainstream of intellectual discourse throughout the literate world. Otherwise they were quite different men, one theoretical and solitary, the other experimental and more gregarious.

Kepler sought in the cosmos the symmetries of mathematics and the harmonies of music. He did little observing of the skies himself and only once or twice used a telescope. His cosmological research was based on data compiled by the Danish astronomer Tycho Brahe, with whom Kepler had worked in Tycho's observatory at Benatek Castle, near Prague. Theirs was a prickly relationship: Tycho, fearing that Kepler would overshadow him, withheld much of his data. The full data became available following Tycho's death (of a burst bladder suffered in a bout of beer drinking), and Kepler eventually discovered the three laws of planetary motion that have since borne his name.

To appreciate the beauty of a scientific law is for most of us an acquired taste, like drinking Scotch or enjoying the music of Alban Berg. And as few become sensitized to science during their formative years, it may well be that connoisseurs of scientific aesthetics are even scarcer than drinkers of MacCallan whisky or devotees of *Wozzeck*. But for those who want to learn how to value science for its beauty as well as for its accuracy, Kepler's laws are a good place to start.

The first law reveals that the orbits of the planets describe, not perfect circles, but ellipses (i.e., ovals), with the sun located at one focus of each ellipse. This masterful demonstration prompted Immanuel Kant to call Kepler the most acute thinker ever born. Like every other cosmologist up to that time, Kepler had assumed that the planetary orbits must be circular. To arrive at the elliptical hypothesis, therefore, he was required to set aside a fundamental aspect of his own intellectual architecture and that of the society to which he belonged. Upon learning of this bold step, his contemporaries reacted with dismay, criticizing not only his hypothesis but his method, which involved intensive application of more sophisticated mathematics than any astronomer of the day employed. Even his old astronomy professor disapproved. (This was Magister Michael Maestlin, who had introduced Kepler to Copernican cosmology and whom Kepler revered.) Kepler's accomplishment was all the more remarkable in that, by the time he resorted to ellipses, he had already earned an estimable reputation and was pushing forty years of age, conditions not normally conducive to mathematical innovation. "I have spent so much pains on it that I could have died ten times," he wrote.[2] The effect was, he recalled, like awakening from sleep to see the light.

The elegance of the first law was not immediately apparent to casual observers, who wondered what difference it made whether planets moved in circles or in rotund ellipses that at a glance did not look all that different from circles. The aesthetic force of Kepler's discoveries emerges more clearly in the second law. It shows that the orbital velocity of each planet increases when it is near the sun and decreases when far away, at just such a rate that the area swept out within its orbit is equal during equal intervals of time. In other words, if one charts the motion of Mars over a period of one month when it is far from the sun, and draws a long, thin triangle connecting the sun with the planet's position at the beginning and end of that month, then draws a fatter triangle inscribing Mars's monthly motion when it is closer to the sun, the areas of the two triangles are equal. The same is true of any orbiting object. This subtle symmetry thrilled Kepler, who compared it to the harmony of contrapuntal music.

Kepler's third law declares that the cube of the semimajor axis (half the long axis) of each planet's orbit is proportional to the square of the planet's orbital period. The third law provided astronomers with a capable tool for mapping the solar system, since it meant that if they knew how long it took a planet to go around the sun—information already available when Kepler was alive—they could deduce the size of its orbit relative to those of the other planets. To measure the actual size of any one planetary orbit is, therefore, to have learned the actual sizes of all the other orbits.[3] Similarly, when one examines planets like Jupiter or Saturn that have many *satellites* (a word coined by Kepler), measuring the size of one satellite's orbit yields the sizes of the other orbits.

While Kepler was doing all this, Galileo was repairing some of the deficiencies in the physics of the Copernican theory. Born in Pisa, Galileo was in many ways a man of the south—more outgoing than Kepler, more at ease in society, more materialistic and more at home with technology. Galileo's most productive years were spent in the Republic of Venice, which ranked among the busiest seaports in the world, and he was familiar with maritime technology in much the same way that many scientists today keep up with goings-on at the Kennedy Space Center. His influential book *Dialogues Concerning Two New Sciences* begins with a straightforward reference to his view that technological progress promotes scientific progress. Says Salviati, one of the three characters in the dialogue:

> The constant activity which you Venetians display in your famous arsenal suggests to the studious mind a large field for investigation, especially that part of the work which involves mechanics; for in this department all types of instruments and machines are constantly being constructed by many artisans, among whom there must be some who, partly by inherited experience and partly by their own observations, have become highly expert and clever in explanation.[4]

Both men were revolutionaries who had managed to shake off the ancient belief that pure thinking is superior to the awkward and often messy business of rolling balls down inclined planes, squinting through primitive telescopes, and otherwise interrogating the material world. Albert Einstein wrote encomiums in their honor, stressing their willingness to look for truth in nature, thus overcoming their culture's traditional preference for abstract thought over empirical observation.[5] Kepler, Einstein noted, "had to recognize that even the most lucidly logical mathematical theory was of itself no guarantee of truth, becoming meaningless unless it was checked against the most exacting observations in natural science."[6] Of Galileo he said: "Pure logical thinking cannot yield us any knowledge of the empirical world; all knowledge of reality starts from experience and ends in it. Propositions arrived at by purely logical means are completely empty as regards reality. Because Galileo saw this, and particularly because he drummed it into the scientific world, he is the father of modern physics—indeed, of modern science altogether."[7]

Galileo's most significant contribution to the physics of cosmology came with his insight into the concept of inertia. Aristotle had assumed, and the Western world had come to believe, that the natural tendency of objects is to remain at rest. This certainly seems to accord with experience—a book or a boulder stays in one place unless one expends energy in moving it—and even today the word *inertia* is commonly taken to mean sluggishness or stasis. Galileo saw that this commonsense assumption was wrong. He pushed wood blocks across a tabletop, then polished the table and the blocks and pushed the blocks again, and pondered the significance of the fact that when there was less friction they traveled farther. He reasoned that if they could be polished perfectly, so that there was *no* friction, they would keep moving forever. Inertia, he concluded, is not just a tendency of bodies at rest to remain at rest, but also of bodies in motion to remain in motion.

Galileo's counterintuitive insight resolved the basic objection to the Copernican assertion that the earth moves. Jumpers don't fly westward nor do easterly gales constantly blow, because the jumpers and the atmosphere are *already moving* with the turning earth, and so tend to remain in motion. Today we have seen enough of the universe to know that motion, not rest, is the ordinary state of matter, and that to be immobile is at most a local trait, measured in terms of a local "inertial rest frame." The farther out one looks, the more one finds that everything, relative to most other things, is moving. The universe was born restless and has never since been still.

Galileo's later years were overshadowed by his futile campaign to persuade the Church to replace the Ptolemaic with the Copernican cosmology. In the end he was forced to make a humiliating recantation on his knees before the Inquisition, and he lived out the remainder of his days under house arrest. But the reason his campaign failed was not solely that the authorities in Rome were unwilling to change their ideas. It was also because Galileo, though armed with many powerful arguments from analogy, was never able to present a quantitative defense of the Copernican cosmology.

That was accomplished by Isaac Newton. Born in England in 1642, the year of Galileo's death, Newton researched everything from alchemy and biblical chronology to optics and mechanics, while publishing virtually nothing. Only with great reluctance and at the behest of the astronomer Edmond Halley did Newton write his *Principia*, the book that inscribed his name in history. The *Principia* presented equations that accurately predicted the motions of the planets and the rate at which objects fall on Earth, revealing both to be caused by a single force, gravity.[8] In so doing, it vindicated the heliocentric, rotating-Earth cosmology of Copernicus, Kepler, and Galileo, while also uniting the physics of heaven and Earth. Newton's research inaugurated two scientific enterprises that have continued ever since—the progress of physics through the investigation of phenomena both on Earth and beyond, and the mapping of a universe that, though vast, is for some reason accessible to human inquiry.

Substantial progress in mapping the solar system was made during the two centuries following the publication, in 1687, of Newton's *Principia*. Explorers armed with telescopes and accurate clocks—marine chronometers, developed to enable navigators to determine their longitude and thus avoid blundering into coastlines at night—observed the transits of Venus across the face of the sun in 1761 and 1769 with results that yielded a fairly accurate value for the size of the earth's orbit. This in turn paved the way for measuring the distances to nearby stars by triangulation (the "parallax" method). The first accurate stellar parallax—that of the star 61 Cygni, 11 light-years from Earth—was measured in 1838.

The basic discovery that must be made by any species if it is to begin mapping the universe at large is that stars belong to galaxies, which are scattered through starless space. This process is complicated by the fact that, when viewed through a telescope, the *nebulae,* which are clouds of gas and dust in our own galaxy, look quite similar to the spiral "nebulae," which are remote galaxies in their own right. Gaseous nebulae typically measure a few tens of light-years in diameter. Spiral "nebulae" are made of billions of stars, typically measure 100,000 light-years in diameter, and lie millions of light-years from Earth. Astronomers had to learn the difference between these two superficially similar classes of objects before their science could mature into something capable of genuine learning about the universe at large.

In a blaze of insight, the philosopher Immanuel Kant in 1755 had proposed that the spiral "nebulae" are galaxies, but not until 1925 did the American astronomer Edwin Hubble, using photographic plates and the giant 100-inch telescope at Mt. Wilson in California, photograph individual stars in a spiral galaxy and thus prove Kant right. Since then, a great many methods have been brought to bear on the problem of the cosmic distance scale, with the result that astronomers today know the distances of thousands of galaxies, extending out for hundreds of millions of light-years, with an accuracy of better than 50 percent.

Meanwhile, astronomy had advanced from its original, taxonomic phase—in which observers classified celestial objects in something like the way naturalists collected dried plants and stuffed birds by the thousands in the days before Darwin—to mature into *astrophysics,* a science that not only reports extraterrestrial phenomena but offers plausible explanations for how they work. The change was rather like watching a play in a foreign language one does not speak, only to have the patterns of behavior become explicable in the second act when a translator begins to whisper explanations of what the actors are saying and how they are motivated. Through astrophysics, it became possible to go beyond describing how the sky looks and to begin learning how it got to be that way.

Essential to the rise of astrophysics was the spectroscope, which breaks down light into its constituent frequencies. These frequencies convey information about stars and other luminous objects from their atoms on up. Light consists of subatomic particles called *photons.* An atom releases a photon when one of its electrons drops from a higher to a lower orbital shell: The atom becomes less energetic, and the excess energy is emitted as the photon. Sunlight is produced in this fashion. So is starlight, and the phosphorescent glow of gaseous nebulae, whose gas is "excited," as the physicists say, by the light of stars embedded in them. By analyzing the spectra of stars and other astronomical objects, it is possible to learn their composition, their temperature, how rapidly they are rotating, and quite a lot else about them.

The first spectrum of the sun was obtained by the German physicist Gustav Kirchhoff, who detected the presence of sodium, calcium, magnesium, iron, and other elements there. (The element helium was to be discovered in the sun in 1868, before it was found on Earth; hence its name, from the Greek *helio,* for "sun.") The English amateur astronomer William Huggins then fitted the telescope at his private observatory in London with a spectroscope, and was able to identify iron, sodium, calcium, magnesium, and bismuth in the bright stars Aldebaran and Betelgeuse. But the most cosmologically significant discovery to be made with the help of the spectroscope came in 1929, when Hubble analyzed displacements in the spectral lines of galaxies and confirmed that most galaxies are rushing away from the Milky Way, and from one another, at rates directly proportional to their distances—the first demonstration that the universe is expanding.

The idea that cosmic space is stretching out, carrying the

galaxies with it, is a twentieth-century innovation—one that was unanticipated, insofar as I can find, in all the prior scientific literature. Yet, curiously, the idea of cosmic expansion emerged in theoretical physics shortly before Hubble found evidence of it in the sky. The groundwork was laid in 1916 by Einstein's general theory of relativity. Researchers studying the theory found that it implied that cosmic space cannot be static but must be either expanding or contracting. Einstein at first resisted this odd idea, but soon found himself obliged to accept the validity of the mathematical reasoning involved. Then in 1929 Hubble, who was not familiar with the theory, independently discovered the expansion of the universe.

The so-called *big bang* model arose from thinking about what an expanding universe would have been like in its infancy. The observable universe today is roughly 15 billion light-years in radius. When its radius was much smaller—only one light-year, say—all the matter in the universe must have been packed together in a lot less space. Any given quantity of matter, compressed to a higher density, gets hotter: That's why a penny, lifted off a railroad track moments after being flattened by a passing train, is hot to the touch, and why compressing air in a bicycle pump heats the air, making the pump warm. So it seems reasonable to imagine that the early universe may have been not only dense, but also hot. *Very* hot: When the universe was one second old, in this scenario, every spoonful of stuff was denser than stone and hotter than the center of the sun. The expansion and resultant cooling of the universe permitted the formation of atoms, molecules, galaxies, and living creatures. What we call *matter* is frozen energy. It froze because the universe, owing to its expansion, cooled.

The big bang theory implied that as the young universe expanded there should have come a time, nowadays reckoned at about five hundred thousand years after the beginning, when the primordial plasma thinned out sufficiently to become transparent to light. Physicists call this event *photon decoupling*, meaning that photons, the particles that constitute light and other forms of electromagnetic energy, were at this point set free. Thereafter they did not often interact with one another, or with matter, but went soaring unhampered through the constantly expanding reaches of cosmic space. Hence most of them should still be around today.

Cosmic expansion would have stretched them out, increasing their wavelengths from those of light to the wavelengths we call microwave radio. In microwave frequencies it is convenient to express energy in terms of temperature—as does, say, the instruction manual that accompanies a microwave oven—so another way to reason through this argument is to say that the universe, having once been hot, should remain a bit warm even today. Physicists theorizing about the existence of this *cosmic microwave background,* or CMB, calculated that it should have a temperature of about three degrees above absolute zero. They also noted that it would display a "black body" spectrum, as is dictated by the relevant quantum physics equations, and that it should be *isotropic,* meaning that any observer, anywhere in the universe, should measure the background as having the same temperature everywhere in the sky. One can think of the CMB as a haze of photons that has permeated space ever since the big bang. As we look far out in space—and, therefore, backward in time, to when the CMB photons were more energetic—we find the haze thickening. At the ultimate distance, where we are peering back into the first million years of time, the haze becomes opaque. Every observer using a microwave radio telescope thus sees the universe as a sphere that is almost transparent nearby but is opaque at its distant and fiery walls.

When this prediction was first made, in the 1940s, it was quickly forgotten. The big bang theory was not yet taken very seriously and there was no such thing as a microwave radio receiver. Then, in 1965, two physicists working with a radio receiver built for communications satellite experiments detected the CMB. Interest mounted as scientists came to appreciate that by studying the CMB they could make direct observation of the universe as it was only half a million years after the beginning of time. In 1989, the American space agency launched a satellite designed to study the CMB from orbit, where its detectors were free from the interference of Earth's atmosphere. Preliminary findings obtained by the COBE (Cosmic Background Explorer) satellite were announced the following year, and turned out to constitute a stunning confirmation of the big bang model. The CMB is indeed isotropic—that is, it has equal intensity all over the sky, as anything genuinely universal must. And, as expected, its temperature is about three degrees

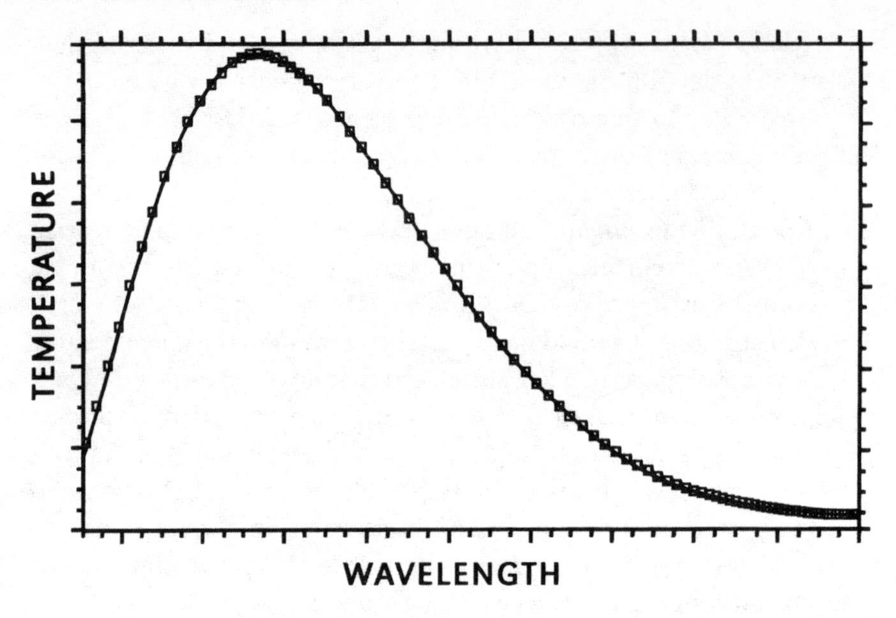

COBE satellite plot of the spectrum of the cosmic microwave background displays *exactly* the curve predicted by the big bang theory.

above absolute zero—2.726 degrees, to be exact. And its spectrum conforms to a black body spectrum: The fit is so precise that the researchers making the announcement had to enlarge the size of the error bars on their diagrams: Otherwise the observational data points would have disappeared into the thin, inked line describing the theoretical prediction.

A final triumph for the COBE scientists came in 1992, when an all-sky map, carefully compiled by repeated observations that pushed the sensitivity of the COBE instruments to their limits, confirmed another important prediction of the big bang theory—that matter, though generally distributed uniformly throughout the cosmos, began fairly early to clump into dense regions from which clusters of galaxies were to form. This was good news for theorists who argued that the vast clusters, superclusters, and bubbles of galaxies we see in the universe today formed by gravitational attraction from *inhomogeneities* in the early universe. The clumps of matter are thought to have originated as *quantum fluctuations,* microscopic departures from the generally homogeneous distribution of matter in the very early universe. Much remains to be stud-

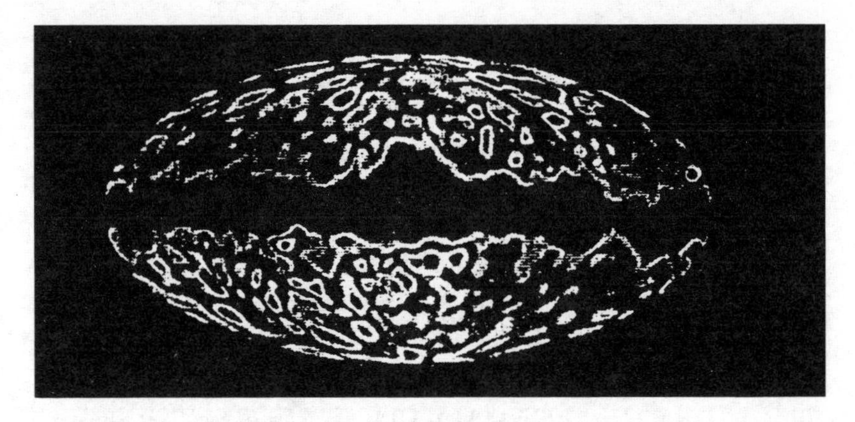

COBE all-sky map shows seeds of structure in the background radiation. The dark belt across the middle is the obscuring disk of our galaxy.

ied about the spectrum and sizes of these inhomogeneities, and how, exactly, they resulted in the large-scale structures we see in the universe today. These findings led most cosmologists to agree that the universe emerged from a hot big bang state.

Several other sorts of evidence support the big bang theory.

One of these is the intriguing fact that the *cosmic element abundance* fits the predictions of the theory. Here the line of reasoning is that as the primordial fireball cooled, protons and neutrons would have joined up to form the nuclei of atoms. The calculations of the nuclear physicists—who have had a lot of experience in this sort of thing, since similar processes occur in the explosions of thermonuclear bombs—indicate that about a quarter of the atom-making stuff should have been converted into helium in the big bang, along with a bit of lithium, while the remainder survived as hydrogen (the simplest atom, whose nucleus in its rudimentary form consists of a single proton). And this is just what we do find: The universe at large is 25 percent helium and 73 percent hydrogen. The theory postulates that all the heavier elements were forged inside stars, notably in *supernovae*—exploding stars, which seed space with clouds of debris, enriched with the heavier elements, from which condensed latter-day stars and planets, the earth and the sun among them. If this theory is correct we should find that older stars are poorer in heavy elements than younger stars are. And this, too, turns out to be the case.

In a big bang universe it ought to be possible to see direct evidence of cosmic evolution by looking out to great distances, since light reaching us from billions of light-years away is billions of years old and so reveals what things were like billions of years ago. Such evidence has indeed been found. A dramatic example is provided by *quasars,* which are the bright cores of galaxies going through a stage in which they emit enormous amounts of energy, enough to make them visible at substantial distances. The nearest quasars lie more than a billion light-years from Earth: the most distant are more than 10 billion light-years away. Surveys of the hundreds of quasars found between these extremes show that they become increasingly numerous as we look deeper into space and farther into the past. Astronomers conclude that the quasar phase of galactic evolution is something galaxies exhibit when they are young, and seldom thereafter.

Distant galaxies are bluer than nearby galaxies. Since bright young stars are blue, this indicates that galaxies billions of years ago formed new stars more profligately than they do today. Observations of remote clusters of galaxies made using the Hubble Space Telescope show that "rich" clusters (those with lots of galaxies relatively close to one another) used to have a higher proportion of spiral galaxies than they do today. So something must have reduced the number of spiral galaxies in such clusters. Nobody is sure what that was. Perhaps the spirals merged to form elliptical galaxies, or perhaps they were torn to pieces to make small irregular galaxies. Much of the bright promise of deep-space astronomy comes from the prospect of directly observing cosmic evolution by using more powerful telescopes as time machines, to look at the universe as it was in the distant past.

The ages of stars fit the age of the universe deduced from its expansion rate—according to some of the data, at least. As we will see, several persuasive sets of observations suggest the universe has been expanding for approximately 15 billion years. This accords with the ages of the oldest known stars, estimated by astrophysicists at about 14 billion years. But there are other observations, which some researchers regard as persuasive, that yield an age for the universe of only about 10 billion years. If these prove to be correct, then something is wrong either with our understanding of the ages of the oldest stars or with some aspect of the big bang theory.

Which leads us to what might be called the *theoretical* proofs of the big bang scenario. It may seem perverse to speak of using one theory to prove another, since theories normally stand or fall on the verdict of observation and experiment. But facts in themselves are as disorderly as cornflakes without a bowl. In practice, science does a lot of pouring cornflakes from box to bowl—checking not just whether facts fit a given theory, but whether the theories work well together. If we ask in what regard the big bang accords with other well-established theories, we find several answers.

General relativity has survived a great many experimental tests and seems to be perfectly accurate insofar as one is concerned with making predictions about the behavior of gravity under conditions that currently prevail throughout most of the universe. And general relativity implies, as we have noted, that the universe must be either expanding or contracting. So the very fact that we find evidence of cosmic expansion in the sky means that a well-established theory, relativity, supports another more hypothetical one, the big bang.

Quantum physics, too, finds a gratifying place within the big bang scheme. Using quantum mechanics, physicists are able to predict the existence and spectrum of the cosmic microwave background, calculate how much of the primordial material was turned into helium in the big bang, and estimate the ages of the oldest stars. Quantum physics makes accurate predictions about events involving three of the four fundamental forces of nature—the weak and strong nuclear forces at work in atoms, and electromagnetism, the force responsible for light and radio energy. But there is not as yet a fully accomplished quantum theory of the fourth force, gravity. This would not matter much were the province of physics limited to the contemporary universe: Gravitation is so weak that it can be disregarded when calculating the interactions of subatomic particles, which have such small mass that their gravitational pull on one another is negligible. But in the high-density early universe, subatomic particles weighed so much that their mutual gravitational influence was comparable to their interactions via the other three forces. To reconstruct events thought to have transpired during the very first fractions of a second of cosmic time will require a quantum account of gravity. Such a theory presumably would lay bare a single principle underlying both quantum mechanics and general relativity, which at our present level of understanding are based on contradictory ways of looking at the world. Candidates include supersymmetry, grand unified theory, and *superstring* theory, in which all the particles of matter and energy found in the universe today are said to be scraps of space, created from shards of a primeval, ten-dimensional geometry that shattered when the expansion of the universe began. Such speculations would be mere abstractions in a static universe, but they take on the many-colored cloak of history in a big bang universe, where they may speak of conditions that actually once prevailed.[9]

The *inflationary* hypothesis has generated considerable interest in cosmology. It proposes that during a dawning moment of cosmic history the expansion of the universe proceeded much faster than had been thought—indeed, at a rate far greater than the velocity of light. For reasons we shall strive to make clear, the inflationary hypothesis not only solves several problems that afflicted earlier versions of the big bang theory but indicates that the universe is extremely large, and flings open a door onto the startling speculation that our universe originated as a microscopic bubble arising from the space of an earlier universe, which may in turn be one among many universes strewn like stars across inaccessible infinities of random spaces and times and sets of natural laws.

To sum up, as the twentieth century draws to a close, the big bang theory looks to be in pretty good shape. It is supported by several solid and more or less independent lines of evidence, and has at present no serious rivals. If one were asked to make a list of the greatest scientific accomplishments of the century, somewhere on that list—along with relativity and quantum theory, the elucidation of the DNA molecule, the eradication of smallpox and the suppression of polio, the discovery of digital computation, and many other worthy attainments—there would be a place for big bang cosmology.

Yet the big bang has its woes. One problem, already mentioned, is that while some observations indicate an age for the universe consistent with other measurements, some do not; if the latter observations hold up, something must be wrong. Another difficult area involves the perplexing question of how, in a generally homogeneous universe, primordial fluctuations produced the vast structures represented by superclusters of galaxies. Quite possibly related to this issue is the riddle of what constitutes the ***dark matter,*** nonluminous material that evidently holds the clusters together. Until these puzzles are resolved we will not be sure that cosmologists are on the right track in working within a big bang context. And almost certainly there will be other vexations to come. The greatest of these surely is the question of how the universe came into being—which is itself a form of the philosophical riddle posed by Leibniz when he asked why there is something rather than nothing.[10]

In this book we will encounter many such questions. For consistency we will address them within the context of the big bang theory, but without pretending that the theory is perfect or assuredly factual, much less "true." We have mountains to climb, and must use the tools available if we are to get somewhere and not just stand and gawk and wait for the high clouds to clear.

So up we go.

2
The Expansion of the Universe

As being is to becoming, so is truth to belief. If then, Socrates, amid the many opinions about the gods and the generation of the universe, we are not able to give notions which are altogether and in every respect exact and consistent with one another, do not be surprised. Enough if we adduce probabilities as likely as any others; for we must remember that I who am the speaker and you who are the judges are only mortal men.

—PLATO[1]

Although the universe is under no obligation to make sense, students in pursuit of the Ph.D. are.

—ROBERT P. KIRSHNER[2]

VITAL WORKS OF ART and science can take on a life of their own, doing things unanticipated by their authors. Beethoven could not have foreseen all the themes that have been discerned in his symphonies, nor Shakespeare have imagined the myriad interpretations mined from his plays; and Jesus of Nazareth and Friedrich Nietzsche and all the other worthy prophets would be astonished at some of the acts committed in their names.[3] This phenomenon, which might be called the creativity of the created work, arises in part because innovations in science and the arts influence not only what we think but how we think. They alter not just the content of ideas but the intellectual landscape in which the ideas comport themselves. To predict where a new work will lead is like betting on the lie of a magical golf ball endowed with the power to alter the slope of the green and the placement of the pin while in flight.

So it was with the curious tale of how Einstein's general theory of relativity predicted the universal expansion that astronomers soon thereafter found written across the sky.

Einstein in composing the general theory was out to create an account of gravitation that would answer to the findings of the special theory of relativity, which deals with light. He knew that such a work would have implications for cosmology, since gravitation is the predominant force on the intergalactic scale. But he had no idea that his theory of gravity would predict that cosmic space expands.

Like virtually all the scientists of his day, Einstein assumed that the universe was static. The stars were known to drift through space, but astronomers had not yet established that stars are gathered into galaxies, much less that the galaxies are moving away from one another. So when general relativity was shown to imply that the universe must be either expanding or contracting, Einstein regarded the implication as a flaw. His response was to modify the equations by introducing a new term—the *cosmological constant*—to make his theoretical universe stand still. Yet even this drastic measure failed: Theorists soon found that, cosmological constant or not, relativity mandated that the universe be dynamic.[4]

The first to see this clearly was Alexander Friedmann, a young Russian scientist who led a short, anonymous, and troubled life that he bore with good cheer. Friedmann's mother married at age sixteen. Four years later she left her rather cold and imperious husband, taking their one-year-old child with her. For this infraction she was convicted in court of "breaking conjugal fidelity," though it was Alexander's father, not his mother, who had in the meantime remarried. She was sentenced to celibacy and ordered to return the child to her former husband. Raised by his paternal grandfather, Alexander was forbidden to see his mother again until he reached adulthood. He studied mathematics and physics at the University of St. Petersburg, was decorated with military honors for his aviation research during the First World War, and by 1920 was lecturing in unheated classrooms at Petrograd University to shivering students huddled in their greatcoats, whose rescue from malnutrition often depended on their professor's ability to rustle them up a few extra food rationing stamps. He sometimes paid research associates out of his own meager salary.

Despite this travail, Friedmann—like Vladimir Nabokov, writing in the toilet of his cramped Paris apartment while his son slept in the main room, or Dmitri Shostakovich, who slept in a hardback chair in the hall outside his flat so that his family would not be involved when the KGB came to arrest him—conducted a career more productive than many colleagues to whom fate had dealt a kinder hand. He published important papers in physics, mathematics, and meteorology, taking a particular interest in cyclones. (A stranger to self-importance, he liked to joke that bad mathematicians become physicists and bad physicists become meteorologists.) At once insouciant and self-critical, Friedmann inspired his students to grapple with the biggest questions they could think of: One of his pupils, George Gamow, emigrated to the United States and there advanced the hot big bang hypothesis that has since become central to modern cosmology.

A born cosmologist, Friedmann taught himself relativity—*terra incognita* to his colleagues at Petrograd—and discovered that if the general theory was correct the universe must either expand or contract. He was thus the first to propose a mathematical model of an expanding universe, one of the genuinely innovative ideas of modern times.[5] He sent word of his findings to Einstein, but Einstein was then at the peak of his celebrity and had to deal with stacks of such letters and was not infallible, and Friedmann was rebuffed. He tried to visit Einstein in Berlin, but Einstein was on vacation in the country, and Friedmann was turned away. There matters stood until Einstein, acceding to the request of another Russian physicist, looked again at Friedmann's paper, and began to change his mind. At first he retreated to the position that Friedmann's result "while mathematically correct is of no physical significance."[6] Then he found an error is his own counterargument. Ultimately he admitted that the unheralded Friedmann was right: The universe of relativity was a dynamic universe.

Further research along these lines was undertaken by the Belgian cosmologist Georges Lemaître, who independently derived Friedmann's results, and by Howard P. Robertson in the United States and Arthur G. Walker in England. Their work led to a fully realized geometric description of a homogeneous, expanding cosmos, known since as the Friedmann-Lemaître-Robertson-Walker model. Friedmann, however, lived to see none of these developments. He died on September 16, 1925, at the age of thirty-seven —of typhoid fever according to some colleagues or, according to Gamow, of an illness brought on by exposure suffered during one of his meteorological balloon excursions.

While relativistic models of the cosmos were blooming in the lofty towers of European theoretical physics, steps in deep-space observations were being taken by American astronomers, who, typically, were familiar with neither relativity nor its cosmological implications. Arthur Stanley Eddington in England was striving to call his colleagues' attention to the literally cosmic importance of Einstein's theory, and Willem de Sitter in Holland was writing influential papers that linked relativity with observational astronomy. But few American astronomers read this work or saw any useful connection between their own research at the telescope and the theorists' musings about whether the universe expands. And yet, in one of the eeriest coincidences in the history of science, these astronomers soon began independently to find evidence of cosmic expansion.

The first glimmerings came when observers took spectra of what were then called "spiral nebulae"—which we today know to be galaxies—and found that their light is shifted toward the red end of the spectrum. This *redshift* occurs when spectral lines are displaced from their normal frequencies toward lower frequencies. Redshifts can be produced by what is called the *Doppler effect,* after Christian Doppler, the Austrian physicist who identified the phenomenon in 1842. Think of a drummer, standing on a railroad car, who hits a kettledrum once every second. Now let the train speed away. As the train accelerates, ever-greater amounts of space intervene between each striking of the drum, so that the intervals you measure back at the station increase to one and a half seconds, two seconds, and so forth. The drummer will seem to be beating more slowly, and the sound will deepen in pitch. Now let the drummer light a torch, and take a spectrum of the light. Compare it with a spectrum taken of an identical torch that is not moving, and you will find that the spectral lines of the drummer's torch are shifted toward the red.[7]

In the expanding universe, redshifts are produced in an analogous but slightly different manner. Classic Doppler shifts arise from motions *through* space. Cosmological redshifts result from the expansion of intergalactic space itself. By keeping this in mind we can avoid lapsing into the parochial notion that galaxies are flying through static space, like shards of a bomb. The universe should not be thought of as expanding "into" preexisting space. All the space the universe has ever had has been in the universe from the beginning, and the space is stretching. This perspective can also help us understand why the special-relativity rule that nothing can be accelerated to a velocity greater than that of light does not apply to galaxies in an expanding universe. That rule is true in static space, but expanding cosmic space can carry galaxies away from one another at velocities greater than that of light. In the cosmological model that forms the centerpiece of this book—a vast, *inflationary* universe of *critical density*—the universe is said to have expanded initially with a velocity much greater than that of light, and as a result most galaxies are so far away that their light has not yet come within reach of our telescopes.

Knowing little of this, the American observers who first detected redshifts in the spectra of galaxies were concerned, not with cosmic expansion, but with resolving the issue of whether the spiral "nebulae" were galaxies of stars or relatively nearby whirlpools of gas. Preliminary evidence was obtained at the Lowell Observatory in Flagstaff, Arizona, a private institution where offbeat ideas flourished. (The place had been founded in 1893 by the Boston Brahmin Percival Lowell, who built it to study the illusory "canals" of Mars.) There the astronomer Vesto Slipher, charged by Lowell with establishing that the spiral nebulae were nascent planetary systems, took spectra of the spirals. Lowell expected him to find redshifts on one side of each tilted spiral and blueshifts on the other. This would show that they were made of gas that, since it was swirling, was receding on one side and approaching on the other. Instead, Slipher found that the spectral lines in most of the spirals he studied displayed only redshifts. By 1922 Slipher had obtained the spectra of 40 spirals, of which 36 exhibited redshifts. But he had no reliable way of estimating the distances of these objects, and by about 1926 he had moved on to other projects.

The startling significance of Slipher's findings became clear three years later, in 1929, when Edwin Hubble established what has been known ever since as the *Hubble law*—that the farther away a galaxy is, the larger the redshift displayed in its spectrum. Using the 100-inch reflecting telescope on Mt. Wilson, near Los Angeles, the largest of its day, Hubble identified individual stars in the spirals, thus revealing that the "nebulae" were in fact galaxies. In the process, he was able to locate Cepheid variable stars in some galaxies. These bright stars brighten and dim over periods that are directly related to their true brightness, a happy characteristic that makes them valuable as distance indicators: If you know how bright a star really is—its *absolute magnitude*—it is a simple matter to measure its distance. All you need to do is measure how bright it appears to be in the sky—its *apparent* magnitude—and apply Newton's finding that the brightness of objects decreases by the square of their distance. The use of Cepheids to chart distances had been pioneered by Hubble's colleague and principal rival at Mt. Wilson, Harlow Shapley. Now Hubble established that he could estimate the distances of galaxies by measuring the brightness of the Cepheids he found there. Combining these data with redshift observations garnered by himself and other astronomers, Hubble put together a list of distances and redshifts for 24 galaxies. He found a linear relationship—the more distant the galaxy, the greater the redshift displayed in its spectral lines. The Hubble law continues to hold up today in studies of thousands of galaxies extending much deeper into space than Hubble ever saw.

A onetime athlete, lawyer, and foot soldier, Hubble was the sort of great man who casts himself as both the author and protagonist of his own life's story. He affected the lordly demeanor of a Picasso or a Toscanini, and took readily to the heroic role that befell

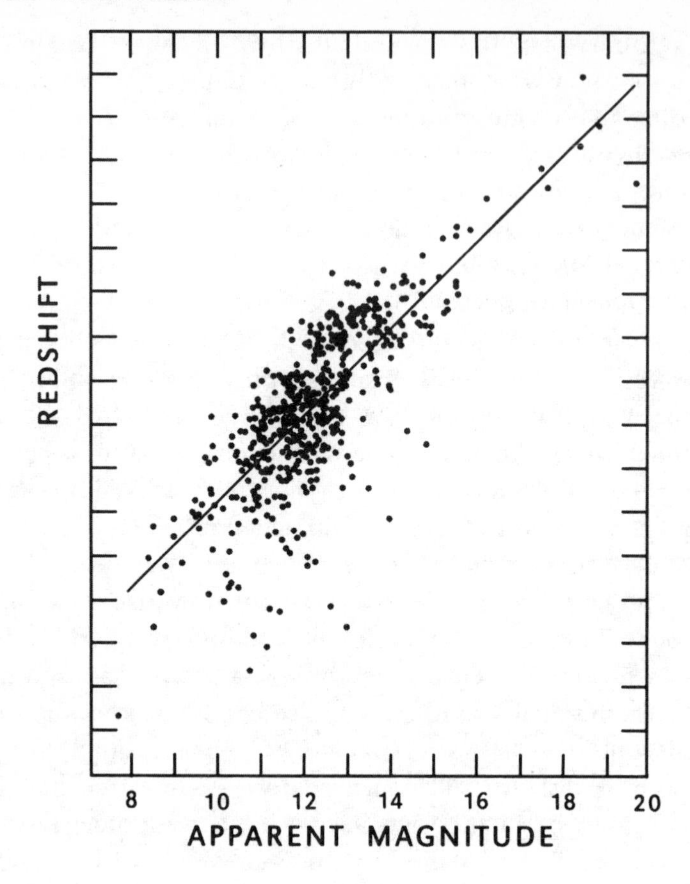

Hubble diagram of redshift (or recession velocity) against apparent magnitude (or approximate distance) of galaxies shows a linear relationship, indicating cosmic expansion. The scatter toward the bottom of the chart is due to the fact that nearby galaxies do not move apart as fast as the Hubble law requires: Their expansion velocity is braked by the gravitational field of the cluster and supercluster to which our galaxy belongs.

him once he became the most esteemed observational astronomer of his generation, his rugged features adorning the cover of *Time* magazine.[8] His enduring fame has not sat well with detractors, who point out that Hubble was not a particularly adept observer, that much of his law had been anticipated by others, and that Hubble understood too little theory to fully appreciate the significance of his own findings.

These points have merit. Hubble was a bit of a showman, and he could be careless about detail. He smoked a pipe in the dome, enjoyed squiring celebrity friends around the mountaintop, and was given to grandiose pronouncements on philosophy and art that struck the night assistants (who have seen them all come and go) as scientifically irrelevant. His subsequent work in charting the Hubble law out to greater distances was accomplished with the considerable assistance of Milton Humason, a former janitor and muleteer at Mt. Wilson, who took over most of the toil of obtaining photographs and spectra at the telescope. And it is true that Hubble's research drew on that of Slipher, Shapley, and others, whose contributions he was not always eager to credit. Nor was Hubble the first to glimpse a possible relationship between the redshifts of the spiral "nebulae" and their distances. As he himself noted, "The possibility of a velocity-distance relation among nebulae has been in the air for years."[9] The mathematician Hermann Weyl, for one, had remarked upon the existence of a redshift-distance relation, in a book published six years before Hubble made his discovery, and the astronomer Carl Wirtz had noticed even earlier, in 1921, that there was a rough correlation between the distances and redshifts of spirals, one that he predicted might have profound implications concerning the dynamics of the universe.[10]

It is also true that at the time of his discovery Hubble understood too little of the relevant theoretical work in general, and of relativity in particular, to appreciate that he had found the cosmic expansion that Einstein's theory predicted. Indeed he seldom spoke of an "expanding" universe at all. He always called what he did a "reconnaissance" and he kept his feet planted in the observational world, where he felt most comfortable. His book *The Realm of the Nebulae,* published in 1936, scarcely mentions cosmological theory.

Yet it seems clear that Hubble's contributions justify his name being attached to the expansion of the universe. Although Shapley had pioneered the use of Cepheid variable stars to establish distances, he failed to carry his campaign into intergalactic space—in part no doubt because he was disinclined to think there *were* any stars beyond the Milky Way. (According to one undocumented story, Humason once showed Shapley evidence of Cepheids in a photograph of a spiral galaxy, but Shapley, convinced that the spirals were whirlpools of gas, wiped off Humason's identifying marks, explaining that there could be no Cepheids there.) And although Slipher had first recorded redshifts in the spectra of the spirals, it was Hubble who came up with distance data accurate and consistent enough to be taken seriously by the scientific community. If Hubble lacked the air of humility cultivated by scientists in the gentlemanly twenties, his broad-brushstroke approach was well suited to the wide canvas of modern cosmology.

The Hubble law brought observation into cosmology, elevating the status of this notoriously speculative subject. Observation put flesh on the bones of the theories, which in turn brought unity and simplicity to the bewildering variety of things the deep-space observers were detecting through their telescopes. As a result, there are today "observational cosmologists," who can both gather data and understand their theoretical implications. And while the data can be maddeningly hard to obtain and interpret, the overarching structure into which they fit is so simple that one observational cosmologist, Hubble's former pupil Allan Sandage, has described cosmology as "a search for three numbers."

The three numbers are the Hubble constant, the deceleration parameter, and the cosmological constant.

The *Hubble constant* denotes the rate at which the universe is expanding. It is symbolized H_0, pronounced "H-nought." Sandage and like-minded cosmologists maintain that H_0 equals about 50 kilometers per second per megaparsec. A *megaparsec* is 3.26 million light-years, so this value would mean that for every 3.26 million light-years one looks out into space, one finds galaxies receding another 50 kilometers per second faster. Other astronomers, using other methods, get a value for H_0 of more like 70. Work on resolving the difference continues, but in the meantime astronomers may take comfort in the thought that they evidently have ascertained how fast the universe expands to within a factor of 2, not 10 or 100.

The *deceleration parameter*—symbolized q_0 and pronounced "Q-nought"—measures the rate at which cosmic expansion is slowing down.[11] Presumably there must be some such braking effect, due to the gravitational attraction that the clusters of galaxies exert on one another. If astronomers knew the correct value for the deceleration parameter, they would know the "mass density" of the cosmos—that is, how much matter there is, on the average, throughout the observable universe. (The mass density of the universe today is thought to be on the order of about one atom per cubic meter of space. That's an average, of course; the actual density is higher in galaxies and lower in the vast voids found between superclusters of galaxies.) That, in turn, would enable them to predict the fate of the universe. If the expansion rate is slowing down rapidly, the universe eventually will stop expanding and then collapse, and all will end in a "big crunch"—the big bang in reverse, with everything being reduced to a hot plasma like the one from which it all once emerged. If, on the other hand, expansion is robust enough to triumph over deceleration, then the universe will continue expanding forever. In that case the energy available to do work will eventually (meaning in about 100 billion years) run out, and the universe will succumb to "heat death"—a state in which black galaxies march endlessly outward in black, endlessly expanding space.

A third possibility is that the universe is perfectly balanced between these two alternatives, so that it will keep expanding forever, endlessly slowing but never quite ceasing to expand. Such a universe is said to have *critical density.* It too may succumb to heat death, but for reasons that I will discuss later it may be one among many universes. In that case we could take comfort (admittedly, a rather cold comfort) in the prospect that while life is dying out in some universes it is being born in others.

For convenience, cosmologists these days combine the Hubble constant and the deceleration parameter into the quantity *omega,* symbolized by the Greek letter Ω. Omega directly indexes the cosmic mass density.[12] As mentioned in the preface, if omega is less than one the universe is destined to expand forever, and if omega is more than one it is destined to collapse. If omega is exactly one, the universe is at critical density and will expand forever, at a rate that approaches but never quite reaches zero.

Theoretically it should be possible to obtain the value of the deceleration parameter by measuring the expansion rate for distant galaxies. Their light, having traveled across billions of light-years of space, brings us news of how fast the universe expanded billions of years ago, when the expansion rate must have been rather faster. By comparing the old, distant expansion rate with the new, local rate —in other words, by comparing the distant value with the local value for the Hubble constant—one could derive the deceleration parameter.[13] In practice it has proved to be very difficult to extract such data. The reason may be that the matter density is indeed critical or very close to it, so that trying to decide whether the universe is destined to end in fire or in ice is like trying to judge a photo-finish horse race. Some cosmologists think that the fact that the observed density looks to be so close to critical constitutes a powerful clue that the density really *is* critical. Critical density is dynamically unstable, like a balancing rock in an Arizona mountain range. Balancing rocks are rare, but we're more likely to encounter one than to happen upon the scene just as the rock has started to fall. An *almost* critical-density universe is a rock just starting to fall. Also, for reasons we will examine later, the critical-density universe is aesthetically appealing. So one can argue that the mass density ought to be critical, rather than merely close to it, inasmuch as nature is beautiful. As the Princeton University cosmologist James Peebles puts it, if omega turned out to be close to one but not exactly one, "I would complain that the Creator got so close to unity yet missed."[14]

Just as omega hints at the fate of the universe, it also promises to reveal the age of the universe. If omega equals one, the age of the universe (in the standard Friedmann-Lemaître-Robertson-Walker model) is two-thirds of the reciprocal of the Hubble constant—meaning that the universe has actually been expanding for two-thirds as long as would be the case had it not decelerated at all. An omega = 1 universe with a Hubble constant of 50 is approximately 15 billion years old.

The quantities we have been discussing—the Hubble constant and the deceleration parameter, and their combination into the shorthand expression omega—are standard tools, the hammer and screwdriver of cosmology today. The third term, the *cosmological constant,* is more speculative. Symbolized by the Greek letter lambda (λ), the cosmological constant represents a repellent force, the opposite of gravitation. Such a force could make the universe expand much faster than its matter density would lead us to expect, which is to say that accurate determinations of the Hubble constant and the deceleration parameter would mislead cosmologists into underestimating how much matter there is in the universe. As we saw, lambda was introduced by Einstein, for reasons that were soon rendered irrelevant, and was thereafter ruefully abandoned by him. There is no established physical mechanism that could produce cosmic antigravity, although some theories envision it. Lambda is cosmology's batty aunt in the attic: Few scientists like it very much, but it keeps turning up.

One reason for lambda's persistence has to do with the inflationary hypothesis, which postulates that the universe in its infancy expanded exponentially, resulting in a cosmos much larger than the part of it we can observe at the present epoch. As we shall see later in this book, there are excellent reasons why cosmologists entertain the inflationary hypothesis: Its physics makes sense, and it solves a host of problems that would otherwise dog the big bang picture. The force that drove inflation would have looked a lot like lambda. So perhaps lambda didn't fall to zero at the end of the inflationary epoch but has continued to function, at a reduced level, ever since.

The cosmic antigravity symbolized by lambda could push omega, the mass density parameter, toward the critical value of one. This might allay suspicions about the otherwise curious fact that we seem to live in a unique epoch in which omega, evolving over time, just happens to be close to unity. Unity is dynamically unstable, a highly unlikely state for the universe to remain in for long. Cosmologists disparage arguments that put us in a unique place (i.e., the "center" of the universe) or time (i.e., just when omega has evolved to a state close to critical density). Cosmic antigravity could counter such objections by driving the universe toward critical density, so that its present value need not be such a remarkable coincidence.[15]

So the expansion of the universe presents cosmologists with a gleaming and challenging goal. If they can measure just three quantities—the expansion rate, the deceleration rate, and the cosmological constant—they can determine the age, size, and fate of the universe. Small wonder that they have devoted years of hard work to this estimable objective, or that societies have equipped them with the telescopes and satellites and other tools required to pursue it.

Unfortunately, the task of arriving at the correct values for the three magic numbers has proved to be as exacting as its object is ennobling. It is not my intention here to recount how hard it is to find Cepheid variable stars in the dusty jungles of distant spiral galaxies, or to sift signal from noise when pushing satellite detectors to their design limits and beyond, or to fix leaks in the plumbing that pumps supercooled helium into the jacket of an electronic imaging system attached to the butt end of a four-meter telescope in the breathless icy darkness of a mountaintop observatory at four o'clock in the morning. But I do want to outline how astronomers today go about mapping the cosmos and charting its motion. This endeavor is important in its own right, as the loftiest chapter yet in the long history of human exploration. And it will help us understand aspects of observational cosmology that will be essential once we go past the basics to explore the frontiers.

Obtaining the redshifts of galaxies is relatively easy. In Hubble's day, it took many hours of exposing a single photographic plate to record the tiny spectrum, the size of a toenail clipping, that an astronomer would then scrutinize with a magnifier to detect the spectral lines. But telescopes now are equipped with charge-coupled devices (CCDs) that do the job in minutes, and some have computer-controlled arrays of fiberoptic tubes that make it possible to take the spectra of dozens of galaxies at once. Thanks to such innovations, the catalogues are growing at an accelerating rate.

Interpreting the data to estimate how much of the redshift is due to cosmic expansion is more problematical. Galaxies, in addition to participating in the expansion of the universe, evidence other sorts of motion as well. Many are binary. Spiral galaxies in particular are often found in pairs—two rather similar-looking galaxies, spinning in opposite directions, presumably because they formed from adjacent whirlpools—and both galaxies in such situations orbit a common center of gravity located at an invisible fulcrum in the intergalactic space between them. Most belong to clusters of galaxies and orbit the cluster's gravitational center at velocities of a few hundred kilometers per second. Many of the clusters belong in turn to superclusters, within which they tug at one another, further complicating their dynamics. And there is evidence that "bulk" motions send whole superclusters sliding along, at rates of some 600 kilometers per second, in directions unrelated to their expansion velocities. Astronomers must correct for all this and more to derive the "pure Hubble flow" that alone is due to cosmic expansion.

Determining the distances of galaxies is even harder. The classic technique involves the *cosmological distance ladder,* a set of overlapping distance-measuring techniques. You obtain the distances for nearby Cepheid variables and other bright stars in our galaxy, identify similar stars in other galaxies, then use whole galaxies to probe still farther into space, and so on. But reliance on each of these "standard candles" is subject to error, and the errors mount up, so that the ladder eventually becomes very precarious indeed. Yet it is precisely at these large removes that measuring the value of the redshift-distance relation becomes most important, since it is only far away—beyond the Virgo Supercluster, to which our galaxy belongs—that local gravitational interference ebbs and pure Hubble flow is exhibited. Additional headaches arise from the fact that galaxies evolve over time, so that when observing galaxies at large distances—which, because of the time it takes their light to reach us, means observing them in the distant past—one is embarking on the risky business of comparing galaxies as they are today with similar systems as they were long ago. And galaxies evolve at different rates, depending on their environments. None of these effects is fully understood. Fortunately the cosmological distance ladder, though rickety, has some overlapping rungs, like those deployed by a hook-and-ladder fire truck. So to some extent, measuring systems can be checked against one another.

Without attempting to describe it all in detail, let me try to sketch the main components of the distance ladder.[16]

Astronomical distances close to home are measured primarily by means of *parallax,* which is simply triangulation. Suppose that you are on one side of a raging river and you want to measure its width. You pick out a tall tree on the opposite bank and line it up with a more distant object, like a mountain peak. You then walk along the bank, carefully measuring the distance you've stepped off. When you stop, you look again at the tree. Owing to your change in perspective, the tree is no longer lined up with the mountain peak. You measure the angle that separates tree from peak. Its value is the same as the angle an observer standing at the base of the tree would measure between the place you stood at the outset and where you are standing now. You can now construct a right triangle for which you have the sizes of all three angles and the length of one side (the side you walked along). This enables you to compute the distance between your starting point and the tree on the other side of the river. You have thus mounted the first step on a distance ladder. The tree stands for the nearby star whose distance the astronomer seeks to know. The mountain peaks in the distance are more distant stars. The river is space.

Parallax has a little ladder of its own. The initial step involves measuring the distances of other planets. This was first accomplished for the planet Mars in 1672, by observers who made parallax observations simultaneously from Paris and Cayenne, in French Guiana. Nowadays astronomers can check the distances of nearby planets by bouncing radar signals off them and clocking the time it takes the radar echoes to return. As we have seen, thanks to Johannes Kepler's discovery of the laws of planetary motion, if you know the radius of one planet's orbit you can derive all the others from their orbital periods. So, having measured the radius of the orbit of Mars, astronomers determined the radius of Earth's orbit. This provided a much longer baseline: By observing a neighboring star at intervals six months apart, when Earth is at opposite points in its orbit, astronomers could measure slight changes in its apparent position against distant stars in the background. The size of this slim angle yielded the approximate distance of the nearby star. In this painstaking fashion, the distances to about one hundred stars had been measured by the year 1900. It has become commonplace to scoff at the nineteenth-century scientists' preoccupation with precision—to quote a few as saying that the future of science lay in making measurements to the next decimal place, that sort of thing—but the determination of parallax demonstrates that attention to exactitude laid the foundations for the more sweeping attainments of twentieth-century science.

Today, accurate parallaxes are available for something approaching ten thousand stars. (The exact number depends on what you choose to call "accurate.") Fortunately, this sample is eclectic: Among the ten thousand are found average stars like the sun, white dwarfs and red giants, binary stars in orbit around each other, stars in clusters, and so forth. After many decades of work in astrophysics —the combining of astronomical data with what can be learned theoretically and in laboratories—astronomers today understand quite a lot about how stars of various masses and chemical compositions behave. Specifically, they often have a pretty good idea of how bright each given kind of star really is.

This makes it possible to add another rung to the ladder. Knowing the intrinsic brightness of a given sort of star, an astronomer can estimate the distances of all other visible stars of the same type, simply by comparing their estimated actual brightness (absolute magnitude) with their observed brightness in the sky (visual magnitude). If one knows, for instance, the absolute magnitude of the giant white star Sirius (spectral class A1), then one can say with some confidence that a similar A1 star that looks one percent as bright as Sirius is ten times farther away, since brightness decreases by the square of the distance. The distance of Sirius, measured by parallax, is 8.6 light-years from Earth, so the distance of the second star is 86 light-years.

A higher rung is provided by stars that pulsate, changing in brightness as they do so. Their rate of change is an index to their absolute magnitude, so by charting variations in their apparent magnitude one can determine how bright they really are. Nearby, this can be accomplished by observing RR Lyrae stars (named after a prototypical example in the northern constellation Lyra, the lyre). Beyond that, one resorts to a brighter class of pulsating variables, the mighty Cepheids.

Cepheid variables are young giant stars that have entered an unstable stage of evolution. The specifics of why they pulsate provide an example of the elegance one can find in something as simple as a star. As a Cepheid contracts, it gets hotter. Heat flowing into the outer portions of the star (its "atmosphere") energizes atoms of singly ionized helium. ("Singly ionized" means that these atoms are missing one of the electrons they would normally have.) The added energy knocks another electron off the helium atoms, making them doubly ionized. Doubly ionized atoms tend to absorb light. As a result, the atmosphere of the star becomes opaque. An opaque atmosphere retains heat, like a blanket; therefore it grows hotter. As it gets hotter it expands. As it expands it cools—naturally enough, since it now is spreading all its energy over a greater area. As helium atoms cool, they return to their singly ionized state. The atmosphere, transparent once again, begins to collapse, and the cycle begins anew. Each cycle typically takes a few weeks.

The beauty of this process from a cosmological point of view is that the rate at which Cepheids pulsate, when combined with a measurement of their color, yields their absolute magnitudes. Bigger Cepheids pulsate more slowly than do smaller Cepheids—just as big gongs produce deeper tones than do small gongs—and the bigger the star, the brighter it shines. So astronomers can infer the distance of any Cepheid for which they have measured a cycle or two of variability. (Polaris, the north star, 466 light-years away, is the nearest Cepheid to Earth.) Cepheids are bright enough to be detected in galaxies as much as 15 million light-years away using ground-based telescopes, and up to roughly 60 million light-years using Edwin Hubble's namesake, the Hubble Space Telescope.[17]

By observing Cepheid variable stars with Hubble and other powerful telescopes it should be possible to map the distances of galaxies from here to the center of the Virgo Supercluster more accurately than ever before. An initial effort along these lines resulted in the announcement in 1994, by a twenty-two-member team using Hubble, that they had located twenty Cepheids in the spiral galaxy M100 and from them had estimated that the Virgo Cluster, core of the Virgo Supercluster, lies closer to our galaxy than had been thought. Comparing M100's redshift with the new estimate of its distance, the team deduced that the Hubble constant has a value of about 80 kilometers per second per megaparsec.

This result startled cosmologists, who had accepted the traditional value of 50 for the Hubble constant. At the heart of their concern was the age of the universe. All else being equal, a rapidly expanding universe must be younger than a slowly expanding one. If the Hubble team was right about M100, the universe is only about 8 billion years old. This is younger than what the astrophysicists believe to be the age of the oldest stars in the Milky Way—about 14 billion years. A universe younger than the stars it contains is obviously absurd. Distressed over this turn of events, some theorists went so far as to invoke the cosmological constant, which by accelerating the expansion velocity over time might make the universe expand more rapidly today than in the past.

The stars most readily observed across vast distances are the exploding stars called supernovae. Supernovae are mind-bogglingly powerful: A supernova can liberate more energy in one minute than is released by all the normal stars in the observable universe during the same amount of time. Only a fraction of this energy—as little as one one-hundredth of one percent, in some cases—is emitted as visible light, but that is enough for the supernova to outshine the entire galaxy it inhabits.

There are two types of supernovae, designated Type I and Type II.

Type I supernovae are thought to arise in binary systems. The supernova candidate is a dwarf star whose orbit carries it close enough to its larger and less dense companion star so that it can, by virtue of its gravitational force, strip gas from the blowsy atmosphere of the companion. As time passes, the dwarf keeps gaining weight in this fashion, until eventually its mass surpasses the *Chandrasekhar limit* (named for the Indian astrophysicist Subrahmanyan Chandrasekhar, who discovered the phenomenon theoretically). At this point, equal to 1.44 times the mass of the sun, the dwarf weighs so much that it collapses even further. Dwarfs are already so dense that normal atoms cannot survive inside them: Their protons, neutrons, and electrons, crushed together cheek to jowl, are kept from merging further by quantum mechanical forces acting principally among the electrons. (This state, called *degenerate* matter, is extremely dense by terrestrial standards: A spoonful of dwarf-star matter set down on Earth would weigh as much as a Rolls Royce limousine.) Yet once a binary dwarf star exceeds the Chandrasekhar limit and collapses further, the weight of matter bearing down on the core smashes its imposing degeneracy structure, and there ensues a titanic nuclear explosion that vaporizes the star. The advantage of Type I—specifically, the subgroup called Type Ia—supernovae to cosmologists is that they all have similar absolute magnitudes. This makes them useful as standard candles. Moreover they are the brightest form of supernovae in the wavelengths of visible light, making them conspicuous to astronomers searching the skies. Preliminary measurements of Type Ia supernovae indicates a value for the Hubble constant of about 50, yielding an age for the cosmos comfortably greater than that of the oldest stars.

While Type I supernovae are dwarfs, Type II supernovae are

giants. They collapse not because they have gained mass, but because they have run out of nuclear fuel at the core. As they run out of fuel they become unstable—there is no longer enough radiative pressure pushing outward to balance the inward pull of gravity—and then they deflate. Since giant stars burn furiously and consequently die young, Type II supernovae are usually found in the arms of spiral galaxies, where the stars originated and from which location they have not had time to venture very far. Type IIs are seldom found in elliptical galaxies, where few new stars form, while Type Is may turn up anywhere there are binary stars, which is to say in all sorts of galaxies. Type IIs are more powerful than Type I supernovae, but they look dimmer—a full magnitude, meaning 2.5 times, dimmer—because they release 99 percent of their energy not as light but in the form of neutrinos. (Astrophysicists, with their customary flair for irony, refer to this process, which takes place at temperatures in excess of 100 billion degrees, as neutrino "refrigeration.") The emerging science of intergalactic neutrino astronomy was baptized in 1987, when neutrinos from a supernova in the Large Magellanic Cloud, a satellite galaxy that orbits the Milky Way at a distance of 165,000 light-years from Earth, were recorded at underground neutrino detectors in Japan and Ohio. Neutrino astronomy has great potential, since neutrinos are plentiful and interact only weakly with matter, meaning that they bring news of events that transpired deep inside the star and not, as is the case with light, only in the star's outer atmosphere.

Type II supernovae can be detected fully a third of the way across the observable universe. But they vary quite a lot in their intrinsic brightness, and this detracts from their usefulness as standard candles. This situation may improve as astronomers get to know them better.

To hone the accuracy of supernovae as distance indicators it will be necessary to observe many more of them, and especially to catch them in their early stages, in the days before the cataclysm reaches maximum brightness. For this and other reasons, finding supernovae has become an urgent enterprise. In professional circles, automated telescopes run by computers are employed to observe scores of galaxies nightly. The computers examine the resulting images and alert their human operators whenever they see a point of light that was not there before. Useful work is also being done by amateur astronomers. Foremost among these is the Reverend Robert Evans, a minister of the Uniting Church in Australia who scans the skies with a telescope on wheels that he trundles out at sunset from his garage into the backyard of his home in New South Wales. The Reverend Evans has an acute visual memory. When looking at a galaxy one sees it through a scattering of foreground stars, some of which are superimposed on the softly glowing disk of the galaxy itself. A supernova in that galaxy looks just like such a star. It can be distinguished from foreground stars only by comparing one's view on a given night with a chart or photograph showing the stars that normally are found there. Evans has memorized the starfields surrounding more than a thousand galaxies, so he can usually tell at a glance if a "new star" (*nova,* the root of the word *supernova*) has appeared. By 1995 he had discovered twenty-seven supernovae, more than any other visual observer in the history of this planet.

The race between professionals with robotic telescopes and amateurs using their eyes (or inexpensive CCD detectors attached to backyard telescopes) may smack of John Henry's triumphant though fatal effort to drive railroad spikes faster than a steam-powered machine. But the robotic campaign has a human side of its own. Through the Internet, people all over the world can order up observations on automated telescopes. In one such educational outreach effort, the Berkeley supernovae searchers made their robotic telescope available part-time to students and amateur astronomers. In 1994 two high-school students using the Internet acquired the first image of a supernova in the face-on spiral galaxy M51. Heather Tartara and Melody Spence, seventeen-year-old juniors at Oil City High School in Pennsylvania, simply wanted to photograph a particularly beautiful galaxy for a science class project. But a few days later, amateur astronomers in Georgia discovered a supernova in the same galaxy, and the image made at Heather and Melody's request turned out to contain an earlier image of the supernova, one that caught it during the critical days before it reached maximum brightness. This was one of the earliest pictures of a supernova ever taken; rarely have the professionals done as well.

Most galaxies are too far away for any of their individual stars to be detected except when one explodes. So their distances are inferred, not by studying their stars, but by using galaxies themselves as standard candles. One way to do this is to identify the brightest galaxy in each major cluster of galaxies. Studies by Allan Sandage and others indicate that there is relatively little disparity in the intrinsic brightness of such galaxies. But astronomers must be especially alert for evolutionary effects. Galaxies in rich clusters are subject to a variety of influences. Close encounters with neighboring galaxies (which happen more commonly in rich clusters) can set off starburst events, in which tidal perturbations caused by gravitational interaction with an interloper galaxy trigger the birth of billions of stars, many giants among them. "Galactic cannibalism"—the gobbling up of small galaxies by big galaxies typically located near the center of the cluster—can temporarily (meaning over a period of, say, a few hundred million years) make the brightest galaxy in a cluster much brighter than it normally would be, prompting the unsuspecting astronomer to underestimate its distance.

A newer way of estimating the intrinsic brightness of galaxies is based on the finding, in 1977, by the American astronomers R. Brent Tully and J. Richard Fisher, that the absolute magnitude of a spiral galaxy is related to a quantity known as its 21-centimeter line width. Twenty-one centimeters is the wavelength of radio noise emitted by the hydrogen atoms that make up most of the interstellar matter in spiral galaxies. This spectral line is blurred—widened—by Doppler shifting, to a degree directly related to the speed at which the galaxy is rotating. The rotation speed, in turn, is related to the galaxy's brightness. Since radio spectra at this wavelength can be obtained for very faint sources, the *Tully-Fisher method* may prove to be usable in estimating the distances of galaxies out to 300 million light-years or more. The Tully-Fisher method typically produces values for the Hubble constant of around 70, although in the hands of some researchers it has yielded values as low as 50.

Recently a few techniques have emerged that measure distances more directly—that is, by skipping over many of the distance-ladder rungs. To date they have produced only approximate results, but they show great promise.

One such direct method is *gravitational lensing.* As we will be discussing in the following chapter, matter distorts the space surrounding it. (What we call gravitational force is simply the result of objects and light beams pursuing the shortest available path through curved space.) Quasars—the bright spots in the center of galaxies—were more common in the violent days when the universe was young; consequently, most quasars, since they belong to the past, are found at great distances. As the light from a quasar travels toward us across billions of light-years of space, it may pass to either side of an intervening cluster of galaxies. The warped space surrounding the cluster can act as a lens, with the result that we see two images of what is (or was) actually one quasar. Unless the cluster is located precisely on a line between Earth and the quasar, the light passing around one side of the lens will have traveled farther than that on the other side. Many quasars vary in brightness, sputtering and flaring over periods of as little as a month. When a variable quasar is lensed, the difference in travel time for the light composing its two images can be measured by observing the images for some time and locating the same incidents of variability in both of them. The difference in arrival time of the same event then reveals how much longer one light beam traveled than the other. Imagine that you are in a sound studio in New York, and that you are recording, on two tape recorders, a live symphony concert being transmitted from Paris, simultaneously over two channels. One radio is receiving the signal via a single satellite over the Atlantic. The other signal is coming in via a longer route, using satellites above Asia, the Pacific, and North America. The second signal therefore arrives a bit later than the first. Once the broadcast is over, you locate a single passage in the music and measure how much time elapsed between the time it was received by the first radio and the time it was received by the other one. Knowing that the signals traveled at the velocity of light—and here we ignore delays introduced by the satellite transponders—you can calculate how much longer was the path through space taken by the second satellite linkup.

If the intervening cluster of galaxies is mapped correctly (one has to make an educated guess as to its center of gravity), simple triangulation then yields the distance of the cluster. Some preliminary gravitational lens results indicate a Hubble constant of 50, others somewhat higher values.

Another direct approach employs what is known as the *Sunyaev-Zeldovich effect,* after the two Russian astrophysicists who pioneered its use. This consists of measuring the intensity of the cosmic microwave background (CMB) through certain clusters of galaxies that emit x-ray radiation. The intergalactic gas in such clusters is relatively warm—that's why it emits X rays—so the CMB photons are heated up when they pass through the cluster. The result is a hot spot in the background radiation. More distant clusters are denser and hotter, and so make hotter spots in the CMB. Hence the temperature yields the distance. The effect is subtle—the hot spot is only a fraction of a percent hotter than the overall background—but it has been observed, by drawing on an arsenal of observational tools. A 1991 measurement of the Sunyaev-Zeldovich effect combined observations made with the *Einstein* and *Ginga* x-ray satellites and those made by a ground-based radio telescope. Early applications of the Sunyaev-Zeldovich method yield a Hubble constant of between 40 and 50. These preliminary data are inconclusive, but ongoing work with specialized instruments like SUZIE (Sunyaev-Zeldovich Interferometer Experiment) could yield important information on the value of the Hubble constant.

Finally there is an ingenious method called the *brightness fluctuation* technique. One points a telescope at the central bulge of a spiral galaxy, or the central part of an elliptical galaxy, and measures the amount of unevenness in its surface brightness from point to point. Since nearby galaxies are more nearly resolved into stars, they will show more unevenness than will distant galaxies, in which the stars merge into a smooth, unresolved blob of light. If, for instance, a narrow-field telescope were pointed at a galaxy so close to us that the field of view contained only one star, we would have the maximum possible surface brightness fluctuation—all black but for one point of light. The same telescope, trained on a more distant galaxy, might capture a hundred stars (less fluctuation) while for a still more distant galaxy the figure would be a thousand stars (even less fluctuation), and so forth. So brightness fluctuation correlates with galaxy distance, everything else being equal. With

Hubble, this emerging technique should be applicable to galaxies at distances of up to half a billion light-years or so.

That's the distance ladder in brief. As we have seen, it produces values for the Hubble constant that differ considerably, with one set of results clustering around 50 (an old universe, age about 15 billion years) and the other around 70 to 85 (a young universe, age about 10 billion years). Amusingly, the astronomers who argue for an old universe are themselves mostly older, and those who argue for a younger universe mostly younger. This is not to say that the scientists are simply projecting their personalities onto the sky; the rules of the game are too stringent for that. What happened, rather, was that a generation of older astronomers and astrophysicists, Sandage prominent among them, put together a grand synthesis that fit the age of the universe with that of the stars. The Young Turks then came along, turned up data that didn't fit, and proceeded to chip away at the imposing edifice. In doing so, they, too, were fulfilling an essential scientific function, which is to challenge the work of the elders and see whether it can be brought down.[18]

Since both sides base their estimates on a variety of methods, for either to be proved wrong will require showing that *all* their methods were somehow in error. So the debate has a richly baroque intricacy. Something of its flavor may be found in the controversy over the Hubble team's having found a young-universe value for H_0 through their study of Cepheid variable stars in the galaxy M100, which lies far enough away that Cepheids could not be seen there until the Space Telescope was launched and its faulty optics corrected by a space shuttle crew.

The Hubble team, led by the young astronomer Wendy Freedman, presented a classic piece of inductive observational science. They located Cepheid variable stars in M100, used them as distance indicators, and found that M100 is closer than the old-universe crowd had maintained. The result, after making various corrections for the local velocity field, indicated that the Hubble constant was more like 80 than 50. Since the universe evidently is expanding more rapidly than had been thought, it is also younger.

Sandage disagreed. Approaching the age of seventy, he was the dean of observational cosmology—a scientist's scientist, with an encyclopedic knowledge of everything from the history and philosophy of science to the arcana of stellar evolution and the intricacies of statistical probability. Sandage had in five decades of research published more papers than most researchers could absorb. As a result, many astronomers pretty much took his word for it that the Hubble constant was about 50. They based this assumption on Sandage's reputations for scientific acuity and personal integrity.[19] Both reputations were well founded, but over the years they became entangled, so when his value for the expansion rate was challenged by younger astronomers, Sandage to some degree took it personally.

He responded to the M100 paper with a bristling arsenal of arguments indicating that the Hubble team's distance for M100 was too small and their value for the expansion rate correspondingly too large.[20] First, he maintained, the Young Turks made errors in their Cepheid measurements: The internal statistics for Cepheids in M100 obtained by the team did not fit generally recognized values, he noted, suggesting that their apparent magnitudes had been measured inaccurately. In addition, M100 may actually lie in the foreground and not in the Virgo Cluster, as his opponents assumed. (It is a spiral galaxy, and spirals are not generally found in the center of clusters like Virgo, where elliptical galaxies predominate.) And since all Virgo galaxies are caught up in the local gravitational field, one must apply corrections before arriving at pure Hubble flow: Here, too, Sandage alleged, the Hubble team erred. A study by the Swiss astronomer Gustav Tammann indicates that Virgo is being pulled away more rapidly than the Hubble team allowed for.[21] This result accords with studies of the cosmic microwave background, the ultimate reference frame to which all studies of galaxy dynamics must refer. If Tammann is right, M100 could have a high recession velocity yet the universe as a whole have a lower Hubble constant.[22]

Wendy Freedman and her colleagues on the Hubble team were not convinced. But they had made the fair-minded concession, in the paper announcing it, that their result was not conclusive. "We do not wish to mislead the reader into believing that the problem of determining H_0 has been solved," they wrote. "It has not."[23] In a conversation soon thereafter, Freedman was asked about the preference shared by many theorists for an omega = 1 universe, which would require a smaller value of H_0 than she had measured. "I think it's a powerful argument," she said. But it's also a powerful reason to go out and measure it. You don't want one without the other. Only a combination of theory and experiment will eventually lead to an understanding of what the universe really is like."[24]

In 1996 Sandage and five colleagues announced the results of a Hubble project of their own. They had identified twenty Cepheids in the galaxy NGC 4639, in which a Type Ia supernova had been observed in 1990. Combining the light curve of this supernova with those of six others, they produced a brutally simple analysis—independent of considerations about local motions and to what cluster the galaxy might belong—that yielded a value for the Hubble constant of 57. For Sandage, that was that. "This marks the end of the Hubble wars," he declared.[25] Freedman and her colleagues were unpersuaded, but predicted that the disparity would be resolved, one way or the other, by the end of the century.

Meanwhile, the controversy over the value of H_0 and the age of the universe prompted a spate of newspaper and magazine stories proclaiming that the big bang theory was, as *Time* magazine put it, "unraveling." So it may be worth noting that the big bang theory has nearly always been in a state of crisis. Indeed, this is true of most living sciences. As long as they remain vital, they are generally as messy as the studios of living artists, which subside into order only when the artist has died and the studio has become a museum. Hubble's original value for the Hubble constant was so high that it yielded an age for the universe of only *two* billion years, less than what geologists knew to be the age of the earth.

That was one of the reasons Hubble shrank from interpreting his discovery in terms of cosmic expansion. Like Darwin—who died before the geophysicists realized that the earth's core is heated not by gravitation but by radioactive minerals, and who therefore could never reconcile prevailing estimates for the age of the earth with the much longer time that evolution would have required—Hubble lived out his life confronted by an expansion rate that paradoxically made the universe out to be younger than the earth. A major step in repairing these errors came in 1952, when Walter Baade at Mt. Wilson found that there are two types of Cepheid variable stars—not one, as had been assumed. This revision nearly tripled the size of the universe, eliminating the conflict between its age and that of the sun and the earth. Another correction was made in 1958, when Sandage showed that some of the bright "stars" Hubble had identified in distant galaxies were HII regions—glowing nebulae lit up by many stars. This doubled the size of the universe.

Astronomers are unlikely to encounter another revolution on the scale of that of the 1950s, when the Hubble constant was revised downward from 500 to 50. As we have seen, the current debate involves values of the Hubble constant that differ by a factor of less than 2, with most research tending to yield values of about 50 or about 70.[26] While less than satisfactory, this is hardly grounds for proclaiming that the big bang cosmology should be discarded. And while work continues on determining the precise rate at which the universe expands, the fact that it *does* expand is today as well established as, say, the fact that biological species arose through the process outlined in Darwin's theory of evolution.

It is, of course, possible to take an altogether different tack, to claim that there was no big bang or that cosmological redshifts are caused by something other than cosmic expansion. At this writing there are at least three such accounts—the "tired light" hypothesis, the steady state theory, and the plasma universe model. "Tired light" is the concept that light as it travels vast distances loses energy, creating a redshift-distance relation that is not due to cosmic expansion at all.[27] But there is no known physical mechanism to cause tired light, and such an account has the disadvantage of decoupling general relativity from cosmology, which at this point serves no evident purpose. The steady state theory maintains that matter is constantly created, by means of a "C-field" that also drives cosmic expansion.[28] But the steady state theory has great difficulty in accounting for the cosmic microwave background. The CMB looks exactly like the signature of the big bang, reducing the status of contrary arguments to something like that of literary scholars who claim, as the old joke has it, that Shakespeare's plays were written not by Shakespeare but by someone else of the same name. The plasma model proposes that parts of the universe expand while others contract, in an ongoing pulsation that occurs when clouds of matter and antimatter collide, generate energy, and are in turn repelled from one another. But this model has not yet made a prediction that can be tested observationally, and so remains too vague to be wrong.[29] There are other dissident theories, but all seem forced in one way or another, rather like the arguments of creationists that God made the geological strata *look* as if the earth were old, when instead it is young. Here it is helpful to wield Occam's razor, the dictum that the simpler of two otherwise comparable hypotheses is to be preferred. The big bang theory explains the Hubble law in an efficient and natural way; and it fits well with the discovery of the cosmic microwave radiation, independent evidence that the universe was once in a high-energy state. So in this book we will proceed on the assumption, shared by most cosmologists, that the universe really is expanding from a big bang origin and has not merely contrived to make itself look that way.

But if cosmic space is expanding, what is it expanding *into*? To explore that question requires us to examine the global geometry of the universe. And that is the subject of the next chapter.

3
The Shape of Space

I think a straight line does not exist. There is no such thing as a straight line in painting.

—WILLEM DE KOONING[1]

How can anyone learn anything new who does not find it a shock?

—JOHN ARCHIBALD WHEELER[2]

COSMOLOGY CHALLENGES US to alter assumptions that have served us well here on Earth but are inadequate when we are called upon to deal with the wider universe. Our subject in this chapter is space. Ordinarily we can make do with a commonsensical sense of space, like the one Werner Heisenberg articulated when he said, "Space is blue and birds fly through it."[3] But if we are to do justice to cosmological space, we shall have to set aside conceptions limited to our terrestrial experience and replace them with the cosmos of modern physics, where space is curved, and on the small scale may dissolve into foam, and in some instances may be torn or pinched off. In return we shall find that the universe is exotically beautiful not only in what it contains but in the very fabric of its spacetime structure. The sign posted above this gateway reads not "Abandon hope" but "Abandon presupposition, all ye who enter here."

The classical physics of Newton (which, for some reason, is still taught to students before they are permitted to study relativity, thus obliging them to recapitulate the provincialisms of their elders) treated cosmic space as if it were but an enlarged version of the space we are familiar with on Earth. Newtonian space is a characterless emptiness, a neutral theater within which events transpire without their being influenced by space. Such a space is known as *flat*. It is the three-dimensional equivalent of a two-dimensional plane, like a tabletop. Just as a ball sent rolling across a flat table goes wherever it is propelled, without being interfered with by the table, a planet floating through flat space follows a path uninfluenced by space. Newtonian space stays out of the dynamical picture.

But Newtonian space when extrapolated to the cosmic scale poses problems, several of which troubled Newton himself. One difficulty has to do with the old riddle of whether there is an edge to the universe. Flat space goes on forever; Newton's equations offer no way to put an end to it. But if infinite numbers of stars occupy infinite space, one can readily calculate that the night sky should be bright as the sun. Instead, the night sky is dark. (This is known as Olbers's paradox, after the nineteenth-century German astronomer Heinrich Wilhelm Olbers, although it occurred to a number of earlier thinkers, Kepler among them.) If, alternatively, the stars are confined to a system of some sort *in* infinite space, one calculates that these stars should long ago have collapsed upon one another, victims of their mutual gravitational attraction. That hasn't happened, either. We might try to counter cosmic collapse by imagining that the star cluster is rotating, but then we encounter the problem of explaining what, if it is surrounded by nothing but space, the cluster is rotating relative *to*. (The Austrian physicist Ernst Mach liked to frame questions of inertia in this manner.)

Another Newtonian problem arises from the operation of *forces*. To account for the fact that the moon and the planets remain in their orbits, Newton postulated the existence of a force of gravity. But neither Newton nor anyone else could understand how gravitational force might propagate through empty space. To answer this riddle, space was said to be suffused with *aether*, a frictionless, invisible substance that carries light and gravity in something like the way that the ocean carries waves. The aether hypothesis survived until the year 1887, when the physicists Albert Michelson and Edward Morley, in experiments conducted at Case Western University in Cleveland, Ohio, demonstrated that there is no aether. (They did this by checking for "aether drift"—a change in the behavior of light waves according to whether the earth was plowing into the waves or passing through them crossways. No such drift was found.) The Michelson-Morley experiment set the stage for the revolution in spacetime physics that Einstein inaugurated with the special theory of relativity in 1905 and culminated, a decade later, with the general theory.

General relativity does away with any need for a force of gravity. It depicts the planets as following paths of least resistance, *geodesics,* through curved space. This dispenses with the (nonexistent) aether. It also resolves the paradox of whether there is an edge to the universe, by raising the prospect that cosmic space has an overall curvature. The universe might, for instance, have a shape comparable with that of a sphere. (It would be a four-dimensional sphere, about which more later.) Three-dimensional space in this model is analogous to the surface of a sphere. Such a universe is finite, since it contains a finite amount of space, but unbounded: One can see forever, or travel indefinitely far, in any direction, without ever coming to an edge.[4]

General relativity may be thought of as simply a way of mapping the universe. Like all scientific theories, relativity is expressed in terms of mathematical equations—mathematics being a codified form of logic that embodies the faith of science that ature works in a rational way. All mathematics, in turn, can be interpreted as mapmaking. Pure mathematics constructs maps of abstract spaces: A mathematician can map the contours of a four-dimensional sphere or a ten-dimensional cube without worrying about whether any such thing actually exists. When applied to physics, mathematics may still be pretty abstract—in quantum mechanics, for instance, theorists often work with imaginary "phase spaces" that exist only in the sense that there is such a thing as "your money" in the bank or an "average voter" in an election—but the equations are supposed to tie in to physical reality in the end. Think of the relevant piece of nature as a terrain and of the equations as laying a gridwork across that terrain, and you have a map.

All maps are imperfect; this is the sadness of maps. They are imperfect in at least two ways. First, since they represent the territory under investigation more economically than does the territory itself, they inevitably contain less information; the resolution of a map is coarse compared to that which it maps. The alternative would be a nightmarish scenario of the sort described by the fabulist Jorge Luis Borges:

> In that Empire, the craft of Cartography attained such Perfection that the Map of a Single province covered the space of an entire City, and the Map of the Empire itself an entire Province. In the course of Time, these Extensive maps were found somehow wanting, and so the College of Cartographers evolved a Map of the Empire that was of the same Scale as the Empire and that coincided with it point for point. Less attentive to the Study of Cartography, succeeding Generations came to judge a map of such Magnitude cumbersome, and, not without Irreverence, they abandoned it to the Rigors of sun and Rain. In the western Deserts, tattered Fragments of the Map are still to be found, Sheltering an occasional Beast or beggar; in the whole Nation, no other relic is left of the Discipline of Geography.[5]

Second, and more to the point where cosmology is concerned, maps introduce distortion. A familiar example of cartographical distortion may be found in maps of Earth drawn on flat paper. Flat maps of small parts of the earth work well, since the curvature of the surface can be disregarded if one is concerned only, say, with getting from downtown Moscow to the suburbs. But such maps become increasingly inaccurate when called upon to take in larger amounts of territory—which is what happens when Newtonian physics is applied to intergalactic space—and they become conspicuously wrong when they are used to depict the entire planet, since the earth is not a plane but a sphere. The various geographic mapping schemes—the Mercator projection, the Hammer projection, the cylindrical projection, and so forth—all contain the same amount of distortion, but each distributes it differently, and each is useful to the extent that it reduces distortion in the arena in which one wants to employ the map. A Mercator map, for instance, is fairly accurate near the equator but inflates the relative sizes of territories near the poles. (Indeed, it must omit the poles altogether: A Mercator map that included the poles would be infinitely large.) The problem shared by all these systems is that they must reduce a three-dimensional plenum, the surface of the earth, to the two dimensions of the map. The only way to eliminate the resulting spatial distortion is to add a dimension. The result is a three-dimensional terrestrial map, a globe. Terrestrial globes reduce resolution—as a child discovers when he squints at his hometown and tries to find his house—but they induce no distortion. The lesson here is that three-dimensional objects like the spherical earth can be mapped accurately in three dimensions, but not fewer.

And this is the story of general relativity as well. The theory indicates that cosmic space can be mapped accurately only by going to four dimensions.

The distortions introduced when three-dimensional maps are employed to chart cosmological space show up in various ways, which Einstein in constructing the general theory of relativity was out to eliminate. One instance is that the motions of the planets—especially of Mercury, the planet closest to the sun—do not quite fit the Newtonian predictions. (The self-educated astronomer Simon Newcomb pointed this out in 1882, and seven years later wondered presciently whether it meant that something in Newton's equations might have to be modified.) Einstein's solution was the same one employed by terrestrial cartographers when they resort to a globe: He added a dimension. The mathematics of general relativity is four-dimensional geometry, with space allotted its customary three dimensions and time assigned the fourth.

Until the nineteenth century the only geometry available to mathematicians was the three-dimensional geometry so beautifully set forth in Euclid's *Elements.* It is for this reason that ordinary, flat, three-dimensional space is called *Euclidean* space. Euclid's monopoly on geometry lasted for two thousand years. It ended on June 10, 1854, when the German mathematician Georg Friedrich Bernhard Riemann, a frail and painfully shy young man who labored in poverty and was to die of tuberculosis at age thirty-nine, gave a talk demonstrating that it is possible to create a four-dimensional geometry that makes just as much sense as Euclid's three-dimensional one. Like most mathematical inventions, Riemannian geometry had no immediately evident application to the real world. But a half-century later Einstein studied Riemann and, in one of the audacious leaps of the imagination that rank him as a creative artist on a par with Beethoven, saw a way to apply Riemann's four-dimensional geometry to the cosmos. "I realized that the foundations of geometry have physical significance" was how Einstein put it.[6] He had hit upon the cosmic equivalent of the terrestrial globe. In much the same way that geographers had to go from two-dimensional ("plane") geometry to three-dimensional ("solid"; specifically spherical) geometry to create an undistorted map of the earth, Einstein, to map the cosmos, went from three-dimensional to four-dimensional (***hyperdimensional,*** also called ***non-Euclidean***) geometry.[7]

The gridlines of Einstein's four-dimensional map are light beams. These he adopts as his fiduciaries, the straight lines of his cartographic scheme. According to the theory, light beams bend when they pass near massive objects. This effect was first observed in 1919, when a British expedition, led by Eddington, pitched tents on Príncipe Island, off western equatorial Africa, and photographed stars near the sun during a total solar eclipse. Light coming from the stars was bent as it passed through the curved space near the sun, making the stars appear to be in a different location in the sky. Curved light beams in themselves were nothing new: Newton's theory contained such curvature, though only half as much of it. But Einstein declared that the light beams are actually straight lines and that space itself is curved. This is not the most intuitive thought in the world; indeed, the curvature of space is probably impossible for humans to visualize.[8] But it can be calculated precisely, through the application of Riemannian geometry, and the resulting predictions have been verified repeatedly by experiment. The theory predicts, among many other things, the precession in the orbit of Mercury that had given Newton such trouble. Indeed, the observational verdict on relativity ranks it among the most accurate theories of physics ever devised.[9]

Relativity also resolves the two aforementioned Newtonian paradoxes—that of whether there is an edge of the universe, and the riddle of how gravitational force propagates through empty space.

As noted earlier, the paradox of the edge of the universe vanishes because curved cosmic space can form a four-dimensional sphere, or *hypersphere*. In this model, three-dimensional space is analogous to the two-dimensional surface of an ordinary sphere: Such a universe is finite yet unbounded: One can travel indefinitely in any direction without reaching an edge. Hence no paradox. As the physicist Max Born exclaimed, "This suggestion of a finite but unbounded space is one of the greatest ideas about the nature of the world which ever has been conceived."[10] The hypersphere is actually one of three classes of cosmic geometry proffered by relativity. The other two possibilities are that cosmic space is flat (meaning Euclidean) or hyperbolic (meaning saddle-shaped). To visualize the difference, imagine drawing a triangle on a flat piece of paper, on a beach ball, and on a saddle. In flat space, the angles of the triangle add up to 360 degrees. In spherical space, they add up to *more* than 360 degrees. In hyperbolic space, they add up to *less* than 360 degrees. The extent to which the total differs from 360 degrees indexes the curvature—positive for a sphere, negative for a hyperbola—of the surface.

As noted in the previous chapter, the question of which of these models accurately depicts cosmic space has to do with the density of matter. If the cosmic matter density is higher than critical, then space is wrapped around the universe like a shell, and the universe is spherical. If the matter density is lower than critical, the universe is hyperbolic. If the density is exactly critical, the geometry is midway between the spherical and hyperbolic cases, and space is flat. In terms of the density parameter omega, the universe, if spherical, has an omega greater than one; if hyperbolic, it has a density less than one; and if it is flat, omega equals one. In the model favored in this book, omega equals one and space locally *appears* flat, although on a gigantic scale—very much larger than the part of the universe we can at the present epoch observe—the cosmic geometry is curved.

The other Newtonian riddle—How does gravitational force cross empty space?—is also resolved, since relativity obviates the need to consider gravitation as such. This is one of the most admirable aspects of Einstein's greatest work of art: It is a theory of gravity that does away with gravitational force. The reason is that in curved space the effects Newton attributes to gravitation are local, not distant. In general relativity, the course of any object describes a "world line," which is a trajectory through space and time. Unless otherwise acted upon, each object follows the path of least resistance in spacetime, a geodesic. Nature does what's easiest—the *least action* principle—and to pursue a geodesic takes no energy at all.[11] An apple falls from the tree because space in the vicinity of the earth is curved: The apple is moving downhill in spacetime. The moon orbits the earth because the earth occupies a basin in spacetime, and the moon rolls around the inner wall of that well like a roulette ball rolling above the roulette wheel. All this behavior is local: In classical physics, gravitation involved action at a distance (which made Newton uneasy), but in relativity, objects simply respond to the contours of space in their immediate vicinity.

To track the trajectories of freely moving objects in space is to chart geodesics in spacetime. The situation is rather as if we were observing, from orbit, a group of blinkered plow horses *sans* drivers moving across a field. The horses, following paths of least resistance, will tend to walk straight when crossing flat fields, and their paths will bend when they encounter hills or depressions. Understanding that the horses are simply responding to local conditions, we can map the contours of the field by charting their paths. In relativity, gravitational "force" is but a consequence of local geometry.

This, by the way, explains Galileo's discovery that in a vacuum all objects fall at the same rate. In terms of common sense—and, for that matter, in terms of Newtonian gravity—it would be perfectly reasonable if cannonballs fell faster than feathers. But in relativity all objects must follow the same geodesics, since the local path of least resistance is the same regardless of who travels it. Big plow horses and small ones follow the same paths, and planets and pebbles pursue identical world lines, because all are tracing out the contours of local space.

To sum up, general relativity demonstrated that gravity, customarily regarded as a force, can instead be interpreted as an effect of hyperdimensionality on the three-dimensional world of our normal experience. By adding a dimension—mapping the universe in four dimensions instead of three—Einstein showed that gravitational force is due to geometry.

This principle can be generalized into an assertion of great potential power, one that will play an important role in our upcoming discussions of supersymmetry and superstring theory. The assertion is that *all "forces" are consequences of geometry.* Riemann raised this prospect. Einstein showed that it could be incorporated into a successful theory of physics, at least insofar as gravity is concerned. Later in this book, we will discuss how, by mapping nature in still higher numbers of dimensions, we may find that all forces are indeed geometrical at the root. Specifically we will see how superstring theories envision the universe as having begun in ten or more dimensions. They are "unified" theories, meaning that they aim to address phenomena currently handled separately by relativity and quantum physics. This, too, is reminiscent of Riemann, who referred to hyperdimensional geometry as the "unification of all physical laws."[12] Superstring theory suggests that all matter and energy—the fact that there *is* anything and that anything *happens* —are due to the fundamentally hyperdimensional geometry of the universe.

Returning to less speculative topics, the curvature of space produces many phenomena that are relevant to cosmology and have already been observed. Among these are the "Einstein lenses" mentioned in the previous chapter—instances where clusters of galaxies in the foreground bend space to create multiple images of distant quasars, or turn their ancient light into graceful blue arcs, so that the far reaches of the observable universe become as dappled as a pond in moonlight. There are also *microlensing* events caused by nearby stars; we will encounter these in the chapter dealing with dark matter. A third prospect is that there are gravitational waves—ripples in space that go rolling outward at the speed of light from binary neutron stars and other gravitationally intense systems. The existence of gravitational waves has been confirmed indirectly and may soon be observed directly.

But our main concern here is with the two relativistic phenomena, significant to cosmology, that lie at opposite spatial extremes. One involves our gaining a better understanding of the overall geometry of the cosmos, which is literally the biggest problem in science. The other is the study of black holes—small-scale objects where the local curvature of space becomes infinite and the theory of relativity breaks down.

First, cosmic geometry. For creatures like ourselves, who evolved in an evidently three-dimensional world, the four-dimensional cosmos can take some getting used to. In the previous chapter we discussed how, since the curvature of cosmic space is dictated by the cosmic matter density, which in turn retards the expansion of the universe, astronomers can in principle infer the shape of space by measuring the extent to which expansion is slowing (the *deceleration parameter*). Here we consider how the curvature of space might be observed directly.

For convenience, we again suppose cosmic space to be spherical, and we model its three spatial dimensions as the two-dimensional surface of the earth. And while we could, of course, choose any system of coordinates, for the sake of clarity we place ourselves at the North Pole. To replicate the bending of light rays in curved space, we imagine that light crawls along the surface of the earth, as if guided by fiberoptic cables laid along the lines of longitude that radiate southward from our post at the Pole. Finally, to give us something to look at, let's decorate the earth with Christmas trees, placed more or less randomly all over the surface. Their glowing lights will represent clusters of galaxies.

What do we see?

First, we notice that the farther we look, the more lights we see in the field of view of any given telescope. Part of this is a straightforward perspective effect found even in flat space. When we look farther away, we see a greater number of galaxies in a given field of view simply because any angle—the long, thin pie-wedge stretching from our telescope out into space—takes in more space the farther one looks. That's why a distant mountain can be obscured by one's index finger. In spherical space, however, the effect is more complicated. As we look outward for some distance, our field of view increases, just as expected. But then—when we probe past the equator in our model Earth—the curvature of space begins to *decrease* our field of view. Think of lines of longitude converging toward the pole: At the South Pole, our field of view becomes zero.

In the big bang model, the South Pole represents the origin of the universe. That's why we and any other observers in the universe see the cosmic microwave background as a distant haze, painted evenly across the sky in every direction. (The background radiation has structure and extent because it was created, not at time zero, in which case it would be infinitesimal, but at the time of photon decoupling, when the universe had already been expanding for half a million years.)

If we imagine that our model Earth has been expanding from a "big bang" origin, and that light moves across its surface slowly enough to represent real light moving across the cosmos, we find that the number of galaxies increases even faster with distance for any given field of view. That is because we are seeing back to an earlier time, when the earth was smaller and our Christmas-tree galaxies therefore were closer together. The point of this exercise is to note that, once we have a good idea of how fast the universe expands and of how much it has been decelerating, it should be possible to measure the curvature of cosmic space by counting galaxy densities at various distances.

Thus far we have been imagining that we could see the entire universe. This, however, is unrealistic. In actuality, we can see only those galaxies that lie close enough to us for their light to have reached us at the present time: These galaxies, the ones at "lookback times" less than the age of the universe, inhabit the *observable* universe. In all feasible expanding-universe models, the observable universe is but a fraction of the whole. An inflationary universe would be incredibly large today, and the observable universe an incredibly small part of it. If the entirety of an inflationary universe were the surface of the earth, the observable part would be smaller than a proton. The inflationary universe might be globally spherical or globally hyperbolic, but to all observers it would look locally flat. So if we do live in an inflationary universe it will be very difficult, perhaps impossible, to determine the overall shape of space, just as it is hard to discern the radius of the earth by measuring only a tiny patch of soil.

Much more could be said about the large-scale curvature of space, and we will return to the subject from time to time. But for now let's move on to black holes.

The black hole is a kind of mandala for twentieth-century physics: a peerless contemplation object, a meeting ground of relativity and quantum physics, a triumphant example of how creative theorizing can open windows on the real world. The existence of black holes is predicted by general relativity, yet black holes also represent the breakdown of general relativity. (Inside black holes the curvature of space becomes infinite, whereupon the relativity equations produce infinite results. Professor Einstein tips his hat and gracefully departs the stage.) To inquire beyond this point will require a theory of quantum gravity. And so it is not surprising that some of the more promising steps toward a unified theory embracing quantum mechanics and general relativity have come from black hole research.

Black hole studies can be divided into two parts, the (comparatively) mundane and the more exotic.

First, the mundane:

The basic concept of a black hole is simple enough to be envisioned without recourse to relativity. That much was accomplished as early as 1783, by a British amateur astronomer, the Reverend John Michell. On this, the "classical" level, black holes are defined as objects with gravitational fields so intense that light cannot escape them. Every massive object has an *escape velocity.* Throw a ball into the air and it comes down. Throw it harder and it takes longer to fall. The speed at which the ball *never* comes down is the escape velocity of the planet on which you are standing. On Earth it amounts to 24,000 miles per hour, the speed that Apollo spacecraft had to exceed to carry astronauts to the moon. The escape velocity is a function of two quantities, the mass of the planet and its radius. The mass determines the overall strength of the gravitational field. The radius determines the distance of a spaceship on its launching pad from the center of the planet, from which the gravitational field seems to emanate. (In relativistic terms, the earth sits in a bowl in space; the closer we are to the center of the bowl, the harder it is to climb out.) You and I are located about 3,963 miles from the center of the earth. At this distance, we feel a gravitational force that we designate as one G. If the earth were put in a giant vise and squeezed down to half its present diameter, our distance from the center would be half what it was, gravity on the surface of the compressed earth would then be four Gs, and the escape velocity would increase by 1.4 times, to 33,600 miles per hour. If we kept squeezing the earth, we would eventually reach a point—when its radius was down to 0.35 inch—at which the escape velocity equaled that of light. The earth would then become a black hole. This critical value is called the *Schwarzschild radius,* after the nineteenth-century German astronomer Karl Schwarzschild, who worked out the first exact solutions to the relativity field equations while serving as an artillery lieutenant on the Russian front (where he contracted a mortal skin disease and died, in 1916, at the age of forty-two). Since escape velocity depends on both mass and radius, the Schwarzschild radius increases with the mass of the object in question: The more massive the object, the larger it can be when it achieves black hole status. The sun, subjected to our giant vise, would become a black hole when squeezed to a radius of 1.9 miles.

The contemplation of imaginary stars that eat their own light remained just a curiosity until Einstein established, in the special theory of relativity, that the velocity of light is not an arbitrary quantity but is deeply woven into the fabric of the universe. This insight is expressed in his famous equation $e = mc^2$, which shows that the amount of energy contained in any chunk of matter is equal to its mass times the velocity of light squared. One consequence of this line of thought is to reveal that nothing can be accelerated to a velocity greater than that of light. Imagine a robust starship rocketing to ever-higher speeds. As it approaches the speed of light, various transformations become evident. The mass of the spacecraft increases. Its length, measured along the axis of its direction of travel, shrinks. The passage of time on board slows. The amount of energy required to make the ship go faster increases steeply. For the ship to attain the velocity of light would require infinite energy. Meanwhile the mass of the ship would become infinite, its length would shrink to zero, and time on board would come to a stop. This is as good a way as I can think of to say that something is impossible, and that is one reason why we say that nothing can be accelerated to the velocity of light.

So the classical picture of black holes—of stars or planets being reduced in size until their escape velocities exceed that of light—turns out, when combined with special relativity, to mean that such objects are in effect cut off from the rest of the universe. Things can be thrown in, but nothing can escape.

To the best of our knowledge, nobody is going around

squeezing planets and stars in vises. But something similar can occur when giant stars collapse. Here gravity acts as the vise.

Every healthy star represents a balance between two opposing forces. Gravity tends to collapse the star. Heat generated by nuclear fusion at the core radiates outward; its tendency is to blow the star apart. Caught in the balance, stars pulsate a bit, owing to the teeter-totter of inward-pulling gravity and outward-pushing radiative heat. The pulsations are modulated by an elegant feedback mechanism. Whenever gravity begins to take over and the density of the core grows, so does its heat. (Any object when compressed becomes hotter.) As the core gets hotter, the rate of thermonuclear fusion increases, because as the subatomic particles there go faster (and that's the definition of heat) they are slammed into one another at higher velocities, and so more of them fuse, releasing more energy. The heat expands the core, which thins it out. The fusion rate then slows down, whereupon the core cools a little bit and starts to contract again, and the cycle recurs.

As long as a star keeps burning steadily at the core, it can remain in this state of equilibrium. For most stars the pulsations are negligible, while others, the variable stars, pulsate dramatically enough to make for detectable changes in brightness. But all survive as long as they have fuel. The end of equilibrium comes when the star runs so low on fuel that its nuclear furnace falters. The question of when this will happen depends on a number of things, including the extent to which the star manages to mix the depleted material at the core with the relatively pristine material farther up toward its surface, but the main determinant is the star's mass: The more massive a star, the faster it burns. The rule of thumb is that the burn rate goes up as the cube of the star's mass. Consequently, a giant star ten times more massive than the sun burns its fuel a thousand times faster, and comes to a quicker end. The sun started out with enough fuel to burn steadily for about 10 billion years—it is now middle-aged, having existed for a little under 5 billion years—whereas a star of ten times the sun's mass runs out of fuel in only 10 *million* years.

Once the furnace sputters, there is no longer enough energy radiating from the core to prop up the star. Gravity's vise now gains the upper hand, and the star collapses. The disaster happens quickly: In less than one second, the star's core shrinks from the size of the earth to the size of an office building. The star sheds much of its material—the outer envelope—while the core continues to contract.

The remainder of the story depends on the mass of the core. A core with a mass less than 1.4 times the sun's (the Chandrasekhar limit) will settle down as a *white dwarf* star. White dwarfs have a *degenerate* core, usually made of carbon and oxygen and surrounded by a dense, gaseous shell about 50 miles thick. They are prevented from collapsing any further by a quantum physics rule called the Pauli exclusion principle (named for the physicist Wolfgang Pauli, a combative theorist who is remembered for his scathing dismissal of ideas that he didn't like as "not even wrong"). The exclusion principle declares that no more than two electrons with opposite spin can occupy the same energy state. It is this mechanism that limits the number of electrons in each shell around the nucleus of an atom, thus making chemistry possible. The "valences" one learns about in chemistry class depend on whether atoms in a molecule have room available in their electron shells, in which case they can link up with neighboring atoms. As electrons flow through white dwarf stars, the exclusion principle limits their density and thus props up the star against further collapse.

If, however, the core of the dying star weighs more than 1.4 solar masses, its gravity overwhelms the exclusion principle and the electrons are smashed into protons, turning the protons into neutrons. The result is a *neutron star.* While white dwarfs are the size of planets, neutron stars are the size of towns, typically measuring less than 10 miles in diameter. A neutron star core consists, as the name implies, primarily of neutrons, which are electrically neutral subatomic particles. The density at the star's core is so great that not even atomic nuclei (which in normal circumstances consist of protons and neutrons) can survive there. Instead we find the neutrons, along with some protons and electrons, in a *superfluid* state, more viscous than our world's most exquisite oils. Surrounding the core is an envelope where the pressure, while still formidable, is low enough to permit the existence of atomic nuclei. The envelope is about a mile thick. It's capped by a crust made of nuclei and electrons. The gravitation at the surface is so powerful, and the crust so malleable, that a neutron star resembles a giant ball bearing. A hill a millimeter high is Everest on a neutron star.

Neutron stars rotate rapidly; some spin more than a thousand times per second. Those with magnetic fields emit intense streams of energy at radio wavelengths from their magnetic poles. When these radio beams happen to be oriented so that they strike the earth, the result is a rapidly beeping radio pulse. Such neutron stars are called *pulsars.*

Neutron stars would seem to be exotic enough to please even the most venturesome thinkers, and indeed for decades many researchers resisted the notion that stars could collapse any further. But it now appears that if a star implodes to a sufficiently small and dense state its gravity will overwhelm all remaining forces that might prop it up, and it will become a black hole.

From the standpoint of general relativity, since curved light rays demark curved space, a black hole is an object wrapped up in a kind of Fabergé egg made of space. Imagine once again that the sun is being compressed in a giant vise, and that we are watching this horrifying experiment from a vantage point on Earth. Just as Eddington observed in the eclipse expedition that first confirmed the general theory of relativity, light rays from distant stars are bent as they pass the sun. We repeat Eddington's observations, but this time no solar eclipse is required, as the sun's light is blocked out by the infernal vise. As the sun shrinks, we see stars that formerly were behind the solar disk. Their apparent position increasingly disagrees with that on the star charts, however, as the light rays are being bent ever more severely. By the time the sun becomes a black hole, the light rays have gone into orbit around it. The sun vanishes, to be replaced by a tiny circle of absolute blackness ringed by a glistening shimmer of warped starlight.

The resulting black hole is wrapped in an oblate zone, the *event horizon,* into which none can venture with any hope of return. Crossing the event horizon of a small black hole would be unpleasant: Tidal effects stretch you into the dimensions of a piece of yarn. The event horizon of a large black hole—one with, say, ten thousand times the sun's mass—is so large and gently curving that you could cross it as comfortably as if in a deck chair on an ocean liner. But you'd still be doomed, for inside the horizon all roads lead to the central singularity, where the curvature of spacetime becomes infinite. From the standpoint of an outside observer, time grinds to a halt at the event horizon. A suicidal astronaut falling into a black hole would notice nothing unusual when crossing the horizon, but well-wishers bidding him good-bye would watch his image grow ever redder and his gesticulations become ever slower until his image finally froze, then faded away. This effect is due to *gravitational redshift,* a displacement of light toward the red end of the spectrum as it loses energy in its attempt to climb out of the pit of curved space surrounding the black hole.

That's the mundane side of black holes. Now for the exotic:

The legendarily counterintuitive qualities of these strange objects crop up as soon as we ask even as straightforward a question as what a black hole *is*. From a quasi-classical or naïve-realist point of view, a black hole is what I've just been describing—a collapsed star or other object with an escape velocity exceeding that of light. But since time freezes at the event horizon, spacetime inside a black hole is denuded of its timelike aspect. The result is *spacetime foam,* a state in which space atomizes and time has no coherent direction. As the relativist Kip S. Thorne puts it, outside a black hole

> spacetime is like a piece of wood impregnated with water. In this analogy, the wood represents space, the water represents time, and the two (wood and water; space and time) are tightly interwoven, unified. The singularity and the laws of quantum gravity that rule it are like a fire into which the water-impregnated wood is thrown. The fire boils the water out of the wood, leaving the wood alone and vulnerable; in the singularity, the laws of quantum gravity destroy time, leaving space alone and vulnerable. The fire then converts the wood into a froth of flakes and ashes; the laws of quantum gravity then convert space into a random, probabilistic froth.[13]

This aspect of the theory cannot be tested. Were I to jump into a black hole—or, as we say in the academic world, were I to send a graduate student in—there would be no way for him to report back to us what he saw. The black hole is cut off from the rest of the universe.

Black holes are nonaggressive: They do not reach out and

swallow innocents at a distance. A black hole with a mass thirty times that of the sun is surrounded by a gravitational field identical in strength to that of a thirty-solar-mass star, and a planet orbiting it at a safe distance simply behaves as if it were orbiting such a star. But since we can neither see inside a black hole nor receive reports back from those who have ventured in, many relativists prefer to think of black holes as consisting of infinitely curved spacetime and nothing else. Hence the name. John Archibald Wheeler, who coined the term "black hole," refers to them as disembodied mass —mass without matter.[14]

The fact that we cannot peer into a black hole means that we cannot observe the singularity—the state of infinite spacetime curvature. To do so would be to behold a gravitational state that relativity cannot define, which to a relativist would be like contemplating a never-never land where one plus one equals three. With a relieved sense of the proprieties being preserved, the relativists say that the singularity is "clothed." They invoke the "cosmic censorship" hypothesis propounded by the English mathematician Roger Penrose, which asserts that there are no "naked" singularities. But some think that if an irregularly shaped object were to implode to black hole status, parts of it might protrude in such a way as to present the outer world with the galling spectacle of a naked singularity. Stephen Hawking, who collaborated with Penrose on research projects in relativity, has speculated that under certain conditions an evaporated black hole might leave a naked singularity behind. Nobody knows for certain whether this can happen, or what you would see if it did and you looked at it.

As an alternative to thinking of black holes as balls of spacetime and nothing else, Thorne and others have constructed what is called the "membrane paradigm." To consider this view, think back to our earlier discussion of the magnetic fields of neutron stars. Neutron stars spin rapidly because their rotation velocity increases as they contract, just as ice skaters spin faster when they pull in their arms. Owing to the superconductivity of the nuclear particles that a neutron star contains, it generates a powerful magnetic field. This field traps electrons and other charged particles, which it induces into rapidly orbiting the neutron star. The larger the orbit of a given bunch of particles, the faster their orbital velocity, and there comes a distance at which their velocity would have to exceed that of light. At this point, as Thorne and his colleague Richard Price write, "Since the particles cannot move faster than light, they resist the [magnetic field] lines' rotation by bending them backward and sliding outward along them at velocities just under the speed of light. The field lines thereby act as levers for transferring the energy of the star's rotation to outflowing plasma."[15]

A black hole, too, may be expected to be rapidly rotating and to have a magnetic field that leverages energy into the surrounding medium. Gas falling into a black hole would form a rapidly rotating disk. It's called an *accretion* disk, because the stuff in it is "accreting" onto the black hole. Its plane is perpendicular to the black hole's axis of rotation. Powerful electromagnetic fields generated by the black hole would turn the whirling disk into an enormous dynamo. (Einstein would have liked this picture; the son of a struggling dynamo maker, he based his special theory of relativity partly on his youthful studies of the electromagnetic fields inside dynamos.) The theorists' calculations indicate that a black hole in this environment would generate energy ten times more efficiently than do the thermonuclear processes that power the stars. Here lie insights into a possible use of black holes as power generators.

For these and other reasons, it is thought that black holes during a feeding frenzy could cause energetic events in their surroundings that would signal their presence to distant observers like ourselves. Material in the accretion disk, swirling toward the hole at great speed and with many attendant collisions among the doomed molecules, would glow white hot, emitting a spectrum of electromagnetic energy ranging all the way up to X rays and perhaps even gamma rays, the most energetic photons of all. The hot disk may also produce glowing jets of plasma. The theorists have come up with several ways the disk might do this. A favorite one, for those of us fond of tornados and violent storms, proposes that the accretion disk is thickened near its center by the intense heat. If we liken the disk to, say, a Kansas plain and the puffed-out parts to an atmosphere, we find that centrifugal force generated by the disk's rotation produces funnel clouds—tornados—above and below the disk. The tornados gather the surrounding atmosphere into thin jets so hot that they go streaking out into space at nearly the velocity of light.

These advertisements—a glowing accretion disk, an outpouring of X rays, a pair of protruding jets—would show up best where the attendant black hole is especially massive and is being fed by infalling matter. So in testing the theory, astronomers look for big black holes in environments where there is something to eat. Where might they find them? At the centers of young galaxies. It is at the cores of galaxies that stars are most densely distributed. The giants among these stars, having rapidly used up their fuel, should have imploded to form black holes early, certainly well before their host galaxy reached its billionth birthday. So the core may be expected to contain many black holes, situated close to one another. Some of these unavoidably would have collided and merged, making bigger black holes which in turn were even likelier to swallow stars and fellow black holes, and to feed on the interstellar gas thought to be abundant in young galaxies. The result would be a galactocentric black hole with a mass on the order of 100 million times that of the sun. When gas is falling into this behemoth, the accretion disk would glow white hot and send off X rays. It might also dispatch plasma jets hundreds of thousands of light-years in length. The jets would vary in brightness, putting on a show when material was falling into the hole and winking out when there was no food for the beast. Compared to this spectacle, the opening of a Hollywood movie is a well-kept secret. It cannot last, however. Over the course of time, as the host galaxy aged and the amount of ambient gas at the center was depleted, the giant central black hole would be deprived of fuel and the jets would disappear. The black hole remains, of course, waiting for the next feeding frenzy. Inherently patient, it can wait forever.

If this scenario is correct, we should expect to find that young galaxies often have bright, sputtering nuclei, while older galaxies generally do not. And that is just what we do find. The bright galactic nuclei are *quasars*.[16] Quasars are abundant at distances of billions of light-years, where we see galaxies as they were billions of years ago. They are uncommon nearby, where we see galaxies as they are today. (The nearest known quasar lies two billion light-years from Earth.) Frequently, quasars are found in galaxies that have been perturbed by interactions with neighboring galaxies: The interaction shakes up the host galaxy and sends some of the interstellar gas in its disk tumbling down into the galactic center, where it awakens the central black hole to renewed violent outbursts. It's icing on the cake that quasars often exhibit jets. Presumably there are more jets than are observed. Jets pointed more or less at Earth cannot be seen against the background quasar, so we see only those that happen to turn up in profile. Trundling into intergalactic space at nearly the speed of light, the jets bespeak an origin in events of enormous power. The inferred mass for galactocentric black holes ranges from about 100 million suns to a billion suns or more. To burn so brightly, the supermassive black holes thought to power quasars must squander a lot of fuel. It is unlikely that even a violently perturbed galaxy could come up with enough fuel to maintain a quasar's lifestyle for long periods of time. It seems more plausible that quasars glow during the relatively brief intervals when the black hole is wolfing down fuel—periods of about 50 million years—then gutter out until feeding time comes around again.

It should be possible to detect the presence of a supermassive black hole even when it is quiescent. Since stars are plentiful at the centers of galaxies, many will have been trapped in orbit around the black hole. Astronomers can determine the orbital velocities of the stars by taking their spectra. Spectral lines of moving objects are blurred, owing to Doppler shifts in their light: The lines of receding stars are shifted toward the red, and those of approaching ones toward the blue, to a degree directly related to their velocities. (Since the orbits are variously seen edge-on, side-on, and in between, one has to measure many stars and average out the velocities, assuming that the planes of their orbits are distributed at random.)

Such observations have produced evidence for the presence of supermassive black holes at the centers of a number of galaxies. Stars in the central 10 light-years of the Andromeda galaxy are arrayed in a disk, and orbit at speeds that suggest they are dancing around an object with a mass of 30 million to 70 million suns. A star or cluster that massive would be bright enough to be seen readily through our telescopes. But nothing is seen. The source of the gravitational force whirling those stars around is massive, small, and black.

Similar evidence for supermassive black holes has been

found in other galaxies. Markarian 315, which has a bright nucleus, evidently contains a black hole recently reactivated by gas dumped into it following a collision with another galaxy. The Sombrero galaxy appears to harbor a central black hole with a mass of almost a billion suns. The giant, violent elliptical galaxy M87, from which opposing jets spit out like bolts of lightning, is an excellent search site in that it is relatively free from interstellar gas, permitting a clear view of the core. A high-resolution study conducted in 1994 with the Hubble Space Telescope confirmed earlier results indicating that M87 has a central black hole with a mass of three billion suns. Velocity studies of the core of the galaxy NGC 4258 indicate that it contains a black hole as massive as 40 million suns. In one particularly spectacular observation, a Hubble image of the center of the galaxy NGC 4261 shows a doughnut of glowing gas and dust with a hellish-looking funnel of white-hot material jetting out of its center—the very signature of a black hole. As Mario Livio of the Space Telescope Science Institute in Baltimore reported at a 1995 meeting at Johns Hopkins University, "We can actually say that, yes, black holes have been observed."[17]

Our own galaxy may harbor a quiescent central black hole. The center of the Milky Way is obscured from our view in visible light by the vast shoals of interstellar gas and dust that lie along the 28,000 light-years of galactic disk between here and there. It can, however, be detected in the more penetrating wavelengths of radio and infrared light. These observations reveal the galactic center to be an exquisitely complex environment, exotic as a tropical coral reef. Its centerpiece is a chalice-like congress of rapidly orbiting stars, a pair of jets about 4 light-years long, a stable ring of orbiting gas clouds and another ring rushing outward, and an array of lacy arcs and threads standing 70 light-years tall. Dynamic studies indicate a mass of nearly a million suns within three-quarters of a light-year of the galactic center, of which only about a hundred thousand solar masses are seen, in the form of stars. The area seems to be enshrouded in a powerful electromagnetic field, like a dynamo. Although much remains to be understood, the picture generally fits well with the presence of a black hole at the center of our galaxy with a mass of a thousand to a million suns.

Just as there are many stars at the centers of galaxies, there may well be more than one massive black hole per galactic center. Black holes sliding down into the gravitational potential well at the galactic center could eventually collide. Relativity theory predicts that colliding black holes would emit gigantic bursts of gravitational waves—ripples in the fabric of spacetime that propagate at the velocity of light. The existence of gravitational waves has already been demonstrated: Joseph Taylor and Russell Hulse of Princeton University won a 1993 Nobel Prize for studies of binary neutron stars showing that the stars are approaching each other at just the rate that relativity predicts as a consequence of energy being carried away from the system by gravitational waves. Work is under way on a pair of gravitational-wave detectors to be built by the LIGO (Laser Interferometer Gravitational-Wave Observatory) project in the United States, and a third detector, VIRGO (named after the Virgo Cluster of galaxies), is to be constructed near Pisa, Galileo's hometown. These observatories will use laser beams to detect slight changes in the geometry of local space as it expands and contracts when gravitational waves go through. Kip Thorne predicts that "gravitational-wave detectors will soon bring us observational maps of black holes, and the symphonic sounds of black holes colliding —symphonies filled with rich, new information about how warped spacetime behaves when wildly vibrating. Supercomputer simulations will attempt to replicate the symphonies and tell us what they mean, and black holes thereby will become objects of detailed experimental scrutiny."[18]

For all the *Sturm und Drang* they can create in their surroundings, and for all the joys they provide as a playground for theoretical astrophysicists, black holes in some sense rank among the blandest macroscopic objects known to science. Normally a black hole reveals to outside observers only three things about itself —its mass, rotation, and electrical charge. Toss in anything you like—encyclopedias, nuclear submarines, whole faculties of social scientists—and the black hole, like a prisoner of war reciting only name, rank, and serial number, will tell you nothing more than its mass, its rotation, and its electrical charge. John Wheeler coined a slogan for black hole blandness. "A black hole has no hair," he said.[19]

Yet black holes may be hairier than had been thought—may

yield, that is, more information than just their mass, charge, and rotation. The basic reason for thinking so has to do with the law of entropy, which states that unless work is done to prevent it, all systems tend toward increasing disorder (meaning greater entropy) over time. If black holes have no hair, one could violate this law by dumping a disorderly system into a black hole, thus reducing the amount of disorder in the observable universe without doing enough work to have earned the right to do so. One might respond by saying, "Fine, black holes violate the entropy law, so what?" And this was the initial reaction of many leading theorists. But the law of entropy is deeply embedded in scientific thought, and efforts to write an exception for black holes tended to raise more problems than they solved. So researchers were left with the possibility that black holes actually do disgorge their hidden entropy, either at present or in the future. Yet nobody could figure out how they might do so. This point turns out to be very interesting, in that entropy is related to information. The less entropy there is in a given system, the more information is available; consequently the entropy issue is called "the information puzzle." It was first elucidated by Hawking, in 1976. To trace its development is to enter into one of the wonderlands of modern theoretical physics.

Wheeler prompted an early inquiry into black hole entropy. Talking one afternoon with his graduate student Jacob Bekenstein, Wheeler was musing about the subject of entropy. Entropy, again, means disorder. The term comes from thermodynamics, the science of heat, the triumph of a nineteenth-century physics formulated when the steam engine, thermodynamics in clanking action, represented the apex of high technology. Physicists in those days established that all work involves a certain irreducible increase in entropy. That's why perpetual-motion machines are impossible. For a machine to do work it must increase entropy. If energy is not imported, the increasing entropy will eventually grind the machine to a halt. It is because of entropy that automobiles need to be serviced and children's rooms to be straightened up. Reducing entropy takes work. Suppose you have two cups of tea, one hot and the other at room temperature. It's easy to mix them together but hard to separate them once they are mixed. Once mixed, the molecules from the cold tea will speed up, while those from the hot one slow down, until all share the same distribution of velocities—a state known as thermal equilibrium. Soon the cup of tea will subside to the ambient temperature of its environment. (This is the basis of a wisecrack made by the intellectually underrated comedian Buddy Hackett, who joked that he was trying to make a glass of water boil by using ESP and although he had not yet done so he had succeeded, after concentrating intently for a half hour or so, in bringing the water up to room temperature.) To reverse the process—to reclaim the molecules that were in the cup of hot tea only minutes ago—is virtually impossible.

More entropy means less information. This was demonstrated by Claude Shannon of Bell Laboratories, and one speaks today of "Shannon entropy" to indicate that one is talking about entropy in terms of information rather than heat. When we had two cups of tea, we had a system with two states. Presenting you with the two cups on a tray, I could say, "Take your pick: This cup is hot tea. The other one is cold." But once the tea has been mixed, the system has only one state: "Here's some tea." The statement of the second law of thermodynamics—that entropy increases unless energy is exerted to decrease it—therefore also means that information tends always to decrease.[20]

This sort of thing was on Wheeler's mind as he talked with Bekenstein. Wheeler said he was bothered by the notion that black holes provide a locker in which one can conceal violations of the law of increasing entropy. "I told him of the concern I always feel when a hot cup of tea exchanges energy with a cold cup of tea," Wheeler writes.[21] "By allowing that transfer of heat I do not alter the energy of the universe, but I do increase its microscopic disorder, its information loss, its entropy. The entropy of the world always increases in an irreversible process like that. The consequences of my crime, Jacob, echo down to the end of time. But if a black hole swims by, and I drop the teacups into it, I conceal from all the world the evidence of my crime."

Bekenstein pondered this conundrum, and came back a few days later with an idea of how to resolve it. "You don't destroy entropy when you drop those teacups into the black hole," he said, as Wheeler recalls their conversation. "The black hole already has entropy, and you only increase it!" Bekenstein proffered an exotic picture, in which black holes grow in size as their entropy increases. One way to think of Bekenstein's idea is to imagine that the horizon of a black hole is made of an enormous number of tiny tiles, fitted tightly together. Each tile represents either a zero or a one—that is, a single bit of information. (A *bit,* short for binary digit, is the smallest possible unit of data. Widely familiar today from computer science, it is used to quantify information in entropy theory.) When you throw a teacup or anything else into a black hole, you increase the black hole's entropy. This adds to the total number of imaginary tiles, which increases the circumference of the event horizon. The size of the event horizon is therefore an index to the amount of information the black hole has swallowed up.

Bekenstein's hypothesis intrigued Wheeler but aroused the ire of Stephen Hawking. Confined to a wheelchair and almost totally paralyzed, Hawking keeps technical papers laid out on tables around his office at Cambridge University, motoring from one to another like a chess champion playing a dozen opponents at once. An assistant comes in periodically and turns the pages. Hawking writes and converses by using his two functioning fingers to manipulate a toggle attached to a computer. His thinking, vividly conceptual, is anchored in strongly held views. He is a fiercely committed intellectual with little of the nice guy about him. "I've given up trying to be rigorous," he once told me. "All I'm concerned about is being right."[22]

Hawking could accept that the size of a black hole event horizon is *analogous* to its entropy. He had himself once worked on a similar idea. He had taken note of research indicating that the size of an event horizon tends to increase slightly over time. (The reason is that since space is never entirely empty, a little matter is always falling into any given black hole.) This, Hawking realized, was similar to the law of entropy. As hot tea is poured into cold tea, the entropy increases rapidly. Leave the system alone and the entropy keeps increasing, albeit ever more gradually. Never, so long as the teacup is left alone, does the entropy decrease—and never does the radius of a black hole event horizon decrease. The similarity was more than intuitive: As Hawking and his colleagues worked on the problem, they found that the laws of black holes and the laws of thermodynamics were functionally identical. Still, Hawking regarded the thermodynamics of black holes as no more than an analogy. Bekenstein wasn't talking in terms of analogies. He was claiming that black holes actually have real entropy. This galled Hawking. To speak literally of black hole thermodynamics was to say that black holes are warm. To be warm is to emit particles. Yet the central idea of black holes had always been that no particles could escape them.

"I was actually very much against this [Bekenstein's hypothesis]," Hawking recalled, in 1983.[23] "I believed that it was really only an analogy. The whole point was that if the black hole really did have an entropy, then it ought to have a temperature, and that would mean that it ought to emit particles. . . . If a black hole does not emit particles, then it cannot really be said to have an entropy or a temperature." Hawking thought *rotating* black holes might emit particles, in which case black hole thermodynamics might mean something literal. A rotating black hole should amplify certain waves, and this could be interpreted relativistically as particle emissions. But since nonrotating black holes do nothing of the sort, Hawking assumed that they could generate no heat.

Hawking is not easily countered, but Bekenstein stuck to his guns. It struck him as unseemly that the laws of thermodynamics could be violated—that entropy could be reduced—simply by throwing lukewarm tea and other entropic systems into a black hole. Nearly everybody else in the field, with the exception of Wheeler, agreed with Hawking. A black hole is virtually cut off from our universe. Throwing lukewarm tea into a black hole is like throwing it out of the universe. Why, therefore, fret about a minor violation of the laws of thermodynamics—which, after all, are statistical laws in the first place?

The controversy deepened in 1972, when Hawking, James Bardeen, and Brandon Carter formulated the laws of black hole mechanics and found them to be identical with the laws of thermodynamics. For instance, the second law—the law of increasing entropy—is customarily stated in words like "A system left to itself will tend toward a state of maximum entropy." As Kip Thorne puts it, "In any region of space, and at any moment of time (as measured in anyone's reference frame), measure the total entropy of everything there. Then wait however long you might wish, and again measure the total entropy. If nothing has moved out through the 'walls' of your region of space between the measurements, then the total entropy cannot have decreased, and it almost always will have increased, at least a little bit."[24] What Hawking and his coworkers had established was that there is an identical law for black holes, known as the area theorem. To state the area theorem, simply take Thorne's definition and for the word *entropy* substitute the words *horizon area*. But the three thought of the relationship as purely metaphorical, which is why in the title of their paper they used the term black hole *mechanics* and not black hole *thermodynamics*. They weren't ready to claim that black holes might actually emit particles.

All is clear in retrospect, and today we may say that the theorists' reluctance to realize that black holes are not so bald was based on an inadequate appreciation of the role quantum mechanics must ultimately play in the physics of space. Space in relativity is portrayed as a continuum, meaning that it is smooth, like silk, not grainy like a beach. Quantum mechanics abhors a continuum. It views everything as divided into discrete units, the quanta. Most relativists agreed, as Wheeler had often reminded them, that the ultimate theory of gravitation would marry relativity to quantum theory, producing a quantized spacetime—a spacetime foam. But no complete theory of quantum spacetime had then been written (nor has one been since, though the goal seems likely to be achieved within another decade or so). So in practice, nearly all black hole theorists worked with smooth, unquantized spacetime.

An exception was the Russian astrophysicist Yakov Borisovich Zeldovich. One night it occurred to him that a rotating black hole ought to emit radiation. Zeldovich liked to think in pictures and work out the mathematics afterward; he based this idea on an intuitive analogy with an ordinary rotating steel sphere. According to quantum mechanics such a sphere should radiate energy, owing to its interactions with quantum space. Zeldovich calculated that the same would apply to a black hole. If rotating, it would emit energy; the emissions would brake the black hole until it stopped rotating; thereafter it would cease to emit anything. A paper expounding this outlandish idea was published in 1971, having got past the referees mainly on the strength of Zeldovich's formidable reputation. It went widely ignored, a time bomb ticking quietly on the library shelves.

Another exception to the smooth-space school was Hawking. Although skeptical about black hole radiation, Hawking was interested in understanding the behavior of quantum fields surrounding black holes. He took note of the Zeldovich proposal but, displeased by the rather *ad hoc* mathematics in which it was couched, undertook to recalculate the problem. The results were not what he had expected. Zeldovich, Hawking concluded, was right—so far as he went. Rotating black holes do radiate, and the radiation can stop their rotation. That much was consistent with the general picture of black holes as permanently closed off from the rest of the universe.

But then—to his "horror," as he recalled the moment—Hawking found that his calculations offered no escape from the conclusion that a *non*-rotating black hole will also emit particles. Bekenstein was right. Black holes don't just behave as if they were warm—they really *are* warm. The black hole laws derived by Hawking, Bardeen, and Carter are not just analogous to thermodynamics, they *are* thermodynamics.

The phenomenological implications were no less provocative. As a black hole emits particles, it loses mass. Unless it is fed matter from the outside, its mass will continue to decrease until, eventually, it is no longer massive enough to keep spacetime wrapped around itself. At that point the black hole will explode.

This idea was sufficiently bizarre that Hawking put a question mark at the end of the title of the paper in which he announced it: "Black Hole Explosions?"[25] But the work has stood up, and today most theorists have learned to live with the extraordinary notion that black holes are not immortal, but are doomed by Hawking radiation to evaporate in a puff of X rays. (An exception is the class called *extreme* black holes, which, if they exist, are as small as subatomic particles.)

The discovery of Hawking radiation enjoys a stature among his works comparable, say, to that of the string quartet Opus 132 in the works of Beethoven. Hawking saw it as an illustration of nature's elegance. "The ideas really guide you in themselves, in the way that they all fit together," he said, years later. "Black hole thermodynamics is an example. It all fitted together so perfectly that it just had to be right. . . . Nature wouldn't have set up anything as elegant as that if it were wrong."[26]

Part of the force of Hawking's theory derives from the fact that it unites relativity, thermodynamics, and quantum mechanics, three distinct fields of physics never before combined in a single set of equations. Relativity gave us the conception of black holes. From thermodynamics came the hypothesis that black holes radiate energy. But it is with the invocation of quantum mechanics that Hawking's theory attains its remarkable result, and in the process snaps into focus as an archetype of a more unified physics of the future.

Central to quantum physics is the Heisenberg indeterminacy (or "uncertainty") principle, discovered by the German physicist Werner Heisenberg in 1927. Heisenberg indeterminacy reveals that subatomic particles do not have definitive positions in spacetime; rather, their locations can be specified only in terms of probabilities. The indeterminacy principle is often discussed as if it represented the difficulty of accurately *measuring* the locations and trajectories of particles. But the point is not that it is hard to find out just where, say, an electron is, but that the electron actually *has* no exact location. Depending on how it's measured, an electron can look specific as a pinpoint or vague as a cumulus cloud.

The effects of Heisenberg indeterminacy had been assumed to be negligible at the event horizon of a black hole, where far more energy is tied up in the gravity fluctuations occasioned by the steeply curved local spacetime than is to be found in the form of subatomic particles. But when Hawking looked at the situation in detail, he found that quantum effects actually count for enough to doom the black hole. According to relativity, particles just inside the horizon of a black hole cannot cross it. But if we bring quantum physics into the picture, we find that the position of a particle is inexact. The odds are that it will stay inside, but it just may quantum leap to a new position, *outside* the horizon. In that case it can escape into the outer universe.

To put the matter a bit more technically: In quantum physics, space is not empty but is roiling with "virtual" particles—ghostly particles that normally exist for but a moment, constantly boiling up out of the vacuum, splitting up, then recombining and subsiding away again. ("Created and annihilated, created and annihilated—what a waste of time," Richard Feynman used to say.[27]) In flat space, virtual particles cannot long survive: They exist by borrowing energy from the vacuum, and normally such energy is in short supply. But the steeply curved spacetime at the black hole event horizon contains enormous amounts of *tidal* energy. This tidal energy is gravitational in origin. It arises from the difference in gravitational force exerted between two points at differing distances from the black hole. Tidal gravity pulls apart pairs of virtual particles and pumps energy into them—so much energy that they can be promoted from virtual (i.e., short-lived) to "real" (meaning permanent) particles. When this happens, one of the newly promoted particles may fall into the black hole while the other escapes. As a result, the black hole is continually radiating particles, and thereby losing mass. Ultimately, unless fed fresh matter, it will evaporate.

Big black holes are cool—the horizon of a black hole weighing three solar masses has a Bekenstein-Hawking temperature of only two hundred millionths of a degree above absolute zero—and the larger the black hole, the cooler it is. So black holes formed from stars radiate very slowly. Consequently they last a long time. A black hole with a mass twice that of the sun will not explode until the universe is 10^{67} years old. But small black holes, if there are any, are hotter, so they don't last as long. (Black hole lifetimes are proportional to the cube of their mass.) If miniature black holes no more massive than hills or mountains had formed when our universe was young, they would have exploded by now. Each explosion would generate as much energy as a million one-megaton hydrogen bombs. Much of this energy would be in the form of gamma rays. Alerted to this possibility, astronomers equipped with gamma-ray telescopes searched for evidence of primordial black holes having exploded in the past. They found none.

"That's a great pity," Hawking said, after the search for primordial black holes had turned up none.

"Why?" he was asked.

"Because had they found them, I would have won the Nobel Prize."[28]

The study of black hole thermodynamics continues today, and points in promising directions. Recall that the concept of entropy is related to that of *information*. What becomes of the information contained in objects that are dropped into black holes? One possibility is that it's lost forever: If the sole manuscript of a first novel is dropped into a black hole, the novel simply vanishes from the universe. That is the outcome of Hawking's formulation.[29] But some theorists are uncomfortable with it. They note that according to the laws of thermodynamics, information may be degraded to the point that it becomes very difficult to restore, but they dislike the assertion that its restoration can be rendered genuinely impossible. Their distaste derives from the consideration that total information loss would violate a law of quantum dynamics, one that we need not delve into here but that is expressed in the evocative phrase "Pure states cannot evolve to mixed states." If we are, in the argot, to "preserve purity," there must be some way of recovering the information seemingly lost when a book is dropped into a black hole.

Two ways of preserving purity have been proposed. The first, put forth by the inventive Dutch researcher Gerard 't Hooft, holds that the information is contained in the particles that Hawking-radiate from the black hole. Deciphering the message would be difficult but perhaps not impossible. The other proposal is wonderfully wild. It postulates that a black hole, when it explodes, does not disappear entirely but leaves behind a remnant, a kind of ash. The remnant would consist of a single particle that the theorists have dubbed the *boltzmon*, after the nineteenth-century thermodynamicist Ludwig Boltzmann. A boltzmon would be tiny —about the size of the Planck-Wheeler area, which at 10^{-66} square centimeters is about as small as anything can be. Bundled inside it would be the sum total of all the information ever consumed by the black hole. Obliged to hold whole libraries in a minuscule volume, each boltzmon would be unique in the universe. Whereas a typical particle has a few states (positive or negative electrical charge, integral or fractional spin, etc.) a boltzmon would have an infinite number of states. As a result, it would be highly unstable. If perturbed, it might very well respond by making a hole in spacetime and vanishing into it, departing from our universe.

Leaving thermodynamics behind, we find that black holes have stimulated other fruitful insights into the potentialities of curved space.

One of these, a favorite among science fiction writers, is the concept of *wormholes.* To envision wormholes, let us first borrow a traditional (if rather unpleasant) popular science image and portray cosmic spacetime as a rubber sheet. The sheet is undulated, with depressions surrounding massive objects. Here and there we find infinitely deep depressions, bottomless pits like the "potholes to China" that motorists complain about: These bottomless pits represent black holes. Each has a fluted stem, like that of a vase meant to hold a single rose. Bend the rubber sheet so that the open, bottom ends of two such stems are joined together. The result is a wormhole—a tunnel connecting two distant points in space. Wormholes could provide the ultimate in efficient travel: A wormhole a mile long might connect two regions of space hundreds of light-years apart. If we could dive into the mouth of a wormhole (it would be a sphere—the three-dimensional equivalent of a porthole) and survive the trip, we would emerge from the other end to find ourselves in a remote sector of the universe, having traveled a vast distance in little or no time. Their putative ability to transport space travelers millions of light-years in an instant has made wormholes a feature of science fiction novels whose plots demand an intergalactic rapid-transit system.

In practice (so to speak), many difficulties would beset a wormhole cosmonaut. First, nobody knows where wormholes exist. Presumably they would have had to be formed in the infancy of the universe—it seems that far too much energy would be required to warp the rubber sheet nowadays—and to have survived ever since; it may well be that neither of these conditions was fulfilled in our universe. Second, a cosmonaut might not survive a wormhole passage. Tidal effects would tend to turn such an adventurer into a long, thin piece of human spaghetti—a forbiddingly unpleasant experience even for those who have flown coach on commercial airliners. Third, it's not clear that a wormhole, if one did pop up, would stay open long enough for anyone to jump into it: Quantum spacetime flux might instantly slam it shut.

Nevertheless, if black holes have taught science anything it is that one should not reject a promising idea solely because it leads to bizarre conclusions. So we may wish to be reticent about rejecting wormhole speculations out of hand, even though they point toward conclusions so strange as to make black holes seem familiar as old shoes.

How strange? Consider the Thorne conjecture.

In the mid-1980s, Kip Thorne proposed that it might be possible to hold wormholes open in such a way as to make them accommodating to travelers. Thorne's recipe was to thread the wormhole with "exotic" material. This is a theoretical form of matter that would, in the wormhole's frame of reference, have negative energy density. The negative energy density would act like antigravity, keeping the wormhole dilated. Exotic matter is not normally found in nature, but research of Hawking's suggests that it may be found on the quantum scale near the horizons of black holes. Thorne speculated that an advanced civilization might be able to gather or manufacture exotic material and use it to open up wormholes for travel. Conceivably, says Thorne, such a civilization might be able to manufacture wormholes from scratch, if they found that none had been left over from the big bang.

The Thorne conjecture raised the possibility that wormholes could be used as time machines, capable of transporting wormhole travelers not only across space, but through time, into the past.[30] This prospect aroused so much attention from the press that Thorne stopped using the phrase "time travel" in his papers, replacing it with a technical synonym, "closed timelike loops," which few journalists understood and few, therefore, wrote about. Traveling into the past of one's own universe violates causality and thus creates severe paradoxes. If you climbed into a wormhole in your living room and returned one minute earlier than you had departed, you would not only have created a copy of yourself but you could stop yourself from climbing into the wormhole, in which case the version of yourself that stopped yourself would not have shown up to intervene—in which case it would! This sort of thing undermines the foundations of science as well as common sense, and many physicists believe time travel to be literally impossible. Among them is Hawking, who has argued on technical grounds that any attempt to construct a time machine will be defeated by the interference of fluctuations in the quantum vacuum. Hawking also offers a pleasing nontechnical argument. He notes that if time travel were ever realized, it would be popular with sightseers and stock market speculators. That this will never happen is demonstrated, he notes, by "the fact that we have not been invaded by hordes of tourists from the future."[31]

There are, however, two arenas in which we can imagine time travel taking place without violating causality. Both involve locations separated from our universe.

One notion is that wormholes connect not one region of our universe to another but a place in our universe to a place in another universe. We will look at this prospect later in this book, when we consider the theory that our universe is one among many.

The other suggests that time travel can indeed take place—but only *inside* black holes. By the early 1990s, several relativists, among them J. Richard Gott III of Princeton University and Alan Guth of MIT, used the Thorne conjecture to conjure up spacetime geometries inside black holes that might whirl a captured cosmonaut into his own past. At a cosmological conference in Irvine, California, in 1992, Gott described one such scenario. "Imagine that you fall into a black hole," he said. "Hoping to survive as long as possible, you steer for a closed timelike loop. There you see, let's say, eleven images of yourself. The first one says to you, 'I've been around once.' The second says, 'I've been around twice,' and so forth. You whirl through the first loop and see yourself in the past, falling in from the black hole horizon. Wanting to be helpful to yourself, you call out, 'I've been around once.' Another loop and you're second in line. 'I've been around twice,' you shout."[32]

Underlying all these developments is a revolutionary change in the way space is portrayed by science. The new vision is arising like the building of an arch, with relativists approaching a unified theory from the large-scale, classical side while particle physicists approach it from the small-scale realm of quantum physics. As we have seen, the seamlessness of space as depicted by relativity evidently is valid only on the macroscopic scale. Theoretically, if we could examine space very closely, we would find that it resolves into the discontinuous, quantum structure called spacetime foam. Early in the big bang, during the *Planck epoch*—a brief but important instant, from time zero to only 10^{-43} second—spacetime foam would have played a dominant role in all events. During the Planck epoch every particle was so energetic (and therefore so massive, since energy equals mass) that its gravitational field, the warping of spacetime around each particle, was powerful enough to influence particle interactions as profoundly as did the normally much stronger electromagnetic and nuclear forces. Full-scale research into the Planck epoch awaits completion of a quantum theory of gravity.

Meanwhile, contributions toward a unified supersymmetry or superstring theory of particles is being made, from their side of the arch, by relativists studying what are called *extreme* black holes. As we will see later in this book, superstring theory depicts subatomic particles as composed of hyperdimensional space. In this scenario the universe began in a higher dimensional state—probably in ten dimensions—and the particles were created from curved space when the extra dimensions collapsed, somewhere near the outset of time. Intriguingly, a subatomic particle as depicted by superstring theory resembles a black hole: Each is made of infinitely curved space. Is it possible that particles in some sense *are* black holes? That's where extreme black holes come in.

Extreme black holes are defined as black holes whose electrical charge is exactly equal to their mass. ("Normal" black holes have more mass than charge. Black holes that had less mass than charge would present naked singularities and thus violate cosmic censorship, so they are thought not to exist.) Extreme black holes are tiny, the size of subatomic particles. They do not Hawking-radiate, so they are stable and don't evaporate. And they have "hair." Indeed they can exhibit a whole panoply of properties, just as one would expect of superstrings. (The Russian physicist Renata Kallosh, now at Stanford University, likes to say that extreme black holes "have superhair.") This abundance of information makes it possible to build a superstringlike theory of subatomic particles out of extreme black holes. As this work continues, with particle physicists brushing up on their general relativity and relativists delving into superstring theory, it is beginning to look as if the two communities may be able to complete an arch, the keystone of which is the long-awaited unified theory that would explain every fundamental interaction by a single set of equations. The point of the theory, one that would have gladdened Einstein's heart, is that everything is made of curved space.

4
Blast from the Past

Matter is energy.

—ALBERT EINSTEIN,
Special Theory of Relativity

Energy is eternal delight.

—WILLIAM BLAKE,
"The Marriage of Heaven and Hell"

ALL ROADS LEAD BACK in time to the big bang, where the chemical evolution of the cosmos began. Every scrap of matter and energy bears traces of its history, if only we can learn how to read it there. Cosmologists search this world for clues to events that happened long ago, though not just far away.

Begin with two facts, hard won by science, both now classroom commonplaces.

The first is that matter is frozen energy. This was revealed in what must be the world's most famous equation, $e = mc^2$: Energy equals mass times the velocity of light squared. The velocity of light is a large number, 300,000 kilometers per second, so this innocent-looking equation tells us that a little mass harbors a lot of energy. It explains, reassuringly, why a star like the sun can burn for billions of years, and, disturbingly, why a bomb the size of an orange can lay waste to a city.

The second fact is that matter is made of atoms. The mass of each atom is concentrated in its nucleus, which consists of protons and (except for simple hydrogen, whose nucleus is a single proton) neutrons. The total number of protons and neutrons in the nucleus determines the ***atomic weight*** of an atom. Protons carry a positive electrical charge; neutrons are electrically neutral. The positive electrical field generated by protons attracts electrons, which carry a negative electrical charge. The electrons snap into place in a series of shells surrounding the nucleus. Each shell has a maximum capacity of electrons, so adding electrons to a shell is rather like putting eggs into an egg carton. The capacity of the innermost shell is 2 electrons, the next shell holds 8 electrons, and the one after that 18.

From this diminutive architecture arises much of the structure of the large-scale world. The periodic table of the elements that hangs on the wall in chemistry classrooms is based on the atomic architecture of matter, as is chemistry itself. The order in which elements appear on the table is that of their atomic weight. The facility with which atoms combine to form molecules is determined by the number and the state of the electrons in the outermost shell. This is the basis of the "periodicities" after which the periodic table is named. Atoms that have vacancies in their outermost shell can combine readily with other atoms; simple hydrogen is such an atom, since it has only one electron and the innermost shell has room for two. Hydrogen consequently is chemically volatile: It combines with, for instance, oxygen, producing what the chemists call rapid oxidation and the rest of us call fire. Ordinary helium, on the other hand, has two protons in the nucleus and so normally carries two electrons, filling the first shell and rendering helium atoms chemically inert. The explosion of the *Hindenburg*—which burst into flames while touching down at Lakehurst, New Jersey, in 1937, resulting in an industry-wide replacement of hydrogen-filled dirigibles by helium blimps—is at the root a tale of the one-electron difference between the valences of hydrogen and helium atoms.

Making matters more elaborate are the chemical *isotopes.* An isotope (from the Greek for "same place") of a given element is an atom with the same number of protons but differing numbers of neutrons. Add a neutron to the one-proton nucleus of a hydrogen atom and you have deuterium. Add a second neutron and you have tritium. The additional neutrons increase the weight of the atom without altering its electrical charge. Isotopes are designated by their weight, measured (arbitrarily) relative to that of the carbon-12 atom. Hence the weight of helium-3 is one-quarter that of carbon-12, and helium-4 weighs a third as much as carbon-12. Some isotopes, like deuterium, are stable. Left alone, they survive indefinitely. Others, like tritium, are unstable (that is, *radioactive*). They decay, emitting particles and losing mass as a result. The moment that a given atom will decay is ruled by quantum uncertainty and cannot therefore be predicted, but as always in quantum physics the uncertainties average out in the aggregate. So the behavior of enormous numbers of atoms lumped together—comprising, say, a tenth of a gram of tritium—is eminently predictable. The half-life of tritium, meaning the time it takes for half the atoms in any given sample of tritium to decay, is 12.3 years.

The point of this disquisition is that neither of these important facts—that matter is frozen energy and that it takes the form of atoms—makes much sense until we take cosmic evolution into account. All things are products of cosmic history. Trivially, that means scientists can explain the world in historical terms, demonstrating that things are as they are because they were as they were. More significantly, it means that the things themselves can be interrogated for information about how the universe evolved. This is true not just of big, astronomical objects like planets and stars but of small, unprepossessing objects all around us. We really can "see a World in a Grain of Sand," as William Blake put it.[1] Literally so: Sand is largely silicon, and the study of silicon continues to absorb astrophysicists interested in such topics as the origin of the solar system and the role of silicon burning in stars.

Our subject in this chapter is the origin of the atoms, and our story is divided into two parts. Part one begins when the universe was one second old, and ends when the universe was about two minutes old. During that brief interval most of the helium and all the deuterium in the universe formed. Part two is concerned with the origin of the heavier elements, which takes place inside stars and continues to this day.

Why is matter frozen energy? Because the universe began in a high-energy state and has been cooling off ever since: Hot energy congealed into cold matter. Why has it cooled? Because the universe is expanding: The same amount of energy in an increasing volume of space means a lower overall energy level. Matter is *frozen* energy.

Why is matter made of atoms? For the light elements deuterium, helium, and lithium, the answer is that during the first minutes of cosmic expansion the ambient energy level fell below that of the strong nuclear force, permitting protons and neutrons to bond together as atomic nuclei. The heavier elements, however, were built inside stars, where light nuclei are fused together to make heavier nuclei. (Nuclear fusion releases energy. That's why stars shine.) Some of the heavier nuclei were then blasted into space when their host stars exploded. Others exited more pacifically, freighted out when their stars grew unstable and shed their atmospheres, or vented as soot and stellar wind.

To heat up matter is to return it to conditions that pertained earlier in cosmic history. Put a scoop of sand in an oven and heat it to 3,000 degrees Kelvin and the electrons will be stripped from the atomic shells. The resulting cloud of naked nuclei and free electrons is called *plasma,* the fourth state of matter (after solid, liquid, and gas). Stars are plasma. So are the jets emerging from the cores of certain violent galaxies. The whole universe was in a plasma state until it was a few hundred thousand years old, at which point it had cooled sufficiently to enable electrons to stick in their shells, forming complete atoms. (It was this event that rendered the plasma transparent to light, liberating the photons of the cosmic microwave background radiation.) Turn up the heat in the oven and you overwhelm the nuclear binding energy. The protons and neutrons in atomic nuclei cannot stay together: You've wrecked basic atomic structure, and resurrected the state of the universe when it was a couple of minutes old. Comparable energy levels are attained in giant particle colliders, which for that reason are sometimes called time machines.

The colliders can go even further, splitting protons and neutrons into the quarks that comprise them. This happens at a temperature of one hundred thousand billion degrees—ten giga-electron-volts, abbreviated 10 GeV. A proton is made of two up quarks and one down quark. A neutron is made of two down quarks and one up quark. We shall have more to say about quarks later on, when we find them, as manifestations of a symmetry group, informative on the role played by symmetry-breaking events in the early universe. For now, the important thing is to understand that matter is mutable—that its current, relatively unchanging state results from encountering it late in cosmic history—and to appreciate that matter is mostly space. A typical atom (I'm thinking here of sodium) is one ten-millionth of a centimeter in diameter. Its nucleus is a hundred thousand times smaller than that. If the nucleus were the size of a golf ball, the outermost electrons would lie *two miles* away. Atoms, like galaxies, are cathedrals of cavernous space. What feels solid about a tabletop is that the electromagnetic fields set up by atoms in the table repel similar fields in your fist. Matter *is* energy. As Werner Heisenberg remarked, in a talk shortly before his death in 1976:

> We ask, "What does a proton consist of? Can an electron be divided or is it indivisible? Is a photon simple or compound?" But all these questions are wrongly put, because words such as "divide" or "consist of" have to a large extent lost their meaning. It must be our task to adapt our thinking and speaking—indeed our scientific philosophy—to the new situation created by the experimental evidence. Unfortunately this is very difficult. Wrong questions and wrong pictures creep automatically into particle physics and lead to developments that do not fit the real situation in nature.[2]

As we proceed to discuss how atoms were constructed in the big bang and in the hearts of stars, we should thus be mindful of the limitations of language and of the received opinions that language embodies. The universe is befuddling not just because it's big but because it challenges our preconceptions. Gathering stars by the bucketful enlarges the bucket, then sets it ablaze.

We start with a bit of human history—a thin slice, culled from the long story of how humans finally learned, in the twentieth century, how stars shine and where atoms come from. A tangled tale, it illustrates that wrongheaded ideas and ill-fated theories can lead, however fitfully, to genuine advances in knowledge.

Edwin Hubble's discovery of the expansion of the universe prompted theorists to consider that if the cosmic matter density is decreasing, then there must have been a time long ago when everything in the universe was as hot and dense as the center of a star—and perhaps even hotter and denser than that. The Belgian cosmologist Georges Lemaître dubbed this original state the "primordial atom," and wondered whether the universe might have begun expanding through a process roughly analogous to the radioactive decay of an unstable atomic nucleus. Lecturing in the early 1930s, in the library of the Mt. Wilson observatory offices in Pasadena to an audience that included Albert Einstein, Lemaître declared: "In the beginning of everything we had fireworks of unimaginable beauty. Then there was the explosion followed by the filling of the heavens with smoke. We come too late to do more than visualize the splendor of creation's birthday."[3] Lemaître's account of genesis was long on oratory and short on specifics—and where it did get specific it was wrong—but Einstein, who understood that style can count for as much as substance, arose at the end of the talk and called it "the most beautiful and satisfying interpretation I have listened to."[4] Neither Lemaître as a physicist nor physics as a discipline was yet up to the task of analyzing the big bang. But by viewing the early universe through the lens of nuclear physics, Lemaître inaugurated what was to become a potent collaboration between cosmologists interested in cosmic evolution and high-energy physicists capable of calculating thermonuclear events in the early universe.

In the late 1940s—by which time nuclear physics had come a long way, owing in part to work done in the Manhattan project on nuclear weapons—the physics of the early universe captured the attention of the protean theorist George Gamow. Born in Odessa in 1904, Gamow worked with Ernest Rutherford in Cambridge, where he laid the theoretical groundwork for the artificial transmutation of elements. (This was achieved in 1932, realizing at last the dreams of the alchemists.) He defected from the Soviet Union in 1933 and emigrated to the United States, where he impressed his American colleagues as a joke-cracking artist of science. He might not concern himself overmuch with where the decimal place belonged, but he had such sound instincts that the mathematicians who cleaned up after him often remarked on the "luck" that made his arithmetical errors cancel out. (To poke fun at his own imprecision he once published a paper that contained an enormous and deliberate error, then submitted an erratum, prepared in advance, advising that one of its equations was off by a factor of 10^{24}—a million billion billion—which, he reassured readers of the journal, "does not affect the result."[5]) Gamow had an extraordinary ability to see to the heart of a question, and by 1948 he had realized that an expanding, big bang universe is evolutionary and that atoms can testify to its history. As he wrote in a *Nature* essay published that year:

> The discovery of the redshift in the spectra of distant stellar galaxies revealed the important fact that our universe is in the state of uniform expansion, and raised an interesting question as to whether the present features of the universe could be understood as the result of its evolutionary development. . . . We conclude first of all that the relative abundances of various atomic species (which were found to be essentially the same all over the observed region of the universe) must represent the most ancient archaeological document pertaining to the history of the universe.[6]

Gamow's work on the early universe incorporated a big right idea and an equally big wrong one. His right idea was that elements were forged in the fires of the big bang. His wrong idea was that *all* the elements were forged there. "These abundances," he wrote, referring to the amounts of various chemical elements found in the universe today (lots of hydrogen, not much gold; that sort of thing), "must have been established during the earliest stages of expansion when the temperature of the primordial matter was still sufficiently high to permit nuclear transformations to run through the entire range of chemical elements."[7] This was an elegantly efficient idea, but wrong. Deuterium, helium, some lithium, and traces of beryllium and boron could have formed in the big bang, but not the heavier elements. The problem is that there are no stable atomic nuclei of the atomic weights 5 and 8. One can imagine deuterium and helium being built in the early universe, by the "capture" of neutrons (i.e., their adhesion to protons), but when the process gets up to atomic weight 5, and again at 8, the nucleus promptly disintegrates, and therefore cannot be employed as the foundation for building any heavier nuclei. Gamow's hopes of finding a way across the gaps were diminished when in 1950 the Italian physicist Enrico Fermi reviewed every conceivable reaction mechanism and concluded that the big bang could have made almost nothing heavier than lithium.

Despite this failure to reach its original goal, Gamow's work was far ahead of its time, and it produced one of the most pregnant predictions in all cosmology: It implied that the universe should be suffused with the photons that make up what we today call the cosmic microwave background radiation. In 1948 Gamow's colleagues Ralph Alpher and Robert Herman calculated the photon flux and concluded that "the temperature in the universe at the present time is found to be about 5 degrees K[elvin]."[8] This was remarkably close to the modern value, measured with the Cosmic Background Explorer (COBE) satellite, of 2.726 Kelvin.[9] And as we have seen, the existence, spectrum, and structure of the microwave background constitute formidable proofs of the big bang theory.

While Gamow and his colleagues were studying the physics of the big bang, other researchers were looking for ways to get rid of the big bang theory altogether. Their motives were in large part scientific: The big bang theory circa 1950 suffered from serious flaws, notably the embarrassing fact that the Hubble constant then estimated yielded an age for the universe less than that arrived at by the geophysicists for the earth. There were also aesthetic reasons for disliking the big bang: Its association with thermonuclear bombs made it seem ugly and abrupt, and it implied an origin to the universe that remained shrouded in mystery if not mysticism. Beginning in 1946, the perceived deficiencies of the big bang scenario were being discussed by the English physicists Fred Hoyle and Thomas Gold with a Viennese colleague, Hermann Bondi, whom Gold had met while Bondi, though classified as an "enemy alien," was working in the British Admiralty during the war. Together the three came up with an alternative, the *steady state* or C-field theory. The steady state theory proposed that matter leaches into existence from out of the vacuum of space. A steady state universe can expand and yet be infinitely old, the newborn matter compensating for the matter density lost through cosmic expansion. To the criticism that a C-field creation of matter seems rather mysterious, the steady state advocates could and did riposte that it was no more mysterious than the alleged origin of all matter at the beginning of time in the big bang model.

All three authors of the original steady state theory went on to have distinguished careers, but Hoyle is the best known today. A gifted popularizer who frequently appears on radio and television programs, Sir Fred combines an admirable grasp of mathematics and theoretical physics with an unusually wide range of scientific interests and a willingness to stump for unpopular ideas.[10] (In recent years he has proposed that the viruses that cause flu epidemics come from space, a notion most epidemiologists dismiss.) In his autobiography, published in 1994, Hoyle recalls that what most offended him about the big bang theory was that it violates the assumption, made by Einstein in general relativity, that the laws of physics hold up in every quarter of spacetime. This assumption is retained in the steady state theory but not in the big bang, which suggests that near the beginning of time the curvature of space was infinite and general relativity therefore inapplicable. Many other laws of physics may not have held sway during the first microsecond of time. It is possible that the laws were dictated by chance, in random "symmetry-breaking" events that occurred during the first moments of cosmic expansion. This prospect Sir Fred refers to with distaste, as "the crude breaking of the physical laws that occurs in big-bang cosmology."[11]

But Hoyle is more than a gadfly. He has made major contributions to cosmology. Nothing illustrates this more clearly than the story of how his pursuit of the steady state theory helped lead to a realized account of the origin of the heavier elements, in stars.

Since Hoyle rejected the big bang, he was obliged to look elsewhere for furnaces where atoms might have been cooked up. Stars were the obvious candidates. A number of nuclear physicists had been piecing together how nuclear fusion—the combining of lighter nuclei to produce heavier ones—produces the energy that powers the stars. Among them were Robert d'Escourt Atkinson, Carl Friedrich von Weizsäcker, Hans Bethe—and Gamow, who in a major step had shown that the uncertainty principle permits protons to make quantum leaps across the force fields set up by their mutual electrical repulsion and thus fuse at higher rates than classical physics could have envisioned.

There are, it turns out, a variety of nuclear chain reactions at work in stars. One important reaction in the sun and similar stars is the *proton-proton* process: Two protons fuse, forming a deuterium nucleus (one proton plus one neutron) and releasing a positron and a neutrino. The neutrino escapes into space; the positron usually hits an electron and annihilates it, releasing a gamma ray. The deuterium then fuses with hydrogen, forming an isotope of helium, helium-3, and releasing more gamma rays. Fusion of two helium-3 nuclei results in a nucleus of helium-4 plus two protons. The end product, two hydrogen nuclei and one helium-4 nucleus, weighs seven-tenths of one percent less than did the original nuclei. That's the fraction of mass that has been converted into energy. The sun, which is thought to produce 98 percent of its energy via the proton-proton reaction, turns 600 million tons of hydrogen into helium every second in this fashion, converting 4 million tons of it into energy. Another reaction, responsible for only about 2 percent of the sun's energy but more important in hotter stars, is the *carbon-nitrogen* cycle. Here carbon acts as a catalyst in a reaction that forges four hydrogen nuclei into a helium isotope, leaving behind as much carbon as there was to begin with but releasing energy. In still hotter stars—giants whose cores can sustain temperatures of above 100 million degrees Kelvin—the *triple alpha* process takes place. Named for helium nuclei, which physicists for historical reasons also call "alpha" particles, triple alpha builds helium into beryllium and carbon.

Hoyle, in the early to mid-1950s, did important work on stellar fusion reactions, producing among other things the first full account of how stars make elements heavier than carbon. Then, in 1957, he published—along with the ebullient Caltech physicist William Fowler and the husband-and-wife team of astronomers E. Margaret Burbidge and Geoffrey Burbidge—a landmark paper that detailed eight fusion processes through which stars turn light elements into heavy ones. These elements, they noted, are recycled into the interstellar medium through stellar winds, ejection of the outer envelopes of red giant stars, and supernovae.[12] The central argument of the paper, that *all* the elements were made in stars—an assertion dear to the heart of Hoyle, since he didn't believe in the big bang—has borne up poorly under observational test. It predicts a cosmic helium abundance of only 1 to 4 percent, rather than the observed value of about 25 percent, and it cannot explain the cosmic deuterium abundance at all, since deuterium is destroyed in stars and not produced there.

Nevertheless the Burbidge, Burbidge, Fowler, and Hoyle paper, when combined with the work begun by Gamow on the big bang, lays out the whole basis for our understanding of the cosmic evolution of the elements. So what we have here is an illustration of fate's penchant for irony. Hoyle and his colleagues, out to undermine the big bang theory by showing that the elements were all made in stars, were unable to do so for deuterium and helium but established that the heavier elements are made that way. Gamow failed to make all the elements in the big bang, but found that the big bang did make the light elements deuterium, helium, and some lithium. Through this free-form dance, the essentials of what Hoyle called "the history of matter" were brought into the realm of human comprehension.

Let's leave history behind, and turn to the current status of *big bang nucleosynthesis,* or BBN, as the study of element production in the early universe is known today. One might think that piecing together how atomic nuclei were fused in the hellish conditions of the big bang would be a formidable task. But the physicists involved don't regard BBN as technically all that challenging. The relevant temperatures were on the order of 10^{10} Kelvin, and the densities less than that of water—conditions in which the possible nuclear reactions are less complicated than they are inside a star. The basic calculations were worked out by about 1950, adeptly enough that Gamow was able to estimate the cosmic helium abundance at almost exactly the modern value. BBN is now by cosmological standards a mature subject. As Gary Steigman of Ohio State University puts it, "Primordial alchemy is conventional physics."[13]

The main characters in the BBN scenario are *baryons, leptons,* and *bosons.* Baryons (from the Greek for "heavy") are the hefty particles, among them protons and neutrons, that constitute most of the mass of the ordinary matter we see and touch and of which we are made. Leptons (Greek for "light") include electrons, which have low mass, and neutrinos, which have zero mass or nearly so. Bosons are force-carrying particles: The bosons that most concern us here are photons, the carriers of light. Photons are insubstantial—they would have zero mass if they ever came to rest, which they don't—but there are plenty of them, a billion photons for each baryon, and the short-wavelength photons found in a high-energy environment like the big bang can pack a punch. (Shorter wavelength means more energy, as you can see if you heat up an iron bar and watch its color mount from red to yellow to blue. That is why ultraviolet light, which is shorter in wavelength than visible light, can cause a nasty sunburn, and why X rays, which are even shorter-wavelength photons, can pass through flesh before colliding with bone.) The role of photons in smashing nuclei, at a declining rate as the universe expands and cools, is a central theme in big bang nucleosynthesis.

As noted at the start of this chapter, the epoch of nucleosynthesis took only a couple of minutes.[14] ("The elements were cooked in less time than it takes to cook a dish of duck and roast potatoes," as Gamow remarked.[15]) Our narrative begins when the universe was one second old. During the first second, atomic nuclei could not exist: Any that formed were instantly blasted apart by the high flux of photons. Also during the first second, protons were constantly turning into neutrons, and vice versa, owing to interactions mediated by the weak nuclear force, the force responsible for radioactive decay.

After one second, the rate of weak interactions became slower than the cosmic expansion rate, at which point the ratio of protons to neutrons "froze out," remaining fixed ever after: There are more protons than neutrons, as neutrons are more massive and so required more energy to construct. Had that been the end of the story, the neutrons would have met an ignoble fate. Left to themselves, free neutrons decay, each becoming a proton, an electron, and an antineutrino, in an average of fourteen minutes and forty-nine seconds. Fortunately for the future of atoms, protons and neutrons continued to be slammed together at high velocities in the heat of the big bang, while the level of destructive interference from photons soon dropped below that of the strong nuclear force. The strong force was then free to bind protons and neutrons together, drawing neutrons into the sanctuary of atomic nuclei, in which environment they cease to decay. This capture process began after the first second and was operating in earnest by about ten seconds. Fusing a proton with a neutron makes a deuterium nucleus. Fusing two deuterium nuclei makes helium-4, and such was the fate of most of the deuterium. Also produced were traces of tritium (one proton and two neutrons) and lithium-7 (three protons, four neutrons). Virtually every neutron was incorporated into helium, mostly helium-4. Parceling out the neutrons and noting the atomic weight of the resulting atomic nuclei, the physicists find that about 25 percent of the mass of the universe emerged from the big bang as helium, while about 0.001 percent wound up as deuterium and even less as lithium.

Tested in a wide variety of observations, this finding has fared well. Confirmation of the standard BBN element-abundance prediction has come from studies of young and old stars; planetary nebulae (which are shells of gas ejected by unstable stars); the glowing gas clouds called HII regions; lunar soil samples that the Apollo astronauts brought to Earth; and particles of solar wind gathered by sheets of aluminum foil that they deployed on the surface of the Moon. The situation is complicated, however, by the fact that most of the stuff in our galaxy and others like it has already been processed through stars once or twice, so that one must make reasonable assumptions about stellar evolution before interpreting what the observations can tell us about the origin of the light elements. Consider helium. Clearly there is much more helium in the universe than could have been made in stars. But since many stars do make helium, their contribution to cosmic evolution must be taken into account before one can say how much of the observed helium dates back to the big bang. The situation with regard to deuterium is similarly subtle. Deuterium nuclei are weakly bound —they have the lowest binding energy of any stable nucleus—so they are readily destroyed in stars. Therefore observations of, say, interstellar clouds today can provide only a lower limit on the primordial deuterium abundance. Lithium, too, can be destroyed in the cores of stars. Lithium has been detected on the surface of old stars, presumably because these stars have not processed much if any of their outer material through their cores. Lithium in greater abundance has been found on the surface of brown dwarfs—stars so low in mass that little or no fusion has taken place in them. (It was by looking for lithium that in 1995 a team led by the astronomer Gibor Basri of Berkeley, using the 10-meter Keck telescope atop Mauna Kea on the island of Hawaii, made the first observation of a brown dwarf star.)

All atoms have a post–big bang history, of course. One way to learn their stories is to look for objects whose histories have been relatively uneventful—like atoms in intergalactic clouds, which if they have always been intergalactic may never have been cycled through stars.[16] Another way is to study distant objects, which are seen as they were when cosmic evolution was less advanced than it is today. One can combine the two approaches by looking for intergalactic clouds at cosmologically significant distances. This is done by taking spectra of distant quasars and searching out *absorption lines,* the spectral signatures of elements in the intervening clouds. The process can be maddeningly difficult: One is looking along a very long line of sight, through many clouds of differing redshifts generating a jumble of spectral lines. (Astronomers call it the "Lyman alpha forest," after a strong line in the hydrogen spectrum that they look for in starting to decipher the redshift of individual clouds.[17])

Trying as it is, after many years the search for distant intergalactic clouds has begun to yield results. Observations conducted in 1994 and 1995 with the Keck telescope indicated that the universe, when one-third its present age, had a deuterium abundance ten times greater than it does today, much as predicted by the standard BBN theory.[18] A particularly dramatic confirmation of the theory came in March 1995, when astronomers using an ultraviolet telescope carried aboard the space shuttle detected helium in intergalactic clouds 10 billion light-years away. The helium abundance approximates the primordial level estimated by the BBN model. "This really shows the helium is primordial," remarked the physicist David Schramm, of the University of Chicago. "This is the definitive measure. It's there. And it's consistent with predictions of the standard model of the big bang theory."[19]

Reassured by the concordance of BBN theory and observation, one can think of the big bang as the ultimate high-energy physics experiment, the outcome of which might reveal things as yet unknown about what the universe is like today. Neutrinos provide an excellent example. Neutrinos are leptons that ignore the electromagnetic and strong nuclear forces, interacting only weakly with ordinary matter. They are of three types, or "families"—the *electron* neutrino, which is produced along with an electron in the weak-interaction event known as beta decay, and the *muon* and *tau* neutrinos, which result from decay events that produce muons and tau particles, heavier partners of the electron.

In the 1970s, when many new sorts of particles were being found in high-energy physics experiments—these were the days when particle physics was being compared to Ptolemaic cosmology and the proliferating particles to the unsavory practice of adding Ptolemaic epicycles to "save the appearances"—there was widespread speculation that additional families of neutrinos might exist. Big bang nucleosynthesis, however, "constrained" the neutrino population: If the BBN scenario was right, theorists noted, the three known neutrino families should be all there are.[20] The BBN account was thus left admirably vulnerable to experimental disproof: Finding a fourth neutrino family would have left it in shambles. So far it has survived. The latest results from collider experiments at CERN (European Organization for Nuclear Research) yield 2.987 neutrino families, plus or minus 0.016. (The number of families is defined in terms of reaction rates and need not be an integer.)

The most intriguing BBN "prediction" has to do with the cosmic matter density—that is, with the value of omega and its implications for the shape and destiny of the expanding universe. The production rate of helium-4 is highly sensitive to the density of matter in the big bang—that's how the number of neutrino families was constrained—and the production of deuterium and helium-3 is sensitive to the density of baryons. Baryons, we recall, are particles of "ordinary" matter, like protons and neutrons.[21] The higher the baryon density in the big bang, the more efficiently helium atoms are made and the less deuterium results. So the BBN equations can be used to estimate how much of the matter in the universe is baryonic—that is, how much of it is made of ordinary matter.

A startling conclusion results: *Most of the matter in the universe is not made of baryons at all!* The exact proportion of baryonic to nonbaryonic matter depends on several uncertain parameters, among them the cosmic expansion rate. But in no reasonable formulation does the baryonic mass of the universe amount to more than 10 percent of the total mass. And if we live in a critical density (i.e., omega = 1) universe, as many theorists believe and as this book argues, then as much as 99 percent of the mass in the universe is nonbaryonic. In other words, if ours is an omega = 1 universe in which light elements were synthesized in the big bang, then all the planets and stars and galaxies we see, all the billions of objects in the astronomical catalogues, constitute but 1 percent of the matter in the cosmos. We are led to conclude that most of the universe is made of dark, nonbaryonic stuff, and that all we have studied to date is but a kind of shadow universe.

This is the dark matter riddle, the subject of our next chapter.

5
The Black Taj

Because all things balance—as on a wheel—
and we cannot see nine-tenths of what is real,
our claims of self-reliance are pieced together
by unpanned gold. The whole system is a game:
the planets are the shells; our earth, the pea.
May there be no moaning of the bar.
Like ships at sunset in a reverie,
we are shadows of what we are.

—F. D. REEVE[1]

Greet the unseen with a cheer!

—ROBERT BROWNING[2]

VISITORS TO the Taj Mahal are told that Shah Jahan, the emperor who in the seventeenth century built the white marble tomb for his wife, Mumtaz Mahal, had intended to erect a second mausoleum, for himself, on the opposite side of the Yamuna River. It was to be identical with the Taj Mahal, but black. Although there is no evidence to support this tale, the myth of the black Taj endures, touching the imagination of all who have heard it.[3]

Imagine what the public reaction would be if scientists were to find evidence that the black Taj actually exists, that it is sitting there just across the river from the white Taj, invisible yet detectable through its gravitational field—and that it is ten to a hundred times as massive as the visible Taj and is made in part of an unknown form of matter. Such a finding would make headlines throughout the world.

Dark matter is the black Taj of modern cosmology. During the past twenty years, theory and observation alike have indicated that at least ninety percent, and perhaps as much as ninety-*nine* percent, of the mass of the universe is dark. It is invisible not because it is far away—most astronomers assume that it is distributed more or less coextensively with the visible galaxies—but because it neither emits nor absorbs light. At least some of the dark matter is made of familiar stuff, but most of it may be of an exotic nature as yet unknown. Until it can be identified, cosmologists will continue to be confronted with the galling thought that the millions of galaxies they have studied are a tiny minority, a sample that may not be representative of the whole. If the dark majority of cosmic mass is different in kind or behavior from the bright minority, then scientific conclusions that ignore the majority may be as distorted as, say, a poll that attempted to predict the outcome of a national election by interviewing only vegetarian monarchists.

The strongest evidence for the existence of dark matter comes from studying the orbital motions of stars and galaxies. We saw earlier how Newton's equations make it possible to determine the mass of the sun by measuring the orbital velocity of any planet whose distance from the sun is known. Since the gravitational force exerted on planets diminishes by the square of their distance from the sun, the inner planets move more rapidly in orbit than do those farther out: The earth, for instance, hustles along at a rate of 30 kilometers per second, while Jupiter, which is five times farther from the sun, proceeds at a stately 13 kilometers per second. This falloff in orbital velocity with increasing orbital size is called *Keplerian,* after Kepler's third law. It is characteristic of all orbital systems where the mass is concentrated inside the orbit, as with planets in the solar system. The sun, which contains 99 percent of the mass of the solar system, could balloon into a red giant or collapse to dwarf size without affecting the orbital velocities of the outer planets: They would continue to behave in Keplerian fashion as long as the sun's mass remained inside their orbits.

But suppose that the sun were surrounded by a heavy halo —a cloud of invisible stuff, massive as the sun and stretching out past the orbit of Pluto. We would now find that the orbital velocities of the planets no longer obeyed a Keplerian distribution. Instead, the outer planets would move as rapidly as the inner ones. The reason is that each planet "feels" the gravitational force exerted by all the mass encompassed within its orbit, as if that force were radiating from a point at the center. The larger the planet's orbit, the more halo mass it encompasses. Therefore Jupiter, though it lies in a weaker part of the *sun's* gravitational field than does Earth, would respond to the stronger gravitational field generated by all the halo mass within its gigantic orbit, and so could move along in its orbit as fast as Earth.

The solar system has no such massive halo, but many spiral galaxies evidently do. Astronomers find that stars located near the edge of the visible disk of such a galaxy are orbiting at velocities comparable with those at the inner edge. Barring some unknown variable in Newtonian gravitation—and one wants to think long and hard before breaking the glass and pulling *that* fire alarm—these galaxies must be embedded in massive halos. If the halo material emitted or absorbed light it could be detected with existing instruments. It is not detected. So it is hypothesized to be made of dark matter.

A similar situation pertains for galaxies orbiting near the outskirts of a galaxy cluster. Their velocities index the total mass of the cluster within each galaxy's orbit. When these measurements are made, they reveal the presence of a much stronger gravitational field than can be accounted for by adding up the mass of all the stars and other bright objects in the cluster. The larger the scale on which we sample the universe, the greater the proportion of dark matter seems to be.

My name is Fritz Zwicky,
I can be kind of prickly,
This song had better start
by giving me priority.
Whatever anybody says,
I said in 1933.
Observe the Coma cluster,
the redshifts of the galaxies
imply some big velocities.

They're moving so fast,
there must be missing mass!
Dark matter.

—DAVID WEINBERG,
"The Dark Matter Rap"[4]

The first clue to the existence of dark matter was garnered by the Dutch astronomer Jan Oort in his studies of the motion of stars in our galaxy. Much of what schoolchildren learn about the Milky Way today was established by Oort: He found that the sun is located roughly 30,000 light-years from the center of the galaxy; he calculated the mass of the galaxy to be about one hundred billion times that of the sun; and, by using radio telescopes to measure the motions of interstellar hydrogen clouds, he first charted the positions of its spiral arms. For all his achievements as a cosmic cartographer, Oort never forgot that there are more things out there than are found on the maps. In 1950, he theorized that comets entering the inner solar system originate in a vast shell of comets, centered on the sun, with a radius much larger than the orbits of the outer planets. Though the Oort cloud has yet to be observed, the theory accounts so well for the distribution of comets' orbits that most astronomers today accept its existence, and postulate that it extends to a distance of approximately two light-years from the sun, tailing off at something like halfway to the nearest star. Oort's studies of the orbital velocity of stars near the sun revealed something equally surprising and much larger in scale —that there is at least twice as much mass in the galactic disk as can be accounted for by adding up all the visible objects it contains.

In 1933, the Swiss-American astronomer Fritz Zwicky discovered that outlying galaxies in the Coma cluster are moving much faster than they would be if its mass were limited to that of the visible galaxies in it. Therefore the Coma cluster must be a good deal more massive than its light would lead us to expect; otherwise these galaxies would have flown out of it long ago. (Since the cluster is estimated to be well over 10 billion years old, it seems highly unlikely that it has been evaporating all the while.) Zwicky calculated that Coma is more than nine-tenths dark matter. Modern estimates arrive at about the same value.

Born in Bulgaria of Swiss parents and educated as a physicist, Zwicky was a memorably stimulating and irritating figure at the California Institute of Technology during the exciting times when the 100-inch Mt. Wilson telescope and later the 200-inch at Palomar were unveiling the intergalactic universe. His protean research went beyond discovering the first evidence for intergalactic dark matter. He proposed that low-mass galaxies exist, then verified his own prediction by discovering the first *dwarf* galaxies, with the 100-inch, within months of the telescope's completion. He and Walter Baade were the first to propose that the burnt-out cores of supernovae can survive as neutron stars. (Zwicky coined the term *supernovae,* and still holds the record for the most supernovae discovered by a single observer.) He predicted that the sky contains compact blue galaxies that might be mistaken for stars, an insight that anticipated the discovery of quasars. He even foresaw—this in 1937—that the dark matter problem could be tackled by studying galaxies that act as gravitational lenses, an approach that did not become feasible until the 1990s. The term *dark matter* is another Zwicky coinage.

Any one of these accomplishments might suffice to elevate an astronomer to a permanent place in scientific history, yet Zwicky remains little known beyond astronomical circles. One reason is that he touted more goofy notions than his colleagues could abide. He took pride in his creativity—"I have a good idea every two years. You name the subject, I bring the idea!"[5] he told Robert Millikan, the president of Caltech. But the bad ideas came at least as frequently. He dreamed up a jet plane that would burrow through the earth. He had a night assistant fire a rifle bullet out of the Palomar dome slit to see if the big telescope could photograph the bullets.

Equally damaging to Zwicky's reputation was his abrasive personal style. In an age when astronomers avoided stepping on one another's toes, Zwicky was a spike-sharpener of the Ty Cobb school. He liked to start arguments by automatically opposing whatever position a colleague most fervently espoused—this on the assumption that one does well to negate everybody's absolute beliefs, since absolute beliefs are almost always wrong. Edwin Hubble was certain that there were no dwarf galaxies, so Zwicky said dwarf galaxies must exist. Zwicky was right. Hubble was annoyed. Anyone who got past these intellectual obstacles to friendship had to contend with Zwicky's belief that the exercise of personal liberty involves flouting the conventions of common courtesy. He railed at his colleagues, calling them "spherical bastards"—meaning they were bastards whichever way you looked at them—and he was physically threatening as well: He liked to show off by doing one-armed push-ups in the Caltech faculty dining room, and he intimidated Baade to the point that Baade ultimately refused to be left alone in a room with his former collaborator. You get back from life what you put into it, and Zwicky's colleagues responded to his disparagement of their ideas by ignoring his. So the study of dark matter that he had pioneered languished until the 1970s.

For nearly forty years,
the dark matter problem sits.
Nobody gets worried 'cause,
"It's only crazy Fritz."
The next step's not 'til
the early nineteen seventies,
Ostriker and Peebles,
dynamics of the galaxies,
cold disk instabilities.
They say: "If the mass
were sitting in the stars,
all those pretty spirals
ought to be bars!
Self-gravitating disks? Uh-uh, oh no.
What those spirals need is a massive halo.
And hey, look over here, check out these observations,
Vera Rubin's optical curves of rotation,
they can provide our needed confirmation:
Those curves aren't falling, they're FLAT!
Dark matter's where it's AT!

—David Weinberg, *"The Dark Matter Rap"*

Interest in the dark matter issue revived in 1973, when Jeremiah Ostriker and James Peebles at Princeton calculated that ordinary spiral galaxies like the Milky Way would be unstable if most of their mass were harbored in their disks. Specifically, they showed that the disk, if it held most of the mass, would long ago have formed what astronomers call a "bar." About one-third of all spiral galaxies do have bars: Their disks, rather than being etched with spiral arms, instead display a pair of opposed tubes a few tens of thousands of light-years in length. In some of these galaxies, the bar terminates in a pair of outlying spiral arms, making the galaxy look S-shaped. In others, the bar is set in a ring of stars, so that the galaxy resembles the Greek letter theta: Θ. Ostriker and Peebles maintained that the two-thirds of spiral galaxies that are *not* barred must be stabilized by a massive galactic halo. The halo would steady the disk in something like the way a pole carried across the chest steadies a tightrope walker.

The Milky Way is known to have a halo that's dotted with old stars. According to the prevailing theory, spiral galaxies originated as spheres of gas that then collapsed into disks. (The same mechanism, working on a smaller scale, flattened the protosolar material from which the sun's planets formed.) As it collapsed, the young galaxy produced its first generation of stars. These remain in randomly inclined circular or elliptical orbits—meaning that their orbits occupy a spherical halo, with halo stars moving at all inclinations, rather like the electron orbits in the old depictions of the atom. Halo stars had been studied for many years; some of the stars in Oort's study were halo stars. But the total mass of halo stars is not nearly enough to satisfy the demand identified by Ostriker and Peebles: Their calculations show that the halo must have at least as much mass as there is in the disk.

This influential work by two eminent cosmologists turned a spotlight on research being conducted by a relatively unheralded astronomer named Vera Rubin. Fascinated by astronomy from the age of twelve, when she would stay awake for hours watching the constellations parade across her bedroom window, Rubin as a graduate student encountered her share of the dismissive attitude toward women that was once ubiquitous in the physical sciences community and continues to blight some of it today: Having written to Princeton asking how she might qualify for a doctoral degree, Rubin received a letter stating that as Princeton did not admit women to the doctoral program, it was biologically impossible for her to qualify. She wound up getting her Ph.D., in 1954, amid a bouquet of Georges—she was enrolled at Georgetown University and her adviser was George Gamow of George Washington University. The iconoclastic Gamow encouraged Rubin's maverick instincts, and early in her career she decided to shun the fashionable subjects favored by savvy young astronomers out to make a name for themselves. "I decided that I was just going to go off and do a problem that nobody would care about while I was doing it," Rubin recalls. "And hopefully when I was finished, I would show them what I had done and everybody would think that it was fine."[6] At a time when the discovery of quasars was drawing heated attention to galactic nuclei, Rubin looked the other way and began studying the outskirts of galaxies.

A typical spiral galaxy consists of a glowing "central bulge" crowded with stars and surrounded by a more thinly populated disk. As anyone who has observed spiral galaxies through a telescope can testify, the bulge is a lot brighter than the disk, and the disk is brighter near the bulge than out toward what appears to be its edge, where it falls off into darkness. If the density of matter in spiral galaxies decreased as does the disk brightness, the stars would obey a Keplerian velocity curve—that is, stars lying toward the edge of the visible disk would orbit much more slowly than those near the central bulge. And that was what nearly everyone assumed was the case: Look at old animated movies explaining astronomy and you will see galaxies rotating in Keplerian fashion.

Rubin found that this assumption was wrong. Working with colleagues at the Carnegie Institution in Washington, D.C., she examined the motions of stars across the disks of spiral galaxies, using an image tube, a relative of the TV camera that (though primitive by the standards of today's CCDs) was more sensitive to light than the photographic emulsions then prevalent in observational astronomy. Her group confirmed that disk stars do orbit more slowly than those in the bulge. But they also found that outlying stars move just as fast as stars in the disk. For the most part there is *no* appreciable decline in the stars' orbital velocity, all the way out to the limits of observation, and in some cases the outer stars actually move faster than those in the main disk. So it must be that galactic mass does not decrease at anything like the rate that the density of visible stars does. "What you see in a spiral galaxy is *not* what you get," Rubin concluded.[7]

". . . If luminosity were a true indicator of mass, most of the mass would be concentrated toward the center. . . . Instead it has been found that the rotational velocity of spiral galaxies in a diverse sample either remains constant with increasing distance from the center or rises slightly out as far as it is possible to make measurements. This unexpected result indicates that the falloff in luminous mass with distance from the center is balanced by an increase in nonluminous mass"—that is, dark matter.[8]

Rubin had struck oil by drilling in what everybody else thought was a dry field. Now fashion swung her way, with many observers gathering data that confirmed her findings. Of particular interest was the work of radio astronomers, who detected 21-centimeter radio emission from hydrogen clouds in galactic disks half again beyond the point where the disk fades away at optical wavelengths. As a result, it is now widely supposed that spiral galaxies are surrounded by massive "dark halos" of unseen matter. Cosmically speaking, there's dark matter right in our front yard.

Lots of dark matter having been found—or, rather, its presence inferred—in the disks and halos of spiral galaxies, astronomers were eager to learn whether the other major types of galaxies, the ellipticals and the dwarfs, also hold dark matter. They were not disappointed.

Ellipticals constitute about a third of all prominent galaxies. They include among their numbers the giant, "cluster-dominating" galaxies found at the centers of spherical galaxy clusters like Coma. Ellipticals display neither disks nor spiral arms. They contain much less interstellar dust and gas than is found in spirals. Some look as spherical as basketballs; others are as ellipsoidal as a circus clown's short, fat cigar. Since a cigar-shaped galaxy seen end-on looks spherical, it is difficult to determine the true shape of any particular elliptical galaxy. Determining its mass is difficult as well. Observing a spiral galaxy that is inclined, say, at a 45-degree angle, one can measure the Doppler shifts of stars receding on one side of the disk and of those approaching on the other side. But elliptical galaxies have little or no net rotation, and their stars orbit in a more complicated fashion: Some, for instance, pass near the center of the galaxy while others do not. So the straightforward application of Newtonian and Keplerian dynamics which works so well for spirals is impractical when it comes to ellipticals. Such preliminary dynamical studies as have been done suggest that ellipticals are at least 70 percent dark matter.

Meanwhile, observations in x-ray wavelengths go further, indicating that elliptical galaxies are about 90 percent dark matter. In the mid-1980s, Christine Jones and her husband, William Forman, both at the Harvard-Smithsonian Center for Astrophysics, studied the coronas of hot, x-ray-emitting gas surrounding giant ellipticals in the centers of clusters and concluded that such galaxies must weigh at least ten times as much as their visible stars would indicate. The reasoning here is as follows: The level of x-ray energy reveals the temperature of the coronal gas. The temperature of the cloud indicates the average velocity of the atoms in it. (That's the definition of temperature, which is why astronomers speak—rather confusingly, I fear—of extremely "hot" clouds that have high-velocity atoms but are so thin that a pond full of the stuff couldn't fry an egg.) The velocities of the atoms put a lower limit on the escape velocity of the galaxy: If the atoms' speed exceeded the escape velocity, the hot gas would have blown clear out of the galaxy long ago. The escape velocity in turn yields a minimum mass for the galaxy. And that mass turns out to be ten times the mass of all the visible stars in the galaxy. A similar estimate was arrived at by David Merritt and Benoit Tremblay at Rutgers, who analyzed the velocities of globular star clusters orbiting the giant elliptical galaxy M87. So elliptical galaxies evidently are mostly dark matter.

Dwarf galaxies are *full* of dark matter. Presumably this has to do with the fact that the dwarfs have low mass—typically only about that of 1 million to 10 million suns, thousands of times less than an average spiral galaxy—and thus have low escape velocities. Material ejected by supernovae and as the stellar wind from red giant stars—stuff that would remain in a major galaxy to be reprocessed into new stars—simply walks out of a dwarf galaxy. Consequently, the stellar mass of a typical dwarf galaxy dwindles as time goes by (perhaps being replenished occasionally when the galaxy collides with an intergalactic gas cloud), while the dark matter—whatever it may be—remains on hand. So dwarfs, poor in stars, are rich in dark matter.

Moving up in scale to groups and clusters of galaxies, astronomers find dark matter in abundance. The smallest aggregations of galaxies consist of pairs, triplets, and groups. The Local Group of some twenty-six galaxies, to which the Milky Way galaxy belongs, is an average-size group. Astronomers examining such groups find evidence of dark matter in the intergalactic space. A 1994 study found that triplets of galaxies dance in stable orbits around their mutual center of gravity at velocities that could be supported only by large quantities of dark matter.[9] An x-ray image taken with the ROSAT satellite of the NGC 2300 group—a little gaggle, dominated by three galaxies, 150 million light-years from Earth, in Cepheus—revealed a dense cloud of gas so hot that it would have blown out of the cluster had the cloud not been held in place by the grip of dark mass amounting to between fifteen and twenty-five times the mass that the visible galaxies contain. Most galaxies belong to such small groups, and it is likely that most have dark matter in their immediate intergalactic neighborhood.

Several sorts of evidence indicate that clusters of galaxies are at least nine-tenths dark matter. X-ray observations of groups like NGC 2300 show that rich clusters contain hot intergalactic clouds that could not remain bound in such clusters if they were quite massive. Dynamical and x-ray examinations of the Coma cluster confirm the maligned Fritz Zwicky's estimate that Coma is at least 90 percent dark matter.

Recently, large masses for clusters of galaxies have been estimated by the gravitational lensing technique.

We previously encountered gravitational lensing as a promising new way of directly measuring the distances of galaxy clusters, so as to measure the Hubble constant and thus determine the expansion rate of the universe. We recall that in several cases multiple images of a distant quasar have been observed to either side of a galaxy cluster, because of gravitational warping of the space surrounding the cluster. Since the strength of the gravitational field reveals how much mass is in the cluster, an astronomer who can map a gravitational lensing field can estimate the mass of the cluster doing the lensing. More gravitational lenses are being detected all the time, and some of them show multiple arcs—slivers of lenses—that permit detailed mapping of the mass of the cluster and of its distribution. A dramatic case was identified in an image taken in 1995 by astronomers using the Hubble Space Telescope. It shows that the rich galaxy cluster Abell 2218 is surrounded by more than a hundred blue-white arcs that resemble a froth of shimmering bubbles. Studies of lensing clusters are particularly useful in dark matter research, because the lens indexes all the mass inside the lens, which typically is found farther from the center of the cluster than are the outermost galaxies used in dynamical analysis and the hot gas clouds studied in x-ray observations. So gravitational lenses, too, suggest that at least some galaxy clusters are more than 90 percent dark matter.

The largest-scale studies also find plenty of dark matter. Particularly provocative is the finding in the late 1980s, by a group of seven astronomers known as the Seven Samurai and led by Alan Dressler, Sandra Faber, and Donald Lynden-Bell, that the Local Group and thousands of neighboring galaxies are sliding off sideways, so to speak, at a velocity of over 630 kilometers per second relative to the universe as a whole. That's pretty fast—nearly three times the velocity of the sun in its orbit around the center of the Milky Way galaxy. The Seven Samurai surmise that the Group is being pulled by a large and as yet unseen concentration of mass they have dubbed "The Great Attractor." The Great Attractor evidently lies 200 million light-years from our galaxy, out beyond the Virgo and Hydra-Centaurus clusters. It appears to be pulling them along, too, at accelerated rates. To throw its weight around on such an enormous scale, its mass must be considerable—the equivalent of fifty thousand galaxies, 90 percent of it dark matter.

The Great Attractor hypothesis is based on statistical studies of galaxies' motions that are vulnerable to many kinds of error, and it involves several theoretical assumptions as well. Given all the variables, some researchers dismiss the project as "scientific mud wrestling." But the Seven Samurai are solid scientists, their work has so far stood up under years of scrutiny by skeptical colleagues, and their morale remains high. "I'm muddy, but unbowed," says Sandra Faber.[10]

The larger the piece of the universe sampled, the greater the percentage of dark matter found within it. This point may be made conveniently by resorting to the cosmic density parameter omega.

As you may recall, if omega equals one the universe is at critical density—meaning that it is "flat" and will expand forever.[11] Observers to date have not found enough matter to yield anything like that high a value for omega. Bright matter amounts to an omega of only 1 percent of the critical value, and adding in all the dark matter inferred on scales up to that of galaxy clusters still gives us an omega of no more than 10 percent. Nevertheless many researchers think omega is one. Some hold this view because they favor inflation theory, and most inflationary theories predict that omega equals one. There are other reputable reasons for holding this opinion: In the standard big bang model, an omega of one is the only density value that does not change as the universe expands. So if omega is *not* one, then we today live in an exceptional epoch, when omega just happens to be close to critical while evolving toward some other value. It *might* be only a coincidence that we have come along just when the cosmic density parameter lies within a factor of 10 of the critical value, but cosmologists are highly skeptical of arguments that put us in a special time or place, since they open the door to many other sorts of special pleading. (Such arguments are said to violate the *cosmological principle.*) As the cosmologist Andrei Linde likes to say, a critical density universe is a "more natural" universe.

For theorists who think this way, omega is a cup that needs to be filled. The observers have supplied one-tenth of the matter needed to fill it—to bring omega to one—and the task now is to find the other 90 percent of cosmic matter. Since a universe less massive than critical density is called "open," finding enough mass to fill the cup is sometimes called "closing" the universe. (This is, of course, just a way of talking. Cosmic geometry is whatever it is, and we are trying to learn about it, not influence it.) Those out to close the universe note with interest that the largest-scale observations find the greatest proportions of dark matter—suggesting that enough may eventually be found to fill the cup.

And so the call goes out for the dark matter candidates:
black holes, snowballs, gas clouds, low mass stars, or planets.
But we quickly hit a snag because galaxy formation
requires too much structure in the background radiation
if there's only baryons and adiabatic fluctuations.
The Russians have an answer: "We can solve the impasse.
Lyubimov has shown that the neutrino has mass."
Zeldovich cries, "Pancakes! The dark matter's HOT."
Carlos Frenk, Simon White, Marc Davis say, "NOT!

Quasars are old, and the pancakes must be young.
Forming from the top down it can't be done."
So neutrinos hit the skids, and the picture's looking black.
But California laid-back Blumenthal & Primack
say, "Don't have a heart attack.
There's lots of other particles.
Just read the physics articles."

—DAVID WEINBERG, *"The Dark Matter Rap"*

Regardless of whether there is enough dark matter to close the universe, there clearly is a lot more dark than luminous matter around. As a paper by one group of researchers notes, "The question has changed from 'Does dark matter exist?' to 'What is this most common of substances?' "[12]

So what *is* the stuff?

One possibility is that all dark matter is made of baryons—which is to say, of protons and neutrons, of ordinary material. One can think of many forms of baryonic dark matter that might reasonably be expected to turn up in space. Leading candidates include dwarf stars too dim to have yet been observed, "Jupiters" (substellar objects with masses about a tenth of one percent that of the sun), black holes, and massive cold gas clouds. Surely at least *some* dark matter is baryonic: After all, if the Oort cloud exists, its hundred billion or so comets qualify as baryonic dark matter, and there almost certainly are a lot of other inconspicuous things out there that are made out of protons and neutrons.

But baryonic dark matter, though it might provide all the invisible mass of the Milky Way and other galaxies, cannot suffice for galaxy clusters and superclusters. The reason is that the calculations of big bang nucleosynthesis—which, as we have seen, correctly predict the cosmic abundances of deuterium and helium—prohibit baryonic mass from amounting to more than about a tenth of the critical value. Otherwise nearly all the deuterium in the big bang would have burned into helium, a possibility ruled out by what David Schramm calls "deuteronomy," the study of the observed cosmic deuterium abundance. So if galaxy clusters weigh more than one tenth of the cosmologically critical mass, at least some of their dark matter must be nonbaryonic in form.

In the swaggering jargon that besets dark matter studies,

baryonic objects are called MACHOs, for MAssive Compact Halo Objects (the reference is to galactic halos), and nonbaryonic objects are called WIMPs, for Weakly Interacting Massive Particles. The "weak" part means that WIMPs interact with other matter only through the weak nuclear force (and through gravitation, which is how scientists detected the presence of dark matter in the first place). WIMPs do not interact via the strong nuclear force or, more to the point, electromagnetically: If they did, we should have identified them already, since they would be sticking to atoms and molecules all over the place. Neutrinos are like that; they "feel" only the weak force. If neutrinos turn out to have mass, then neutrinos are WIMPs. We will have a look at the fascinating question of neutrino mass and consider why neutrinos probably are not the best candidate for WIMPdom. But first let's look at the observational evidence for the existence of MACHOs.

When it was discovered that our galaxy has a massive dark halo, there was no immediate way to determine what the dark halo objects might be. Not until the end of the 1980s did the sun rise on the prospect of searching out MACHOs. In October 1989, the astrophysicist Charles Alcock, of the Lawrence Livermore National Laboratory, gave a talk at Berkeley's Center for Particle Astrophysics in which he outlined a search strategy that he and other researchers envisioned—one that would use charge-coupled devices, or CCDs, to detect *microlensing* events caused by dark objects in the galactic halo.

CCDs are the light-sensitive silicon chips that have revolutionized professional and amateur astronomy alike. Their most obvious virtue is that they are much more sensitive than photographic emulsions. An amateur's backyard telescope, if equipped with a CCD camera, can perform at the level of the Palomar telescope of old, and CCDs attached to big telescopes like Palomar and Keck more than double the distance to which these telescopes can penetrate in a given exposure time. In his Berkeley talk, however, Alcock focused on another aspect of CCD technology. A CCD chip consists of an array of light-sensitive spots called pixels. Each pixel reports digitally, in real time, to a computer. This means that CCDs can make what amount to motion pictures. Observing an object bright enough to be recorded in five minutes, one can make a movie in which each frame consumes five minutes. And since a CCD chip containing millions of pixels can image thousands of stars in a telescopic field of view simultaneously, astronomers using CCDs can look for variations in the brightness of all those stars at once.

Alcock noted that the Large Magellanic Cloud, the most massive satellite galaxy of the Milky Way, presents millions of stars within a single wide-angle telescopic field of view. If such a telescope, equipped with a CCD, were kept pointed at the cloud, he reasoned, the stars in the cloud could be monitored for changes in brightness caused when a dark halo object—a MACHO—in the halo of our galaxy passed through the line of sight leading from the star to the telescope. This is the microlensing part. What happens is that the MACHO—which might be, say, a dwarf star with half the mass of the sun—bends space enough to focus light from the distant star, with the result that the star appears to brighten when the MACHO passes in front of it. There aren't enough observers in the world to monitor all the visible stars of the Magellanic Cloud, even were they to volunteer for such arduous duty. But by using computers to keep track of the stars, Alcock concluded, it ought to be possible to seek out MACHOs in the Milky Way's halo.

The Berkeley astronomers reacted with enthusiasm, and soon a team of scientists from Livermore, Berkeley, and the Australian National University—predictably dubbed MACHO—had renovated an obsolescent 50-inch telescope at Mt. Stromlo Observatory near Canberra, Australia, fitted it with CCDs, and trained it on the Large Magellanic Cloud. Each clear night, the computers monitored the brightness of nearly two million stars in the cloud. Sifting through the data was, as expected, a stringent business: Every five minutes, the instruments generated seventy-five megabytes of data, equal to the content of a stack of books five feet high. The astronomers had to screen out thousands of previously uncatalogued variable stars, which could be eliminated once they were seen to change brightness regularly rather than on just the one significant occasion. (In the process, the team discovered several new kinds of variables.) But the method worked. By mid-1995, the MACHO team had observed eight microlensing events in the cloud. (Two other groups inaugurated similar projects. In the prevailing temper of the nomenclature they were called EROS, for Expérience de Recherche d'Objets Sombres, and OGLE, for Optical Gravitational Lens Experiment.[13] They, too, detected microlensing events.)

In each of these instances, it seems that a small dark object had drifted between the telescope and a Magellanic cloud star, briefly brightening the star's image. From the period of brightening, typically about a month, it was possible to estimate the diameters of these MACHOs. The results indicate that MACHOs are mostly dwarf stars—either white dwarfs, which are the cinders of dead stars, or brown dwarfs, which have too little mass for nuclear fusion to commence at the core.

The MACHO team then trained their telescope on the central bulge of the Milky Way galaxy, an egg-yolk-yellow sphere of stars that looms above the black clouds of the foreground disk, in the Southern Hemisphere constellation Sagittarius, high in the skies above Mt. Stromlo. There they found many more microlensing events than had been expected. They surmised that either the Milky Way is a barred spiral and the bar happens to point our way, so that our view toward the bulge contains more material than we had assumed, or else the halo is flattened, so that its density increases as one looks closer to the disk. In any event, this ongoing work has provided the first direct evidence that MACHOs exist, has indicated what at least some of them are, and has confirmed that our galaxy is mostly dark.

Microlensing is still in its infancy. As data accumulate and techniques improve, microlensing should make it possible to map the galactic halo in greater detail. At present, it is not known how far out the halo extends. If the halos of major galaxies like the Milky Way typically stretch out to a radius of, say, three or four million light-years, a distance comparable to the average distances separating such galaxies, then baryonic matter in the halos might account for the dark matter inferred to exist in clusters of galaxies. At that point, we might simply stop and declare that the question has been resolved in favor of purely baryonic dark matter—that the dark matter consists of nothing but MACHOs. If the halos fall short, there still may be plenty of baryonic dark matter out there—perhaps in the form of "dark galaxies," which certainly exist and may be as numerous as bright galaxies, though astronomers don't yet know how many there are, since their low surface brightness makes them hard to find. So maybe baryons alone can do the job. And some scientists do take this minimalist position.

But the constraints of big bang nucleosynthesis make it unlikely that baryons will provide enough dark matter to explain the dynamics of galaxy clusters. And baryons certainly won't suffice to close the universe. Those who advocate a critical density cosmos must look beyond MACHOs and hope for WIMPs. As the University of Chicago physicist Michael Turner puts it, "The grander—and more radical—view is that . . . we live in a universe dominated by nonbaryonic dark matter. From a theoretical perspective this is the most attractive scenario—and it may even be true!"[14] Now, to propose that there are two true explanations for one phenomenon—that cosmic dark matter is made of (10 percent) baryons and (90 percent) some exotic stuff not yet known to exist—is a step few scientists take eagerly. Some refuse to take it at all. As the cosmologist Joseph Silk of Berkeley says, "I personally think the organizer of the universe would really be playing a very mean trick on us if he had several different kinds of dark matter around."[15]

Yet there are hints that WIMPs—nonbaryonic particles—do exist.

One clue is that it does not seem possible for galaxies and galaxy clusters to have assembled themselves gravitationally if baryonic matter was all they had to work with. There are at present *no* viable theories of how these structures could have formed if the universe contained only baryons. Another clue, even more provocative, is that a wide variety of particle physics theories predict the existence of particles with just the characteristics required by cosmologists searching for nonbaryonic dark matter. The theoretical WIMP emerged from physics concerns that have nothing to do with cosmology. As one group writes, "This striking coincidence suggests that if there is [such a kind of particle] it *is* the dark matter."[16]

The existence of such a WIMP is predicted by supersymmetry theory. A leading candidate goes by the name LSP, for Lightest Supersymmetric Particle. Its precise characteristics are predicted differently by different versions of supersymmetry—which is why nobody has yet seen fit to give it a proper name—but in virtually all cases it fills the bill as far as dark matter is concerned. Later we will discuss the case for supersymmetry (SUSY, for short) as a prospective unified theory, while allowing that SUSY remains purely theoretical, and incomplete at that. No particles envisioned by supersymmetry have yet been found in nature, nor will experimenters be able to search for them until higher-energy accelerators are built. But before we dismiss the LSP as just a lot of theoretical hand waving, consider that we already know of one weakly interacting nonbaryonic particle that definitely exists. It is the neutrino.

Neutrinos interact with ordinary matter only via the weak nuclear force, so they are Weakly Interacting Particles, or WIPs. The question is whether they have mass, which would make them WIMPs. Lots of neutrinos were released in the big bang: The cosmic neutrino background should contain roughly as many relic neutrinos as the microwave background does photons. So if just one kind of neutrino has just a little mass, that in itself could be enough to close the universe.

The story of how neutrinos were discovered serves as a cautionary tale for those who might too readily dismiss WIMPs as "merely" theoretical. In the 1920s, physicists investigating the radioactive decay of unstable atoms were troubled to find that these events seemed to violate the law of conservation of energy. When atomic nuclei decay, they emit electrons and lose mass. Experimenters monitored radioactive material as it decayed, recording the electrons produced and weighing the samples to see how much mass had been lost. They found that more mass was lost than was carried off by the electrons. This violates the conservation laws, which require that all the mass (or energy; they are the same thing) in any transformation must be accounted for. Furthermore, if one considered electrons alone, which were the only particles seen to be emitted, radioactive decay also violated the conservation of momentum and angular momentum.

In 1930, Wolfgang Pauli theorized that the missing energy was being carried off by an undetected particle. This particle had to have very low mass—Pauli proposed that its rest mass was zero—and no electrical charge; otherwise it would have been noticed by the experimenters. To solve a problem by conjuring up a new particle for which there was no experimental evidence might have seemed *ad hoc*—in the spirit of the physicist Isidor Rabi, who said of the muon, "Who ordered that?" But Pauli's theory was artfully realized, and Enrico Fermi was sufficiently impressed by it to give the imaginary particle a name—*neutrino*, Italian for "little neutral one." Twenty-six years passed before the detection of evidence that neutrinos are real, in a subtle experiment conducted in 1956 by Frederick Reines and Clyde Cowan at the Savannah River nuclear reactor in South Carolina.

Today neutrinos are major players on the physics scene. Scientists understand that they were produced in the big bang, are being generated in thermonuclear reactions that power stars today, and that neutrino "refrigeration" almost certainly carries off the lion's share of the titanic energies produced in supernovae. Yet Pauli's assumption that neutrinos have no mass has never been proved, and neither any experimental evidence nor any symmetry principle of physics requires it to be true. If anything, theory is moving in the opposite direction, since several versions of the emerging unified theories imply a nonzero neutrino mass. So the physicists, spurred on by cosmologists looking for an exotic particle to close the universe, have been pursuing the question of whether neutrinos have mass.[17]

Of the three families of neutrinos—electron, muon, and tau—the electron neutrino seems the best dark matter candidate, and several studies of radioactive decay have suggested that the electron neutrino might have some mass. But not much: Various observations have placed stringent upper limits on neutrino mass. Notable among these was the detection of neutrinos released by supernova 1987A in the Large Magellanic Cloud. Supernovae release neutrinos in a sudden burst: Neutrinos interact so weakly with other particles that they fly out of the core of the doomed star even faster than the light of the explosion does. The heavier the neutrino, the slower it goes; so if, say, electron neutrinos have mass, they will travel through space more slowly than will the zero-rest-mass neutrinos that were also released by the supernova. Therefore the total amount of time it takes for the burst of supernova neutrinos to pass through a detector—zero-mass neutrinos first, then the massive ones, if any, trundling along behind—places an upper limit on the mass of the heavier neutrinos. The shorter the burst, the less the dispersion in neutrino mass can be. Neutrinos from supernova 1987A struck the earth on February 23, 1987, at 11:19 P.M. Greenwich Mean Time—about six hours before the exploding star's light was first noticed, by an astronomer on a mountaintop in Chile—and were recorded by several underground detectors in the United States, Japan, and elsewhere. Some of the data are questionable, but depending upon which events one chooses to accept, the burst lasted some nine to twelve seconds. That's a pretty tight pattern. It indicates that the heaviest neutrinos from supernova 1987A had little if any mass. Which is fine with the cosmologists: Since neutrinos are so numerous, only a modest neutrino mass is required to close the universe.

A particularly troubling hint that science does not fully understand the neutrino shines down from the sky all day—and up *through* Earth all night. This is the solar neutrino problem.

Astrophysicists think they know the sun quite well. In particular, they believe that they understand the major thermonuclear reactions at work in the solar core and that they know the temperature of the core, which dictates the reaction rate, to an accuracy of within 5 percent. The reaction rate in turn yields the rate at which the sun should release neutrinos. And solar neutrinos are indeed detected, indicating that the astrophysicists are on the right track. The trouble is that their theories predict twice the number of solar neutrinos detected on Earth.

For more than twenty-five years, solar neutrinos have been monitored at an observatory consisting of a 100,000-gallon tank of cleaning fluid installed 4,850 feet underground in the Homestake gold mine in Lead, South Dakota. (The underground location reduces "noise" produced by cosmic rays, which are stopped by rock that neutrinos fly through.) Occasionally a neutrino collides with a chlorine atom in the cleaning fluid, transforming it into an argon atom. Every couple of months the experimenters siphon off the cleaning fluid and filter out the argon. The quantity of argon indicates the number of neutrino collisions, which is then extrapolated to find the total number of neutrinos that have passed through the tank.

Solar neutrino detectors rank among the strangest observatories in astronomical history, and the scientists who study them have had to put up with a certain amount of public puzzlement about the nature of their work.[18] John Bahcall, of the Institute for Advanced Study in Princeton, who is a contributor to the standard model of solar fusion that has been challenged by the detector results, was once consoled by a mine worker who noted that the weather lately had been unusually cloudy. And shortly after ordering ten railroad tank cars full of cleaning fluid to fill his detector, Ray Davis, Jr., who directs the Homestake experiment, found himself the target of mailings from a coat-hanger salesman.

Still, the detector physics is well understood by physicists, if not by everyone else, and the results are clear: The tanks catch only about half as many electron neutrinos as theory predicts the sun should produce. Vigorous efforts have been made to explain the discrepancy, usually by lowering estimates of the core temperature (the "cool sun" theories). But that would make the sun shine less brightly than it does. As Thomas Bowles of Los Alamos puts it, with the martini-dry understatement fashionable in the field, "All of these models have run into problems in trying to reproduce other measured parameters (e.g., the luminosity) of the Sun."[19]

In the early 1990s, two new and more sensitive neutrino detectors came on line. In keeping with the esoteric tincture of neutrino studies, they use gallium, a rare metal so expensive that loading a gallium detector can cost $20 million. (The metal can be recovered and sold when the detector is decommissioned, which makes each such project a speculation in gallium futures.) Both are located beneath mountains—GALLEX (GALLium EXperiment) in a tunnel that runs under the Gran Sasso mountains of central Italy, and SAGE (Soviet-American Gallium Experiment) at the Baksan Neutrino Observatory under Mt. Andyrchi in the Caucasus. Data from the gallium detectors confirm the paucity of solar neutrinos, and pretty much rule out astrophysical explanations like the ones that argue for a reduced or intermittent fusion rate in the sun. So it seems that new physics are called for.

One plausible explanation is that neutrinos "oscillate"—quantum mechanics lingo meaning that one sort of neutrino turns into another on its way out of the sun. Oscillation could transform electron neutrinos—the kind captured by the gallium and cleaning-fluid detectors—into muon neutrinos, which the detectors do not trap. The theory of neutrino oscillations is elegant and straightforward—and, most important, it can be tested, in the large neutrino detectors that have recently been coming on line. Preliminary results from a 1996 experiment at the Los Alamos National Laboratory indicated that neutrino oscillations are real. Whatever the final verdict, the best evidence for neutrino mass, relevant to cosmological arenas long ago and far away, shall have come from the nearby sun.

> *Who's right? It's hard to know, 'til observation or experiment*
> *gives overwhelming evidence that relieves our predicament.*
> *The search is getting popular as many realize*
> *that the detector of dark matter may well win the Nobel Prize.*
>
> —DAVID WEINBERG, *"The Dark Matter Rap"*

But while the experimentalists hunt down neutrino mass, the detection of which would in any event be of fundamental importance to particle physics, neutrinos have fallen out of fashion among cosmologists looking for the missing mass needed to close the universe. The reason is that neutrinos do poorly at one of the essential tasks required of nonbaryonic dark matter, that of forming galaxies and galaxy clusters. Computer simulations using neutrinos as the nonbaryonic dark matter either form galaxies too slowly or wind up with more structure (i.e., a clumpier universe) than is observed. And so the spotlight has shifted to the lightest supersymmetric particle, or LSP.

In most current theories, from conservative ones involving minimal extensions of the standard model to shining castles in the distance (in the air, some would say) like superstring theory, the ranking LSP candidate is the *neutralino*. Readers who are marking their scorecards will want to note that the neutralino is a linear combination of the supersymmetric partners of the photon, of an early-universe boson called the Z^0 and of the theoretical Higgs boson. We will get to know these and other inmates of the supersymmetric zoo later on, when we discuss unified theories of physics. For now the important point is that most versions of supersymmetry theory predict the existence of a stable particle with the mass and abundance required to close the universe. The evolutionary history of such particles is similar to what we've seen for matter-antimatter annihilation: In the furnace of the big bang, LSPs would have been annihilated until the primordial plasma thinned out, leaving the universe with a constituency of relic LSPs amounting to 90 percent of its mass.

The mass (i.e., energy) of the LSP puts it almost within reach of the particle accelerators at CERN and elsewhere, so it should soon be possible to test this prediction of supersymmetry theory. Until then, it seems premature to dismiss talk of exotic dark matter as "merely" theoretical. One seldom sees what one is not looking for, and theory tells experimenters where to look: Sometimes they find what they were seeking; sometimes they don't; sometimes they find something unanticipated. Making a model of the universe is like trying to pitch a tent on a moonless night in a howling Arctic wind. The tent is theory. The wind is experiment. When one gets to the precipice, where the secure lands of the known have been left behind and the dark canyons of the unknown fill one's field of view, it becomes very difficult to guess just where to set the tent pegs and to predict which ones will hold once the wind comes up. But there are signs of a beautiful structure lying just over the epistemological horizon, and we shall not get to it if we prematurely dismiss the possibility that it exists.[20] As the Harvard physicist Percy Bridgman remarked back in 1927, "Whatever may be one's opinion as to the simplicity of either the laws or the material structures of Nature, there can be no question that the possessors of such a conviction have a real advantage in the race for physical discovery. Doubtless, there are many simple connections still to be discovered, and he who has a strong conviction of the existence of these simple connections is much more likely to find them than he who is not at all sure that they are there."[21]

Meanwhile we might pause a moment to enjoy the view. Whatever happens to the current crop of conjectures, we have learned that most of the matter in the observable universe has not (yet) been observed. Ahead lies a new challenge to cosmic cartographers: To chart the black Taj as they have charted the bright one, to learn its characteristics and to see to what degree it resembles the bright side. It may be, for instance, that dark matter does not "trace" bright matter—that it is not gathered in galaxies and galaxy clusters, as bright matter is. Theories that postulate a different distribution for dark matter include what is called "biasing," and theorists tend to be rather apologetic about them, saying of them what Robert Frost said of poetry that doesn't rhyme, that it is like playing tennis without a net. But just as the universe is under no obligation to make life easy for scientists, neither need it have gone out of its way to make their lives difficult, and dark matter may not cluster as bright matter does. Central to that question is the subject of the large-scale structure of the universe.

6
The Large-Scale Structure of the Universe

We emerged to see—once more—the stars.

—Dante, *Inferno*[1]

I saw that in its depths there are enclosed,
Bound up with love in one eternal book,
The scattered leaves of all the universe—
Substance, and accidents, and their relations,
As though together fused in such a way
That what I speak of is a single light.
The universal form of this commingling . . .

—Dante, *Paradiso*[2]

Scientists assume that on the "cosmologically significant" scale, where distances are measured in billions of light-years, the universe is isotropic, meaning that it looks the same in all directions, and homogeneous, meaning that matter is distributed evenly throughout space. The assumption of global isotropy and homogeneity plays a central role in modern cosmology, making possible simplified models that treat the universe as if it were a gas and the galaxies molecules. And it seems to be borne out observationally: The cosmic microwave background glows with almost exactly the same intensity all over the sky, indicating that the early universe was homogeneous and isotropic. Radio maps probing a billion light-years into space show galaxies distributed as evenly as sifted sand, indicating that the universe at large remains isotropic and homogeneous today.

But on smaller scales—below, say, a few hundred million light-years—there's lots of *structure*. At this level, visible matter is not distributed evenly at all, but is clumped and clustered together. There are galaxies, clusters of galaxies, superclusters, and enormous bubble walls surrounding low-density voids that yawn 300 million light-years wide. So the universe as a whole may be isotropic and homogeneous, but on smaller scales these characteristics break down.

In one sense the cosmologists have long understood this. Were the universe *perfectly* homogeneous there would be no such things as stars, planets, and galaxies, since all these objects are much denser than average. Were it *perfectly* isotropic space exploration would be pointless, since everything would be the same everywhere. The question is just where the does local structure fade away into a homogeneous and isotropic distribution? What, in other words, *is* a "cosmologically significant" scale?

This question can be quantified by asking how big a volume of space must be sampled before all the samples start looking alike. Historically, structure studies have involved observing ever-larger volumes, like building a set of nested Chinese boxes, each larger than the one before. Boxes the size of the solar system aren't big enough to reliably enclose a homogeneous sample. Neither are boxes the size of galaxies, or of groups, clusters, or superclusters of galaxies. In recent years, the boxes have grown to be hundreds of millions of light-years on a side, yet these, too, have not been big enough to overwhelm the level of structure. So some cosmologists have begun worrying about whether the assumption of isotropy and homogeneity is valid. Cosmology has assumed the posture of the princess in the fable whose rest is disturbed by the pea. The cosmological mattress is made of sponge. All sponges have holes in them—they possess local structure—and sponge will make a decent mattress only if the holes are not too large relative to the size and sensitivity of the princess who reclines on the mattress. (That's why the fable concerns a princess rather than a giant, who is less likely to be disturbed by a pea.) Various cosmological theories can tolerate various sizes of holes in the sponge. But past a certain point—just about the point that the observations have now reached—all will start to toss and turn.

Large-scale structure is a subject of intense scientific interest. The Russian astrophysicist Yakov Zeldovich saw it as "the most important feature of the physical world around us."[3] His colleague Igor Novikov went so far as to say that "producing a thorough analysis of the deviations from the ideal picture of a homogeneous, isotropic universe is the most important task confronting modern cosmology."[4] There are three principal reasons for this.

One is that by studying structure, astronomers make better maps of cosmic space. Margaret Geller of Harvard, who has contributed significantly to this field, remarks, "When people on airplanes ask me what I do, I used to say I was a physicist, which ended the discussion. I once said I was a cosmologist, but they started asking me about makeup, and the title 'astronomer' gets confused with astrologer. Now I say I make maps."[5] Obviously it would be useful if astronomers could identify all the major galaxies within, say, a few hundred million light-years from Earth, ascertaining both where they are in space and how they are moving. The local universe is the part we can explore in greatest detail, and the bigger the sample the less parochial and the more representative it will be.

A second motive derives from the fact that local concentrations of matter perturb the velocity field produced by the expansion of the universe. Measuring the rate at which the universe expands (i.e., obtaining the value of the Hubble constant) is complicated by the fact that our galaxy is embedded in a supercluster, whose local gravitational field retards the expansion velocity of all galaxies less than about 100 million light-years from us. Until astronomers understand these effects, there will continue to be troubling uncertainties in basing estimates of the cosmic expansion rate on the velocities of nearby galaxies.

The third reason to investigate large-scale structure is to learn about cosmic history. The big bang theory declares that the vast structures we see around us today originated in minute density fluctuations that arose in the very early universe. The structures are themselves relics, and to understand how they formed would be to learn a great deal about processes that took place when the universe was young.

In this chapter we consider what the mapmakers have learned about cosmic structure, then see how structure affects the motions of galaxies (and is revealed by them), and, finally, discuss some of the leading theories and observations bearing on the question of how visible matter got to be distributed the way it is.

First, cartography.

Cosmic structure is a continuum, with the boundaries between various levels of hierarchy not always clear. Discriminating between, say, a large *supercluster* and a small *supercluster complex* can be a bit arbitrary, like sorting eggs according to which are "large" and which "jumbo." The task is sometimes compared to making maps of Earth, but it is actually more difficult than that. The existence of the oceans at least permits a straightforward distinction between land and sea, whereas ordering the distribution of galaxies in space is more like trying to categorize clouds in the sky. Until we have a proper theory of structure formation, every taxonomic scheme will to some extent be a matter of subjective impression. Anyway, what follows is meant to approximate the consensus view.

Astronomers have identified five levels of structure larger than galaxies—groups, clusters, clouds, superclusters, and the vast aggregations called supercluster complexes or walls. The walls enclose voids—cosmic bubbles, if you will, with few galaxies to be seen within. Similarities have been found among these systems, the significance of which is not well understood. Superclusters, for instance, contain groups of galaxies, in something like the way that an individual galaxy contains clusters of stars. Some superclusters evidently are flat in shape, as spiral galaxies are, but what this means is unclear. According to the standard big bang model, all these structures result from gravitational effects—they trace the shape of local space—but until the mechanisms that produced them have been worked out quantitatively, a certain amount of arbitrariness and confusion will endure. Nevertheless the practice of dividing structure into these five bins seems reasonable, if only as a way station toward a fuller account.

The smallest units in this scheme are *groups* of galaxies. Groups typically are a few million light-years wide, and consist of three to six conspicuous galaxies plus a dozen or so smaller and dimmer ones. Many of the larger galaxies are found in pairs, and sometimes in trios. Many of the smaller galaxies are satellites of the dominant ones. So one might say that groups are themselves built from such lesser associations of galaxies.

Clusters measure 10 million to 20 million light-years in diameter and contain hundreds to thousands of galaxies. They are much "richer"—meaning denser—than groups; groups represent enhancements of only a few times more mass per cubic volume of space than that of the universe at large, while rich clusters are at least ten to twenty times the average cosmic density, and at the cluster center the ratio can go as high as ten thousand times. Clusters come in two forms. Irregular clusters have a loose, somewhat scattered shape. Spherical clusters resemble giant elliptical galaxies. Irregular clusters have lots of spiral galaxies, while elliptical and SO galaxies (SOs are spirals whose disks display no spiral arms) predominate in spherical clusters.

A sharp distinction can be drawn between clusters and all larger structures, in that clusters, like groups, are gravitationally bound, meaning that their galaxies are held together by gravity, not drawn apart by the expansion of the universe. Clusters are the largest gravitationally bound systems in the universe; everything larger is being stretched by cosmic expansion. The distinction between clusters and groups is that the orbital velocities of galaxies in groups are low—100 or 200 kilometers per second—while in clusters the velocities are ten times faster, about 1,000 kilometers per second. This is because clusters are higher in density than groups: Galaxies in clusters orbit faster for the same reason that, say, Saturn's satellite Dione, caught in the powerful gravitational field of that massive planet, completes an orbit comparable in radius to that of our moon's in only a tenth the time (2.7 days for Dione, 27.3 days for the moon).

On larger scales than clusters, since we have now left gravitationally bound systems behind and are dealing simply with various degrees of density enhancement, classification becomes somewhat more subjective. Some astronomers identify what they call *clouds,* associations measuring 30 million light-years or so in diameter and often linked together in strings called *filaments* and *spurs.*[6] Others prefer to skip directly to the term *supercluster.*

Superclusters typically measure 100 million light-years or more in diameter and contain something like ten thousand galaxies each. Alan Dressler aptly describes them as "extensive plateaus of modestly enhanced galaxy density . . . that reach like vast, lacy bridges to neighboring clusters."[7] Once thought to be the largest structures in nature, superclusters are now understood to be subordinate to enormous walls or sheets, sometimes called supercluster complexes, that can span a billion light-years in length. That's more than 5 percent of the radius of the observable universe. Efforts are under way to chart these gigantic structures, and preliminary results suggest that they form curved walls enclosing vast voids or bubbles, within which galaxies are scarce and superclusters nonexistent. The voids measure approximately 300 million light-years in diameter. Dense superclusters are found at points where bubble walls intersect. If the bubble picture proves accurate, the distribution of bright matter in the universe resembles the structure of a Swiss cheese or a sponge.

Looking out from our perch in the Milky Way galaxy, we can map our cosmic environment in terms of this five-part hierarchical scheme. A good place to start is by looking toward the constellation Virgo, home of the richest concentration of bright galaxies in Earth's skies. There, some 60 million light-years from us, stands the Virgo Cluster, an irregular aggregation of 1,170 prominent galaxies—and, no doubt, of thousands of dwarf galaxies too dim to have yet been detected. The Virgo Cluster is centered on a pair of giant elliptical galaxies that fattened themselves by swallowing spiral galaxies that fell into the center of the cluster. The path leading from the Virgo Cluster to the Milky Way is a Louvre gallery of picturesque galaxies. Along it are to be found landmarks like Centaurus A, a giant elliptical caught in the act of eating a spiral galaxy; the graceful spiral M81, sailing past its companion, M82, which is going through a fiery storm of star formation set off by its having been hit by M81's gravitational wake; and the clotted spiral M101, with its bright attendant galaxies M51 and NGC 5055.

Why such scenic splendor along the road to Virgo? Because the Virgo Cluster has a vast halo of galaxies—analogous to the halos of stars and globular clusters that surround major galaxies—that in turn merges into a supercluster, one piece of which stretches out to our location and a bit beyond. This is the Virgo Supercluster. It has a radius of more than 100 million light-years. Its shape is somewhat flattened, rather like that of an enormous galaxy whose nucleus is the Virgo Cluster. Most of its galaxies belong to clouds, among them the Virgo II and III clouds, the Leo II cloud, and the Canes Venatici cloud and spur. (Like most such structures, these are named according to the constellations through which we see them.) The clouds in turn contain many groups. In all, the Virgo Supercluster contains at least eleven clusters or clouds at its core, and perhaps fifty more in the halo.

Pausing for a moment to take in our immediate neighborhood, we find that the Milky Way belongs to an ordinary group of galaxies known to earthlings as the Local Group. Like other such entities, the Local Group has a dominant spiral (the Andromeda galaxy, M31) paired with a second-ranking spiral (the Milky Way). There is also a dimmer but still impressive spiral called M33. The other Local Group galaxies are mostly inconspicuous dwarf spirals, irregulars, and ellipticals. In addition to its twenty-six known galaxies the group probably harbors others, hidden and as yet undiscovered behind the coal-black dust clouds that clutter the disk of the Milky Way. Bound together gravitationally, the galaxies of the Local Group form a genuine association, like a school of fish or a flock of birds; they don't just happen to be in the same place at the same time. The Andromeda spiral and the Milky Way are currently approaching each other, at a velocity of 300 kilometers per second. They will perform a do-si-do, a couple of billion years from now, then move apart as they continue to pursue their orbits around the Local Group's center of gravity.

The Local Group stands toward the outer edge of a large cloud, variously called the Local or Coma-Sculptor cloud, which itself lies toward the edge of the Virgo Supercluster. So once again we find ourselves on the fringe of things. Not only does the earth not occupy the center of the solar system (as Copernicus asserted, and before him Aristarchus of Samos), and the solar system reside far from the center of the Milky Way galaxy (as Harlow Shapley ascertained), and the Milky Way galaxy take second place to Andromeda as the dominant spiral of the Local Group, but the Local Group is situated toward the outer limits of the Virgo Supercluster. Cosmography is discomfiting to anthropomorphists.

The largest known structures—multiple superclusters arrayed like the walls of bubbles, enclosing cosmic voids—begin to emerge when charts are drawn on scales approaching a billion light-years in diameter. These maps are crude: They extrapolate the distances of millions of galaxies from data for only a few thousand, and unquestionably contain many errors that will bemuse astronomers of the future. But they do show that superclusters are in many instances connected to neighboring superclusters by bridges and tendrils, forming supercluster complexes. One example is the Hydra-Centaurus-Pavo Supercluster, which stands off to the left of the constellation Virgo from the standpoint of an observer in the Northern Hemisphere of Earth, is situated a couple of hundred million light-years away from us, and appears to be linked with the Virgo Supercluster. Similarly, the Perseus-Pisces Supercluster forms a continuum with the Sculptor complex. The Leo and Leo-Coma superclusters, too, may be linked. The hypothesis that supercluster complexes like these are bubble walls encompassing voids is reinforced by consideration of rich streamers like the Great Wall and the Cetus Wall, supercluster complexes more than a billion light-years long that appear where the walls of several bubbles intersect.

Most of the surveys on which the large-scale maps are based derive distances for galaxies based on nothing more than their redshifts. The redshifts yield approximate distances in an expanding universe, but since the galaxies also have velocities produced by the gravitational fields of the structures to which they belong, such maps are distorted. When surveying a cluster, for instance, astronomers using redshifts alone will overestimate the distance of galaxies whose orbits in the clusters are carrying them away from us, and underestimate the distance of those that happen to be on the approaching side of their orbits. As a result, clusters in redshift maps appear drastically elongated on an axis pointing to Earth, a phenomenon known, for some reason, as "fingers of God." Similarly, galaxies on the near and far sides of walls are drawn by gravity toward the middle of the wall: Those nearest us therefore will be mapped as too far away (since they are accelerating away from us, toward the wall), and those on the far side will seem too near, with the result that the wall will map out as thinner than it actually is. These distortions can be compensated for, however, and the maps themselves still represent quite an achievement. Given that the very existence of galaxies was not established until 1925, we can say that in less than seventy-five years the reach of human cartography has expanded ten thousand times—from the scale of the Milky Way galaxy to that of bubble walls. Since another leap of ten to twenty times outward would exhaust the observable universe, this great leap may forever represent the most ambitious cartographic advance of all time.

The revolution in intergalactic mapmaking was slow in starting, for want of both data and an adequate theoretical framework within which to interpret the data. The most obvious sign of extragalactic structure is that bright galaxies prefer a swath of sky that describes what is now understood to be the plane of the Virgo Supercluster. This was noticed as early as 1784, when the perspicacious English astronomer William Herschel remarked that most of the hundreds of "nebulae," meaning fuzzy objects, that he observed in the sky apart from the Milky Way were found in a broad belt—a "stratum," Herschel called it—perpendicular to the Milky Way. (Our galaxy happens to be oriented so that it presents an open face to the Virgo Cluster, which must make it a nice sight for anybody in Virgo with a telescope.) The Swedish astronomer Knut Lundmark in the 1920s similarly noted that bright spiral galaxies seem "to crowd around a belt perpendicular to the Milky Way."[8] Strong evidence of hierarchical structure showed up in galaxy catalogues compiled by Harlow Shapley, by Fritz Zwicky—and by George Abell, whose classification of "Abell clusters" of galaxies has become increasingly helpful with the passage of time. (Abell, like Zwicky, was somewhat underestimated in his day, though for opposite reasons: Rather than putting on fireworks shows of inspiration he worked steadily on a few good ideas, and his cheerful, good-natured accessibility—he was an adept science popularizer, and one of the most popular teachers among undergraduates at UCLA—led some colleagues who hadn't read his papers to take him for a lightweight.) Abell's study of nearly two thousand galaxy groups showed evidence of superclustering, and when Neta Bahcall of Princeton examined Abell clusters in the 1980s, her findings showed that clusters are even more likely to be found in superclusters than galaxies are in clusters—that is, isolated clusters of galaxies are statistically rarer than isolated galaxies.

Research that led to identification of the Virgo Supercluster was begun, midway through the twentieth century, by the French-born American astronomer Gerard Henri de Vaucouleurs. Born in Paris, de Vaucouleurs worked at Mt. Stromlo in Australia and later at the University of Texas at Austin, where he died in 1995 at the age of seventy-seven. His mixed cultural background was reflected in his costume, which in public almost invariably combined gray Parisian dress gloves with a ten-gallon hat. He had an encyclopedic knowledge of the astronomical journals, and he marshaled data gleaned from a great many papers, as well as his own research, to support his argument that galaxies are clumped together in clusters and superclusters.

For years this work was received with "resounding silence," as de Vaucouleurs recalled it.[9] The cosmologists—concerned with the big picture, in which the universe was (and is) presumed to be homogeneous—regarded local inhomogeneities as parochial at best and an irritation at worst. They were building railroads, and did not care to be told that here and there a few ties were crooked. Also, there were not then enough data on galaxies to prove much. Astronomers had redshifts for only about one hundred galaxies by 1950, and perhaps two thousand by 1970, and independent distance estimates had been obtained for only a fraction of those. Today redshifts are now on file for more than one hundred thousand galaxies, and the data rate is increasing exponentially. One project alone, the Sloan Digital Sky Survey, is expected to collect the spectra of one million galaxies and thirty-five hundred rich clusters. The result of all this work has been to exonerate the claims of Herschel, Lundmark, Zwicky, Abell, and de Vaucouleurs that the *in*homogeneity of the universe—the organization of matter into coherent structures—extends to larger scales than most theorists had imagined.

Watershed observations were conducted in the 1980s by a graduate student at the Harvard-Smithsonian Center for Astrophysics, Valérie de Lapparent, under the guidance of the astronomers John Huchra and Margaret Geller. Huchra and a few others, Marc Davis of Berkeley among them, had become interested enough in large-scale structure to conduct a survey that obtained redshifts for twenty-four hundred galaxies. This search hinted at structure, so a second, more systematic project was undertaken, this time to map the redshifts of a thousand galaxies across a specific strip of the northern sky. (Since any angle of sight expands as it is extended into space, the strip describes a three-dimensional wedge.) Huchra and Geller had not really expected to find anything interesting. "I recognized that if there was structure there we would see it," Geller recalled. "But I didn't expect to see it, because the party line was that it wasn't there."[10] Consequently, they were in no hurry to have the data reduced. When de Lapparent finally presented her results, the presence of huge bubbles and walls made of galaxies stood out with startling clarity. Huchra suspected that a mistake had been made. Geller was quicker to accept the result as genuine, thus abandoning the "party line" that there were no structures that large. "I have a strongly held skepticism about any strongly held beliefs, especially my own," she said recently.[11]

The original Geller-Huchra wedge has become one of the iconic images of contemporary cosmology. It displays the "fingers of God" and other distortions characteristic of redshift maps, but the Great Wall shows up distinctly, as do at least two of the bubbles whose walls it helps define. The Great Wall is a billion light-years long yet only a few tens of millions of light-years thick. Any future version of the big bang theory that is to be taken seriously as advancing the state of the art will have to account for the formation of the 300-million-light-year voids and their skins made of galaxies.

Suddenly finding themselves surfing one of the biggest waves in science, Geller and Huchra set to work mapping wedges adjacent to their first survey. Soon they were able to graduate from what Geller refers to as the "crummy" telescope they had originally used to a more appropriate instrument, and by the late 1990s they expected to be observing with a giant 6.5-meter telescope equipped with a robotic, fiberoptic "octopus" capable of acquiring more than fifteen hundred galaxy redshifts per night.

As such surveys continue, it's possible that even larger levels of hierarchy will be found. Certainly nature displays a self-similarity on many spatial scales. As the physicist Philip Morrison of MIT points out, if we span the universe from nuclear to intergalactic dimensions, we encounter a two-beat rhythm: Areas of emptiness—the voids between the nucleus and the electron shell of the atom, between stars and their planets, between galaxy clusters and voids, and so forth—are interspersed with high-density regions, like atomic nuclei, molecules in crystals, and galaxy clusters and walls.[12]

Self-similarities like these lead some theorists to maintain that the universe has a *fractal* geometry. Whereas familiar geometries are based on whole numbers like 2 or 3, fractal geometries are based on fractions. One can compute fractals in, say, 1.2 dimensions. Structures built from fractal geometry are self-similar, displaying repeating features on various scales. The forking of big branches near the base of an oak tree, for instance, is of much the same form as the little forks of twigs near the tips of its branches and the forking veins in its leaves. The mathematician Benoit Mandelbrot created a stir in cosmology in the 1970s by arguing that galaxies on the superstructure scale are organized as if generated by a fractal geometry of 1.23 dimensions, while on larger scales (where homogeneity takes over) they resolve into a classical 3D picture. For Mandelbrot, Alan Dressler's "vast, lacy bridges" made of galaxies resemble the meteorologists' cirrus clouds, which in his view are also fractals.[13] Few cosmologists have been able to do much with this suggestion, though James Peebles incorporated it into some of his research. If the universe turns out to be built on a fractal geometry, or if for some other reason it consists of ever-larger structures replicating the features of smaller ones, then the assumption of homogeneity is false and the large-scale structure problem is probably going to be very difficult to solve.

It seems more likely, however, that the great walls and giant voids that Geller and Huchra and others have been mapping will prove to be either the largest or at least the penultimate structures in the observable universe—with the top level perhaps consisting of an occasional "superbubble" in which several bubbles have nested and merged. The cosmic microwave background suggests that this is the case: While its inhomogeneities are very interesting, the CMB is generally homogeneous, a situation that would seem to preclude ever-greater hierarchies of cosmic structure. Studies of remote quasars point to the same conclusion: They display a degree of clustering in the young universe not greatly different from what the maps are showing around us today.

Another body of evidence suggesting that astronomers are approaching the upper limits of structure hierarchy comes from "pencil beam" surveys. In the pencil beam approach, observers limit themselves to a small piece of sky—typically about half the diameter of the full moon—and obtain redshifts of every galaxy they can find there, from bright, nearby systems to dim smudges at the very limits of detection. If Geller's and Huchra's technique is like digging a trench, pencil beam observations are like digging a well. From the pencil beam data, one makes a chart in which increasing redshift is the horizontal axis and the numbers of galaxies found at various redshifts are plotted as peaks. The technique is fraught with sophisticated statistical perplexities. Since the long, thin cone of the pencil beam line of sight takes in ever-larger volumes of space as it extends ever deeper into intergalactic space, it is more likely to miss nearby structures—clusters, say, that the beam happens to poke right through without hitting a galaxy—than structures farther away, where the beam is larger and the odds of hitting a galaxy correspondingly greater. On the other hand, the farther one looks, the brighter a given galaxy has to be for the survey equipment to record it at all. The beam might intersect a dozen major galaxies in a distant cluster but fail to detect any of them, yet record an even more distant galaxy that happens to be abnormally bright. Furthermore, in an expanding universe the density of galaxies increases with lookback time—the effect of which is to *increase* one's chances of hitting more distant galaxies. The ranks of the pencil beam data analysts include some of the more sophisticated mathematicians to have lent their talents to astronomy. They roam the globe, going from conference to conference, mulling over thin cones that they speak of as "probing" space, as did those ancient Greeks who thought of vision as a beam projecting from the eye to illuminate a benighted world.

The farthest-reaching pencil beam surveys conducted to date produce a chart of galaxy distribution against redshift that looks like a picket fence in which the pickets stand tall where the beam crosses a wall or rich supercluster and short where it hits less dense concentrations of galaxies. The separation of the pickets, though, is fairly regular.[14] The regularities indicate concentrations of galaxies spaced at intervals of about 300 million light-years, with a few possible 600-million-light-year superbubbles showing up here and there. If the data are to be believed, the largest recurring structures in the universe are superbubbles with diameters no more than twice that of the bubbles already being mapped locally. The Johns Hopkins astrophysicist Alexander Szalay and his colleagues, who obtained these results, concede that they "are tentative at present, and possibly unappealing."[15] Still, if larger structures existed, one would expect them to have shown up in these pencil beam searches, since the searches go so deep. Putting together two of the beams end-to-end, one pointing out each galactic pole, we find that the astronomers have probed *six billion* light-years of cosmic space, turning up dozens of concentrations of galaxies. So it may be they have finally found the limits to cosmic structure. "I think we have discovered a characteristic scale in the universe," Szalay asserts. "It's not a random distribution of little blobs, big blobs, bigger blobs. It's a characteristic size."[16]

So much for large-scale mapmaking. Now for dynamics—the motions of galaxies imposed by these vast structures and superimposed on the expansion of the universe.

Were matter distributed homogeneously throughout the universe on, say, all scales larger than galaxies, then cosmic expansion—the stretching of space itself—would be the only important large-scale motion, and its rate, the Hubble constant, could be measured with relative ease. Cosmologists would need to obtain the distances and velocities of only a few dozen galaxies, at distances on the order of a few tens to a hundred million light-years, and the answer would be in hand.

The existence of large-scale structure complicates this task considerably. Each concentration of mass generates a gravitational field that influences the motions of galaxies within it and, to a lesser extent, of those beyond. Groups and dense clusters exert enough gravitational force to keep them gravitationally bound: The clusters participate in cosmic expansion, but do not themselves expand. Superclusters are stretched by cosmic expansion, but their gravity retards the local expansion rate. The Local Group, for instance, is moving away from the Virgo Cluster at less than two-thirds of the rate one would expect from cosmic expansion, owing to the braking effect exerted by Virgo. Supercluster complexes tug at galaxies by the thousands. It would be difficult to interpret all these contending motions even if astronomers had accurate data on their distances, which they do not, and could see all the galaxies responsible—which they cannot, owing to obscuration by the disk of the Milky Way. (Such obscuration hides lots of galaxies, and efforts are under way to peek through the dust clouds to improve our mapping of this covert slice of the sky. Astronomers who conducted a single search in which they studied plates made in optical wavelengths—which penetrate dust poorly—announced in 1994 that they had "detected over eight thousand previously unknown galaxies" in the clouded skies along the southern Milky Way.)[17] And one must also keep in mind the gravitational influence of dark matter, which may or may not be distributed in the same way as the visible galaxies are. So the study of how galaxies move owing to the various hierarchies of large-scale structure promises to keep researchers busy for decades to come.

Fortunately there is more than one way to skin this particular cat. The cosmic microwave background provides a universal backdrop against which our motion can be charted on an absolute scale—the *Machian rest frame* of the universe as a whole. There is a hot spot in the CMB caused by our motion in one direction relative to the rest frame of the universe, and a corresponding cold spot on the opposite side of the sky. These reference points can be used to set the framework for studies of our motion and, by inference, that of other galaxies. Our own motion is complicated—the earth orbits the sun at a velocity of 30 kilometers per second, the sun orbits the galactic center at 220 kilometers per second, and the galaxy traces an orbit in the Local Group, which in turn is involved in a tug of war among cosmic expansion and the gravitational attraction of the Virgo Cluster and structures beyond. Such motions can in some instances be attributed to the presence of known structures and in other instances may imply the presence of structures that have not yet been seen. The history of extragalactic motion research thus cuts both ways, with the existence of known structures being used to account for peculiarities of motion (like Virgo's retardation of the Local Group's expansion velocity) and motions being used to tip astronomers off to the existence of previously unmapped mass concentrations (like the Great Attractor). Earlier in this book we envisioned the curved space of general relativity by imagining how one might trace the contours of a landscape by looking down at plow horses walking across it. Pangalactic structure dictates the curvature of space in our part of the universe, and we can map this terrain if we understand how the galaxies within it are moving.

Evidence for large-scale galaxy motions not due to the expansion of the universe began to crop up in 1976, when Vera Rubin, Norbert Thonnard, and W. Kent Ford collected data on 96 spiral galaxies beyond Virgo and found that they all appeared to be flying toward Perseus at nearly 500 kilometers per second. A few big-name astronomers paid heed to the "Rubin-Ford anomaly," as it came to be called. One was Fred Hoyle, who welcomed the anomaly as potentially undermining the big bang theory. Another was Allan Sandage, whose data on some southern galaxies had been cited by Rubin and her colleagues in support of their findings. Sandage, a big bang traditionalist, was unenthusiastic about having his research enlisted in support of so exotic a result, and he suggested that the Rubin-Ford effect might be caused by a sampling bias, but he did not rule out the disquieting possibility that it was real. Most astronomers, however, ignored the finding. They shared the basic assumption of large-scale homogeneity, and if questioned pointed to the cosmic microwave background, which since its discovery in 1964 had been measured by rocket- and balloon-borne instruments and was known to be highly isotropic, just as one would expect if matter was distributed smoothly throughout the universe.

But then, less than a year after the Rubin-Ford paper was published, George Smoot, Marc Gorenstein, and Richard Muller announced that high-altitude observations conducted from a U-2 aircraft under their direction had detected the CMB anisotropy—the hot spot indicating that the Local Group is moving rapidly relative to the cosmic reference frame. The vector of our motion implied by the CMB result was not the same as that specified in the Rubin-Ford paper—directions are a bugaboo of large-scale dynamics studies—but with two entirely independent studies producing high velocities for the Local Group, opposition to the notion that we're all going *somewhere* promptly evaporated.

Rubin, whom we encountered previously as doing the lonely and unheralded galaxy-rotation work that proved central to the study of dark matter, reacted to the turnabout with her customary aplomb. "We got this result, and not a single person who looked at it thought I could really have this large a motion," she recalled. "Then, six months later, the microwave dipole was discovered, and showed almost the same velocity. After that nobody said we couldn't be moving that fast. Instead, everybody said it [i.e., her result] doesn't agree with [the CMB] direction.

"I never really doubted that our sample showed what it did," she added cheerfully. "Astronomy is still an observational science. We're in for a lot of surprises, and you don't detect them except by observing. Even the relatively nearby universe is complicated, and our very simple models of uniform expansion may have to be modified. . . . It's sort of nice, I think. It's nice to be doing cosmology relatively nearby. You don't always have to be observing things near the edge of the universe."[18]

Having helped to redeem another previously disreputable field of research, Rubin could watch with pleasure as it flourished. In 1981 de Vaucouleurs and colleagues published a study of several hundred galaxies indicating that the entire Virgo Supercluster is moving, apart from cosmic expansion, at a rate of nearly 500 kilometers per second. In 1985 Sandage and Gustav Tammann found that the Virgo Cluster has a proper motion of over 600 kilometers per second, toward Hydra. And then, the following year, came the Seven Samurai, with word that the Local Group is part of a flotilla of thousands of galaxies that share a bulk motion of over 630 kilometers per second, toward an as-yet-unmapped mass concentration—the Great Attractor.

The term occurred to Dressler spontaneously, during a press conference at an American Physical Society meeting in Washington, D.C., in May 1987. "While trying to explain the enormity of a supercluster mass capable of pulling in galaxies on a cosmic scale, the name 'Great Attractor' slipped out, as I waved my hands groping for words grand enough to describe the universe," Dressler recalled.[19] The equally provocative moniker "Seven Samurai" to describe Dressler and his six collaborators was coined by Amos Yahil during a 1986 Santa Cruz workshop on nearby galaxies: "What are we to do with these seven, these, these . . . these Seven Samurai," growled Yahil, who was not entirely pleased with their findings.[20] Neither term sits well with Dressler, who notes that "Great Attractor," in attracting publicity and ridicule for its grandiosity, "gave us both wings and an albatross, and my fellow Samurai were happy to pass on both the credit and the blame."[21] But the work itself has stood up, and provides evidence for the largest-scale complexities of cosmic structure yet divined by measuring galaxy motions. Presented to what Dressler describes as a stunned audience of eminent scientists at an Aspen, Colorado, conference in 1986, the Seven Samurai's study indicated that all the galaxies in our cosmic neighborhood are being pulled in the direction of the constellation Centaurus. The direction of this gigantic drift is not far in the sky from the CMB hot spot.

What's doing the pulling? The Centaurus part of the Hydra-Centaurus-Pavo Supercluster lies that way but cannot be responsible, as it, too, evidently is being pulled in the same direction. So the Samurai speculate that the Great Attractor, presumably an extremely rich supercluster complex, lies farther out, at a distance of perhaps 200 million light-years. In obedience to the military adage that decisive battles are always fought on terrain located at the intersection of two inaccurate maps, the hypothetical Great Attractor is hidden behind the dark clouds of the Milky Way's disk, so investigation of it will prove difficult. Further study could, however, provide direct evidence of its mass, preliminarily estimated at roughly that of a few hundred thousand galaxies the size of the Milky Way. That information would be extremely useful in determining the overall shape of space and the expansion rate of the universe.

A map in which mass density is plotted vertically shows the Virgo Supercluster and its neighboring superclusters as hills, separated from the towering mountain of the Great Attractor by a gently undulating valley. On one level, this is just another reminder that we are not at the center of the cosmic action. But in terms of big bang theory it is powerfully suggestive. Since these structures are presumed to have begun as density fluctuations in the big bang itself, they could provide important clues in finding out how structure formation got started.

Which leads us back to the beginnings of things. The standard big bang model assumes that gravity alone formed cosmic structures ("because," as James Peebles irreverently puts it, "no one has been able to think of any other reasonably effective force on such large scales"[22]). According to this *gravitational instability* picture, some regions of the primordial plasma were denser than others, and their slightly stronger gravitational fields thus attracted mass from surrounding regions. The denser regions became galaxies, clusters, and superclusters—in what order they did so is a bone of contention—while the less dense regions became voids. The growth of clusters and voids that began in the big bang continues today, as we see from galaxies drawing toward structures like the Great Wall and away from the voids that separate them.

The gravitational instability picture dates from early in the twentieth century, when the British astrophysicist Sir James Jeans calculated what is now known as the *Jeans mass,* the density and temperature at which a gas cloud of a given mass will collapse under the influence of its own gravity. Had the universe begun as a nonexpanding pond of gas, random motions would eventually have produced regions dense enough to attain the Jeans mass and collapse. But in the expanding plasma of the big bang, the overall matter density dropped too rapidly for that. So we need seeds of some sort—high-density areas robust enough to build to the Jeans mass even as the overall density drops due to cosmic expansion. Since the presence of matter dictates the contours of space, the problem can be viewed in terms of pure spacetime geometry. As Peebles puts it, "In the search for a theory of the origin of large-scale structure, we are looking for something that caused ripples in the geometry of the universe."[23] These seeds need not be too big —indeed they cannot be very big; otherwise neither the cosmic microwave background nor the universe today would be as homogeneous and isotropic as it is—but they have to be dense enough to get gravitational collapse going, and they have to have come from somewhere. One can of course simply write them in as "initial conditions," and in practice that is often done, but ultimately one wants a physical explanation and not just a *deus ex machina.*

The most promising way to account for initial density perturbations is to ascribe them to quantum flux—the random occurrence of high-density regions here and there in the primordial material. (Since random variations are often defined as "noise" in science and engineering, quantum flux is also called *quantum noise.*) Quantum physics not only permits but indeed requires that such regions occur. The quantum approach was stressed by John Archibald Wheeler. We have encountered Wheeler in his role as a leading theorist in black hole research. He worked with Einstein and knows relativity inside out, but, as Einstein was not, he is equally happy working with quantum mechanics. The quantum principle, he likes to say, is *the* principle of nature. Wheeler has since retired, but when he was in the lecture hall he embodied Niels Bohr's dictum "Learn by teaching." He would put knotty problems from his own research up on the blackboard and try to solve them in the classroom with the help of his students, guiding them through roller coasters of free association ranging across broad landscapes of physics, mathematics, philosophy, and common sense. He educated pretty much an entire generation of American physicists, and part of his generous intellectual legacy is the application of quantum precepts to cosmology in general and to structure formation in particular.

Physicists do not yet have an adequate account of how the universe got from quantum flux to large-scale structure, in part because they do not yet have a theory that enables them to calculate events transpiring during the Planck epoch—the first tiny fraction of time (the first 10^{-43} second, to be exact). Encouraging developments have come in quantum cosmology, unified theory, and the inflationary hypothesis. We will report this good news in chapters to come. For the present, let's assume that quantum flux provided the seeds for structure formation. And, like most theorists, let's assume that these fluctuations were random ("Gaussian," in the jargon). This is the simplest assumption, and it avoids the God-like invoking of initial conditions designed to make the equations come out the way we want them to by putting in what we want at the start.

What, then, might have followed? There are basically two approaches to the theory of structure formation via gravitational instability—bottom-up and top-down.

Bottom-up theories assert that the primordial seeds formed protogalaxies, with masses only about a million times that of the sun, and that these objects subsequently got together to form galaxies, groups, clusters, and so on. Important work was done in this area in the 1960s and early 1970s by Peebles and his colleagues at Princeton, among others. But the bottom-up approach, though it could with some stretching indicate how galaxies might have formed, has great difficulty accounting for the vast superclusters, supercluster complexes, and voids that have since been observed.

Top-down theories claim that the largest structures formed first, thereafter subdividing into clusters, groups, and galaxies. The top-down approach was explored in the early 1970s, independently by Edward Harrison of the University of Massachusetts at Amherst and the seemingly ubiquitous astrophysicist Yakov Zeldovich of the University of Moscow.[24] Consequently, its prediction of what the primordial density spectrum would look like is known as the Harrison-Zeldovich spectrum.

Harrison, a polymathic scholar, went on to write a series of books in which he addressed such simple but profound riddles as why the sky is dark at night. His answer is that the total energy of the universe—that is, the energy that would be released were all matter converted into energy—is insufficient to light up the sky. The distant fires of the big bang, envisioned as the poet Edgar Allan Poe's "continuous golden walls of the universe" and perceived today as the cosmic microwave background, are dim, and the aggregate stars don't shine brightly enough to do the job.[25]

Zeldovich, who died of a heart attack on December 2, 1987, at the age of seventy-three, ranked among the brightest stars in the dazzling constellation of twentieth-century physics. An unfettered thinker with powerful appetites for ideas and the pleasures of mortal life, he was a vodka connoisseur, who when going out drinking in Moscow would wear his many state medals, the better to intimidate police officers charged with arresting drunks. He read extensively in German and French literature (his mother, a member of the Union of Soviet Writers, translated Proust into Russian) and was fond of Proust's remark that "the highest praise to God is the unbelief of a scholar who is sure that the perfection of the world makes the existence of gods unnecessary."[26] He knew how to laugh and, like most people who have a sense of humor, was deeply serious at the root. As the physicist Andrei Sakharov recalled, Zeldovich "was almost childishly delighted when he had managed to achieve some important piece of work, or had overcome a methodological difficulty by an elegant method, and felt failures and errors keenly."[27] He brought to his theoretical work an impeccable instinct for how nature actually works. In structure formation he is remembered particularly for having shown how, in a top-down model, galaxies would be organized into vast, flattened sheets—"Zeldovich pancakes"—that look a lot like the Great Wall and similar membranes now turning up at the intersections of cosmic bubbles.

Zeldovich died knowing that the cosmic microwave background should contain a historical record of just how matter was clumped when the universe was half a million years old—a record that he expected would conform to the Harrison-Zeldovich spectrum. The background formed, we recall, when primordial matter thinned out sufficiently to become transparent to light—the era of ***photon decoupling***. It's a kind of gigantic Jackson Pollock canvas covering the entire sky, full of coded information about the transpirations of early cosmic history. For our present subject, what counts are the codes that reveal how matter was distributed during the big bang. Photons coming from denser regions had to climb out of the deeper gravitational potential wells formed by stronger gravitational fields. Therefore light coming from dense regions is a bit dimmer than light from surrounding areas. It was to scrutinize this tapestry that the COBE satellite was launched. It gathered data for two years, mapping the microwave background across great swaths of sky, looking for variations in its brightness. Scientists around the world eagerly awaited the results. "You could say we're close to a crisis, but the truth is, we're getting down to the point where we should see fluctuations," said Dressler. "We are now positioned to see them—and boy, we'd better see them, or otherwise these models are wrong."[28]

On April 23, 1992, George Smoot of Berkeley and his colleagues on the COBE project presented the long-awaited map they had pieced together from the COBE satellite data. It showed huge structures painted across the primordial sky. If Harrison and Zeldovich were right, the difference in temperature of the hot and cold regions would be about one part in 10^{-5}—one one-hundred-thousandths of a degree. The difference detected by COBE was 5.5 times 10^{-6}, very close to the Harrison-Zeldovich value. "What a triumph for the gravitational instability picture!" exalted J. Richard Gott III of Princeton.[29] Later he added, "I always said this was the test: COBE must yield a Zeldovich spectrum and random Gaussian noise. It did."[30] Other physicists were similarly unrestrained in their enthusiasm for the COBE result. Stephen Hawking called it "the scientific discovery of the century—if not all time." Joel Primack of the University of California at Santa Cruz ranked it as "one of the major discoveries of the century. In fact, it's one of the major discoveries of science." "They have found the Holy Grail of cosmology," declared the University of Chicago's Michael Turner.[31] Whatever one may make of these effusions—physics by now sports more "holy grails" than the Vatican Museum—certainly the COBE result provided powerful reassurance that the theorists were on track in proposing that cosmic structure formation, which has created the largest objects in nature, resulted from random quantum flux, the smallest phenomenon in nature.[32]

But a great deal remains to be understood. A central part of the problem is the riddle, discussed in the previous chapter, of what cosmic dark matter is made of. Even if it may be said that humans now know that the universe built superclusters from seeds born of quantum flux, we don't yet understand what recipe it used. Basically there are two sorts of cookbooks, using two kinds of particles, hot and cold. "Hot" means particles that, at the time of photon decoupling, were moving at velocities close to that of light. "Cold" particles would have been moving more slowly. The two theoretical approaches are known as "hot dark matter" and "cold dark matter."

The leading hot (i.e., fast-moving) dark matter candidate is the neutrino. Neutrinos interact with ordinary matter only via the weak nuclear force, which is both weak and short in range. Consequently, they are impressively aloof. Billions pass through your body and mine every second, yet do no harm—as far as they are concerned, we might as well not be here—and as many hit us from below as from above, having flown unscathed through the earth. As we've noted, lots of neutrinos were released as products of nuclear reactions in the big bang. But neutrinos with mass do a poor job of forming cosmic structure. In computer simulations, they tend either to make galaxies too slowly or to promote more large-scale structure than has been observed.

Cold dark matter models, in which the dark matter is slow-moving particles, perhaps of a supersymmetric variety, seem more promising. But they, too, have problems. At the present state of the art they can be made to fit either today's cosmic structures or those seen by COBE in the microwave background, but not both. The cold and hot models make a number of differing predictions, which should make it possible to choose between them by observation. One important distinction is that they give differing answers to the puzzle of whether dark matter and bright matter are distributed similarly—whether, for instance, the cosmic voids are empty or full of dark matter.

There are also "mixed" dark matter theories, which invoke both hot and cold particles. "It looks like it could be we need both," said Marc Davis. "The last several years have seen accumulating evidence that the simplest dark matter models were just too simple." The economy of nature usually militates against theories that ascribe one result to two different agencies. As Davis put it, "We come to this conclusion reluctantly. Nobody likes to make their theories more complicated."[33] But this may be the odd case where a dualistic explanation is the right one. After all, it is possible to be *too* conservative when it comes to declaring what sort of matter might exist across the wide and varied universe. A century ago, nobody took seriously photons, neutrinos, quarks, or any of a hundred other varieties of subatomic particles that today are known to exist and have repeatedly been observed in physics experiments. Given, then, that the 1 to 10 percent of the universe that is made of bright matter is fairly complicated, it may not be too farfetched to imagine that dark matter, which makes up the great majority, might be composed of more than one sort of particle. One auspicious mixed model, cooked up by Davis with his Berkeley colleagues F. J. Summers and David Schlegel at about the same time that the COBE results were announced, uses a recipe of 30 percent hot and 70 percent cold dark matter.

Researchers are busily examining the cosmic microwave background in greater detail. Some fly detectors aboard high-altitude balloons lofted from the frozen plains of Antarctica where the dry, stable air is favorable to microwave astronomy. Others continue to gather data from the COBE satellite; and a second, much more sensitive COBE-style satellite has been approved by the European Space Agency for launch in the year 2004. It would map perturbations on smaller scales. (Those shown on the COBE map were all larger than the horizon of the observable universe at the time the background radiation blossomed into existence.) Some want to use the CMB to check on the density and expansion rate of the universe. And, of course, everyone is intrigued by the fact that if the perturbations did indeed originate as quantum flux events, then the CMB offers a window onto the quantum behavior of the very early universe. Until COBE, writes Joel Primack, "we were building cosmological theories on sand. We had scant information about the initial conditions and the events of the crucial first second, and we had only speculative theories about the origin of the galaxies and their large-scale distribution in the universe. The new COBE data has finally allowed us to begin building cosmological theory on solid ground."[34]

Lurking behind much of this work is the possibility that we inhabit a critical density universe—one in which cosmic space on the observable scale is nearly flat. The cold dark matter theories imply as much; so do the mixed models. The dynamics associated with the Great Attractor make sense only if the universe is at critical density. (Interestingly, as Alan Dressler pointed out prior to the COBE announcement, the Great Attractor requires about the same degrees of temperature variation that COBE observed. "Basically, photons lose this much energy as they climb out of the gravitational potential well of [an] embryonic Great Attractor," Dressler notes.[35])

This brings us to the event that would have rendered cosmic space apparently flat, and amplified primordial density fluctuations in just the fashion predicted by theory and confirmed by the COBE results. Spatially it would have been the biggest thing that ever happened—in our universe, at least. It goes by the name of inflation. But before considering inflation, we need to investigate cosmic evolution and the prospect of unified theories that not only predict the existence of particles that could constitute dark matter but also imply that inflation really happened.

7
Cosmic Evolution

Creation is not the work of a moment. When it has once made a beginning with the production of an infinity of substances and matter, it continues in operation through the whole succession of eternity with ever increasing degree of fruitfulness. Millions and whole myriads of millions of centuries will flow on, during which always new worlds and systems of worlds will be formed after each other in the distant regions away from the center of nature, and will attain to perfection.

—IMMANUEL KANT[1]

Why stay we on the earth unless to grow?

—ROBERT BROWNING[2]

IF WE WERE TO EXPRESS in a single word the principal liability of the prescientific philosophies of nature—if, to put it another way, we were to name the one among their shortcomings that science has done the most to repair—I think it would be that they presumed that the universe was *static*. Many thinkers dismissed change as an illusion. In this view, the manifestations of time—"the moving image of eternity," as Plato called it—are noise, while what matters, the signal, is the invariant hum of eternal stasis. Others admitted that change occurs, but regarded it as trivial. They held that time moves in cycles, so that events, though they may seem unique and important from our limited point of view, are in the long run destined to repeat themselves, tracing out endless orbits of fatalistic destiny. The grinding of these eternal wheels, music to the ears of many a sophisticate, sounds through the works of Hesiod, Pythagoras, Plato, Aristotle, his student Eudemus, and many who followed them. Even the vast timescapes of Hindu belief, often invoked by science writers as foreshadowing today's astronomical figures—and the numbers *are* big; a thousand *mahayugas,* each lasting four billion years, make up but a single Brahma day—amount to little more than fitting larger cogs to the same old mechanism of eternal return.[3] One reads in the ancient books of *re*volution, but almost never of *ev*olution. Change occurs *in* the universe, but in the end amounts to nothing important, and the overall picture remains the same. The gods are said to look pityingly on humans naive enough to imagine that there is such a thing as progress or that anything is really altered by the fall of a city or the composition of a poem. Indeed it is the allegedly Olympian stature of such fatalism that has made it so popular with generations of philosophers and philosophy professors, and among undergraduates longing to be unhappy so that they may appear to be profound.

The major exceptions to the prevalent denigration of time were found in Judeo-Christian theology. The Jews of old obstinately refused to dismiss time as an illusion. On the contrary, they made much of it. The Hebrew account of Genesis is wholly time-bound. It ticks like a clock from the seven-day creation of the world to the unfolding of human generations. Its tradition has persisted, as in the remark of the revered rabbi who, asked by an incredulous student what could possibly be learned from modern innovations like railroads, replied that missing a train can teach you that a moment makes all the difference. The Kabbalah overtly embraces evolution, albeit as a progressive concept, declaring, "Evolution follows a path of ascent and thus provides the world with a basis for optimism. How can one despair, seeing that everything evolves and ascends?"[4]

Christians adopted a similar outlook, depicting history as a *story* in which historical events such as the birth of Jesus made important and lasting changes in the world. This was so marked a departure from Greek thought that in the year 1215, when the works of Aristotle had risen over Europe like a second sun, the Fourth Lateran Council found it advisable to denounce Aristotle's belief that the universe is infinitely old and to affirm that, for Christians, the universe had a beginning in time and was moving through a series of unique—*not* eternally recurring—events. Modern scholars miss the point when they poke fun at James Ussher, the seventeenth-century Irish bishop who counted up the "begats" and concluded that "the beginning of time . . . fell on the beginning of the night which preceded the 23rd day of October, in the year . . . 4004 B.C." What matters is not that Ussher's figures were wrong —after all, he was off by only a factor of a million, which was not all that bad by cosmological standards—but that he thought of time as having *had* a beginning, and of the world as unfolding in unexpected ways, like a play. The influence of Christianity on scientific thought constitutes an embarrassment for scientists who recall that Christians persecuted Galileo and scoffed at Darwin, and who are understandably reluctant to entangle science and religion today. But it deserves to be kept in mind, if only as an antidote to the vulgar and vigorous tendency to paint history in stark monotones of heroism and villainy.

With these notable exceptions, very few prescientific philosophies took seriously the notion that time had a beginning and perhaps will have an end, and that the universe is constantly changing as it careens through time. Less popular still was the view that the universe is *evolving*, by which I mean that it used to be much less complex and varied than it is today. Consequently the concept of cosmic evolution was largely neglected until science summoned it up.[5] Even now, evolution remains problematic. The word comes from the Latin for *unroll* or *unfold*, implying that evolution merely brings to light what had been concealed, as when a rug merchant unrolls a carpet. Yet one of the most compelling attributes of evolution is that its products are largely unpredictable. Nor do we yet have an adequate means of quantifying just what is meant by saying that nature today exhibits more "variety" and "complexity" than it did, say, when all was quark soup.

Still, no useful investigation can proceed if we must first

scotch every pertinent ambiguity, and in this chapter I will argue that evolution is at work on many scales. Not only do galaxies evolve chemically, their stars brewing hydrogen and helium into heavier elements so that old galaxies are more chemically complicated and varied than young ones, but galactic evolution results from processes operating across entire clusters of galaxies. Molecules are built within the interstellar thunderheads known as giant molecular clouds and on planets. Terrestrial life may be viewed as a baroque extrapolation of cosmic molecular evolution—an outlook that adds poignancy to the notion of some biologists that a chicken is an egg's way of making another egg and that all living creatures are but mechanisms that DNA molecules employ to make more (and more varied) DNA molecules. The very laws of nature seem to have evolved from simpler, original laws, which in turn may have arisen from a state of primordial lawlessness. Evolution may be as efficacious a concept as the quantum principle, in which case all thought that purports to anchor itself in material reality will need either to be dynamic or to encompass dynamics within some new and more sophisticated version of stasis. Dynamics and stasis will tug at one another throughout much of the remainder of this book, an essential tension. In this chapter, we examine how evolution entered into scientific thought, then inquire into evolutionary forces at work on the cosmic scale.

We begin with Charles Darwin, a man so habitually underestimated by his contemporaries that he found it necessary to defend himself against the popular slander that he had come up with nothing new. (This was an example of Alexander von Humboldt's dictum that there are three stages in the popular attitude toward a great discovery: First, people doubt its existence; then they deny its importance; and finally they give the credit to the wrong person.) It is true that Darwin had many immediate antecedents. Among them were progressivists and perfectionists like his grandfather Erasmus Darwin, who thought of evolution as a construction project through which God wrought such alterations in living beings as were required in order to arrive at a perfect end product (amazingly, *us*). But Charles Darwin's contribution was not to put forth a case for evolution as such; indeed, until the public linked his name to the term, he seldom spoke of "evolution" at all. As far as I can tell, the word *evolve* appears only once in Darwin's *On the Origin of Species,* and there it is literally the last word, concluding the book's famous final sentence: "There is grandeur in this view of life, with its several powers, having been originally breathed into a few forms or into one; and that, whilst this planet has gone cycling on according to the fixed law of gravity, from so simple a beginning endless forms most beautiful and most wonderful have been, and are being, evolved."[6] Rather, what Darwin did was to identify a mechanism through which the innate characteristics of species might change and new species thus appear. His theory was precise though not brittle, and vulnerable to experimental test. It was in short a *scientific* theory, and with it the concept of evolution ceased to be solely a philosophical issue and edged into the arena of science.

Darwinism was premature. Those who would have us reject theories that fail to explain everything from the get-go would do well to consider that Darwin's account in its initial form lacked several elements essential to its eventual success. Most important it lacked an account of *genes,* which by quantizing heritable information prevent mutations from being diluted out of existence before they can become widely established. Darwin admitted that he was unable to remedy this embarrassing inadequacy, just as he had no answer to the physicists who, for want of an understanding of how radioactive isotopes generate heat inside the earth, concluded that since the planet is still warm inside, it could not be old enough for evolution to have produced woodpeckers and whales, much less the wonderful us. Notwithstanding these difficulties, Darwin stubbornly maintained that his theory was valid, and rightly so.

Darwinism has three implications that are particularly pertinent to our concern with cosmic evolution.

First, by depicting every living creature as a product of its ancestry, Darwin helped sweep aside static conceptions of the world and set the axis of history at the center of things. As the historian Bert James Loewenberg writes, "He succeeded in putting the *whole* of past life into *every* aspect of *every* form of present life."[7]

Second, his theory invoked not just time, but *lots* of time. For a single original life-form to have diversified into the millions of species found on Earth today would have required hundreds of millions of years—so much time that Darwin, who was not mathematically inclined, sometimes threw up his hands and proposed that the past was infinitely long. If this was extravagant, the impulse behind it was sound. History is revealed by evolution to be not only central, but *big*.

Third, having erected a maypole of history and found it to be unexpectedly tall, Darwin pointed it at the stars. It is no accident that he refers in the concluding sentence of the *Origin* to Earth as a planet that has long "gone cycling on." For once we realize that the earth is old, and that everything on it has been fashioned through lengthy and interrelated historical processes, we can begin to see that our own story is intertwined with an even older tale of cosmic evolution.

In the house of cosmic evolution are many mansions. I will touch on just a few, in ascending order of scale.

On the terrestrial level, *massive extinctions*—events in which the majority of species suddenly disappeared—punctuate the fossil record. Recently they have been ascribed to the dolorous environmental effects of comets or asteroids striking Earth, touching off earthquakes, floods, and fires, the soot from which blocked out sunlight for decades. Seven global dieouts have been found to coincide with the age-dated impacts of comets or asteroids—and the same may prove to be true of the others. One of the most dramatic and best-studied mass extinctions occasioned the break between the Cretaceous and Tertiary periods, 65 million years ago, when the dinosaurs and most of their contemporaries suddenly expired. Outlines of the "smoking gun" impact crater have been mapped gravitationally in the Gulf of Mexico, off Yucatán, initially by oil geologists who knew they were onto something big but didn't know quite what to make of it, later by academic scientists bearing the lamp of the impact hypothesis. (Buried beneath newer strata, the crater is more than 100 miles in diameter.) Such catastrophes had dramatic consequences for biological evolution, by clearing the way for the emergence of new species that would otherwise have had little chance of displacing senior life-forms from ecological niches to which they had become well adapted. As we are descended from one such beneficiary—a little lemurlike creature that cowered in trees as the dinosaurs thundered by—we may owe our existence to the comet that made a mess of things 65,000 millennia ago.

More benign—though far more remote in time, and less well established scientifically—is Earth's bombardment by the icy comets thought to have abounded in the infant solar system. Primordial comets could have formed the oceans and rained down the amino acids from which life originated here. Evidence for cometary cornucopias of life-brewing water and amino acids may be found in the existence of complex carbon molecules detected in the spectra of modern comets and of satellites like Saturn's moon Titan, and in the giant molecular clouds of the Milky Way, where new stars and planets are forming today.

An evolutionary viewpoint invites us to seek out evidence of historical events in the modern-day appearance of just about everything. The solar system is particularly rich in clues relating to how the sun and its planets formed. Consider the relative sizes of the planets. Near the sun orbit the rocky worlds Mercury, Venus, Earth, and Mars. Farther out lie the gas giants, Jupiter, Saturn, Uranus, and Neptune. Each giant consists of a rocky core wrapped in ice and cold gas. Then, beyond Pluto—a maverick planet whose eccentric orbit carries it both inside that of Neptune (where it is now) and beyond—is the realm of the comets, which are made of ice and snow and rocky debris.[8] So if we place the solar system's constituents side by side, as in a police lineup, we find first a zone of small, rocky planets; then a domain of cold, gas-enshrouded giants; then a still colder sphere of icy comets. What's the evolutionary message?

The answer preferred by astronomers today consists of an updated version of the *nebular hypothesis,* in which the planets are said to have coagulated from a disk of dust and gas that originally circled the young sun. This account explains why the planets all orbit the sun in the same direction—a fact that puzzled Newton, who ascribed it to the will of God, and fascinated the French mathematician Pierre-Simon de Laplace, who found it "very remarkable."[9] To de Laplace, as to Immanuel Kant, this and several other regularities in solar system dynamics suggested that the planets had emerged from a whirling disk or whirlpool. The Kant-Laplace nebular hypothesis ruled nineteenth-century cosmogony (the science of cosmic origins) and formed part of an evolutionary outlook that portrayed the solar system as striving, in Kant's words, "to evolve itself out of the crude state of chaos."[10] Psychologically affable (something in us loves a vortex), it has considerable charm —or so it seems to me, perhaps because my first encounter with it, at about the age of six, kindled a lifelong fascination with science. But while focusing on regularities in the solar system, the nebular hypothesis in its original form failed to explain several *ir*regularities. Why, for instance, do some planets (and some of their satellites) revolve the wrong way on their axes? Why do some planets have a tilted axis of rotation? These exceptions to the rule loomed larger as astronomers learned more about the planets in detail, and an updated and strenuously revised nebular account resulted.

The modern version retains the original disk of gas and dust, but emphasizes the role of the dust. Here is one of nature's more dramatic examples of great things arising from humble beginnings. As the theory has it (and it is by far the most rigorous account of planetary formation yet devised), every planet, including Earth, began as a microscopic grain of dust in the nebula from which the sun formed. By the time the protosolar disk formed, this grain had been joined by others, stuck together by electrostatic force into a clump the size of a medicine ball. Differences in the orbital speeds of such objects in the disk resulted in gentle collisions that built the ball up to about a mile in diameter. It now ranked as one among millions of *planetesimals,* objects massive enough to begin attracting one another gravitationally rather than meeting by chance. There ensued an epoch of titanic collisions, during which planetesimals that were struck at high velocities shattered, while those subjected to gentler impacts survived and grew. The rule of thumb is that a planetesimal hit at a speed faster than its escape velocity will be destroyed, the resulting debris flying away too fast for the local gravitational field to halt its escape, while those that happened to be hit at lower velocities will capture their assailants and thus gain mass. This in turn endows them with a higher escape velocity, making them more likely to survive—an instance of the rich getting richer (or, if you like, of a Darwinian law of the jungle).

This all took place quickly, by astronomical standards. The first three acts—from micron-size dust grains to medicine balls;

from there to mile-wide planetesimals; and from planetesimals to protoplanets a few hundred miles in diameter—took only about ten thousand years each. It was, however, efficient, and by the end of Act III the protoplanets' orbits had been swept clean. The fourth act, which would merge these objects into full-fledged planets, required that some be tugged by gravitational interaction into orbits that were more elliptical, promoting collisions. This took longer—more than 10 million years. The results were worth waiting for: The construction of the solar system climaxed in a dramatic epoch when worlds collided, some of them shattering into debris, others combining to form the sun's permanent planets.

With lots of leftover scraps of rock and ice around, episodes of bombardment persisted for hundreds of millions of years thereafter, scarring the planets but seldom threatening their existence. The interplanetary hailstorm climaxed with the "late terminal bombardment," when leftover planetesimals, their orbital velocities braked by the solar wind, came spiraling down to pelt the planets of the inner solar system. On Earth, where the oldest surface rocks date back only 3.8 billion years, the great majority of the resulting craters were erased long ago by erosion, but craters on the airless and geologically inert moon still bear witness to the troubled times of old. The moon was hit by so many meteors that its crust remains littered with a layer of shattered rock 16 miles deep. One can distinguish the older terrain of the lunar highlands, pocked with craters, from the younger terrain of the relatively unbroken "seas" formed when late-bombardment impacts sent molten lava flowing over craters that had been there before. When *Apollo* astronauts collected "young" moon rocks from the seas and "old" ones from the highlands, they were gathering the tangible evidence of bombardment history.

Much of all the smashing and bashing transpired in a wider context of equally spectacular violence, with the furiously burning sun ejecting white plasma jets from its poles and sending gale-force solar winds screaming across a disk peppered with lightning bolts. This suggests why the outer planets retain lots of ice and light gas while the inner planets do not. Near the young sun, the heat and solar winds melted ice and stripped away hydrogen and other light gases. Out at Jupiter, where the sun casts only one twenty-fifth as much warmth as at the earth, the planets could retain their primordial gas—ice, too, as we see with the ice palaces of the Jovian moons. Jupiter's satellite Callisto is as large as Mercury, yet less than a third as massive; the reason is that while Mercury is made of iron and rock, Callisto is half ice. Ice becomes ever more dominant out to Pluto and into the realm of the comets, where the sun is so distant that it looks like not much more than a particularly brilliant star. Had comets not ferried ice to Earth—our "baked" little planet, as Newton called it—we might have had no oceans. And without organic molecules contributed by the comets, the earth might have remained devoid of life. So comets may well have given Earth life, as subsequently they were to take so much of it away.

But the beauty of an evolutionary perspective resides less in its facility for spinning pleasing history tales than in the framework it provides for further research. Darwin saw this clearly, and in his later years emphasized that his theory had not completed the biological sciences but rather had reinvigorated them, bringing them to life like rain on a parched garden. "How wide and rich a field for study has been opened up through the principle of evolution," he wrote. "And such fields, without the light shed on them by this principle, would for long or for ever have remained barren."[11] Much the same thing happened in the middle of the twentieth century, when the slumbering science of geology was awakened by the theory of plate tectonics, based on the realization that heat from Earth's core pushes the continents around. Mars is too small to retain enough heat to do the same; consequently, Mars is geologically inert, its evolution having stalled when the red planet was only about a billion years old. The riddle of what, if anything, this development may have had to do with the rivers and lakes of the young Mars having frozen solid back then is the sort of question addressed today by the discipline called *comparative planetology.* Cosmic evolution is not yet as accomplished a paradigm as Darwinian evolution, but it has already enlivened planetology and the other sciences that investigate the solar system. Information garnered through the use of space probes such as the manned *Apollo* missions to the moon and the unmanned missions—*Viking* to Mars, *Magellan* to Venus, and *Pioneer* and *Voyager* to the outer solar system—can be fitted into an evolutionary scheme that permits scientists to piece together the history of the sun and its planets as a solar *system*, not just a clutter of oddities.

The planetesimal model cracked the puzzle of how the sun's planets rotate. Most, the earth included, spin on their axes in the same direction—counterclockwise as viewed from the north. But Venus spins clockwise. So does Pluto. And Uranus is unique: Its north pole is tilted more than 90 degrees, right through the plane of its orbit. The planetesimal theory accounts for these anomalies by proposing that Venus, Pluto, and Uranus were struck, during the late-bombardment epoch, by objects that knocked them over or reversed their directions of rotation. This portrayal gains support from the strange case of Miranda, a satellite of Uranus. Its surface, imaged by *Voyager 2* in 1986, is a remarkable mix of old cratered terrain, young smooth terrain, and several features unlike anything else seen in the solar system. Theory has it that Miranda was hit so hard it was nearly destroyed. Its remains flew apart, then fell back together to form a strangely jumbled world. If so, Miranda is late-bombardment roadkill.

Earth's moon is a mystery. The moon is too big: It's more than a quarter of the diameter of Earth. Only Pluto has a satellite so large in relation to itself. The distinction is sufficiently striking that planetologists classify Earth-Moon and Pluto-Charon as "double planets." Moreover, the moon is made of the wrong stuff. While the earth has a massive iron core, the moon contains virtually no iron. Its density, about 3.3 grams per cubic centimeter, resembles that of Earth's mantle but not Earth's core. Nor does the moon have much in the way of volatiles—like water—with which the earth is well endowed. And yet, to further confuse matters, there are similarities between the two objects: Many minerals are found on both, and their relative abundances of various isotopes, such as those of oxygen (which in the moon is bound up in rock), are much the same. Finally, the moon's orbit is all wrong. Every other major satellite in the solar system orbits above its planet's equator. This is true even for the satellites of tipped-over Uranus. But while the earth's axis is tipped relative to the plane of its orbit by 27 degrees, the moon orbits along Earth's orbital plane, not its equator.

Taken together, these considerations seem to rule out two otherwise promising theories of how the Earth-Moon system originated. The first of these proposed that they formed together, from an eddy in the vast disk of material that surrounded the newborn sun. But were this the case, both bodies would have about the same chemical composition, so that the moon would have, say, as much iron as Earth does. Yet the moon is poor in iron. The other theory held that the moon formed elsewhere in the solar system and was subsequently captured by Earth. But studies of orbital dynamics show that such a capture would have required the unlikely intervention of a third planetary body. This is a long-odds proposition, and anyway no such object has been found. And capture theory fails to account for the similarities in composition between Earth and Moon revealed in their isotopic ratios.

Recently a new account, based on the planetesimal model and dubbed the "big splash," has provided possible answers to the riddle of where the moon came from. According to this account the young Earth, still red-hot beneath its thin new crust, was struck by a massive, contending planet approximately the size of Mars. (The interloper was one of the elliptically orbiting bodies that lit up Act IV of the planet-formation drama.) Most of the material of the wrecked invading planet was incorporated into the liquefied Earth, adding to the mass of our home planet. But vapor generated by the intense heat when the two objects collided was squirted out into space in a matter of minutes. There the vapor settled into orbit and condensed as the moon. Because this material came mostly from Earth's crust rather than its core, the moon today lacks iron and so is lower in density than Earth. But because the two objects shared a commingled origin, some chemical similarities survived. Computer simulations suggest that the infant Earth was indeed hit by at least one planet as big as Mars and also by two or three lesser but still hefty planetesimals. A similar scenario could explain why Mercury, the innermost planet, has a much larger iron core than would be expected in so small a planet. A smaller planet hitting Mercury could have vaporized much of its rocky crust, leaving behind only the core of what had been a larger planet—and, perhaps, one farther from the sun, the crash having knocked Mercury down into its current orbit. Plenty of work remains to be done before these questions are fully resolved. Meanwhile it is salutary to keep in mind, as we talk knowingly of the origin of the universe and of the far-flung galaxies, that we remain uncertain about the origin of the nearest object to our home planet. The plump, romantic moon, as it climbs in the east tonight and "releases / Twig by twig the night-entangled trees," as Archibald MacLeish put it, still wears its tattered old cloak of mystery.[12]

While the planetesimal model has made progress, thanks mainly to the growing mountains of data gathered by space probes and to computer simulations constrained by those data, the evolution of the solar system may not be fully understood until it can be compared with that of other planetary systems. The good news here is that since the Milky Way continues to make new stars, astronomers should in principle be able to observe extrasolar planetary systems at various stages in their development, from newborn stars with protoplanetary disks to mature systems like our own. The bad news is that finding them isn't easy. Stars form inside dark clouds that block their visible light. (Fortunately they can be detected in infrared light, the long waves of which can often penetrate the clouds.) Mature stars have drifted out of the clouds and can be seen more clearly, but by that time their protoplanetary disks may be gone, and planets are too dim and too close to their stars to be resolved with existing telescopes. If we lived on a planet circling Alpha Centauri, the nearest star to the sun, we should not yet have observed any of the sun's planets—not giant Jupiter, and certainly not the small inner planet Earth. Nevertheless some remarkable observations conducted in recent years suggest the presence of extrasolar planets and thus indicate that planetary evolution may compose a major part of the cosmic picture. Gains have been made in observing planetary systems at all three stages—inside molecular clouds, as mature systems in open space, and at an intermediate phase where the star has emerged from a cloud but its planets have not yet congealed.

Molecular clouds are intriguing objects—huge, ink-black bags and globs and tendrils that dot and line the galactic disk. Some are easily overlooked by the inexperienced observer, who may mistake their massed blackness for empty space. Others, silhouetted against background fields of stars, emerge as grand, grape-shaped spheroids, haughty chimneys, and rambling rivers of ink. There are about six thousand giant molecular clouds in our galaxy. Most are ranged along the spiral arms, like thunderstorms stacked up against a mountain range. They are "giant" in that they are more than 100 light-years in diameter, typically weighing up to a million times the mass of the sun—and "molecular" in that they harbor not just the free atoms of hydrogen and helium common to interstellar space, but also a variety of molecules, including those of carbon monoxide, formaldehyde, alcohol, and water. They rank among the coldest things known, with temperatures ranging from a few degrees to about 100 degrees Kelvin. But they are star-making machines, and when stars form they heat up their part of the cloud, sweeping it clean and exciting the remaining gas to glow. The resulting blisters form a class of *emission nebulae* called HII regions. ("HII" refers to hydrogen atoms in their ionized state, meaning that their electrons have been knocked off, leaving them with a positive electrical charge. "HI" designates nonionized hydrogen, which is electrically neutral.) The Orion nebula is an HII region. It presents a spectacular, blossomlike array, glowing with the red light of ionized hydrogen and the blues and key-lime-greens of ionized oxygen and nitrogen. It is dotted with embryonic star-forming regions that astronomers call "evaporating gaseous globules," or EGGs. It sits on the near side of a giant molecular cloud complex that embraces much of the constellation Orion—a patch that, even at its distance of 1,500 light-years from Earth, takes up more of the sky than is covered by an outstretched hand.

Scholars of stellar origins have calculated that newborn stars should exhibit both protoplanetary disks and *bipolar jets,* which are spurts of plasma ejected by the star from its poles. (Both result from the dynamic processes that cause a collapsing gas cloud to rotate, flatten, and eject material to dissipate angular momentum. The same thing happens, though more violently, around black holes.) Remarkably, the disks and jets of protostars have both been observed. By 1994, astronomers using the Hubble Space Telescope had found disks associated with 56 of the 110 stars they imaged in the Orion nebula. And in 1995, Hubble obtained high-resolution images of bipolar jets being ejected by young stars. The jets stretch out for billions of miles, and contain knots and bulges that constitute a historical record of when different amounts of gas were ejected. "The jets' clumpy structure is like a stockbroker's ticker tape," said Jon Morse, of the Space Telescope Science Institute, in Baltimore, shortly after the discovery.[13] "They represent a recorded history of events that occurred close to the star. The spacing of the clumps in the jet reveals that variations are occurring on several time scales," with small knots being pumped out every few years, and big ones every century or so.

Thanks to Hubble and other high-resolution instruments, we are entering an era when stars within a couple of thousand light-years can be examined for disks, jets, and other evidence of the processes thought to have produced the sun's planets. Within that range lie the vast star-birthing clouds found along the Orion and Sagittarius spiral arms of our galaxy, so there should be ample systems to study. William Herschel, the greatest astronomer of the eighteenth century, was also one of the first to appreciate cosmic evolution as a fruitful, if tentative, way of understanding astronomical phenomena. "This method of viewing the heavens seems to throw them into a new kind of light," he wrote.

> They are now seen to resemble a luxuriant garden, which contains the greatest variety of productions, in different flourishing beds; and one advantage we may at least reap from it is, that we can, as it were, extend the range of our experience to an immense duration. For, to continue the simile I have borrowed from the vegetable kingdom, is it not almost the same thing, whether we live successively to witness the germination, blooming, foliage, fecundity, fading, withering and corruption of a plant, or whether a vast number of specimens, selected from every stage through which the plant passes in the course of its existence, be brought at once to our view?[14]

Herschel's dream is coming true today. Besides seeing new protostars in Orion and elsewhere, astronomers have found intermediate cases of youthful stars that still retain their protoplanetary disks. The signal discovery came in 1983, when two astronomers who were calibrating a telescope on the Infrared Astronomical Satellite (IRAS) found a disk surrounding the bright young star Vega, located 26 light-years from Earth. The discovery spurred interest in other nearby stars that IRAS showed to be surrounded by an excess of infrared light. In 1984, observers working at Las Campanas Observatory in Chile managed to photograph a protoplanetary disk around Beta Pictoris, a star in the southern sky, 50 light-years from Earth. Subsequent observations of Vega and Beta Pictoris identified gaps in their disks, such as would have been swept clear by freshly minted planets.

Confirming the existence of mature planets is highly difficult, since astronomers do not yet have telescopes with sufficient resolution to image them directly. A time-honored, though exacting, technique for seeking out fully formed extrasolar planets consists of looking for evidence of their gravitational interaction with the stars they orbit. As a massive planet like Jupiter orbits, it pulls its host star toward it, tugging the star this way and that as the planet goes around. This effect should be observable in nearby stars. When planetary systems are oriented face-on to the earth, it should be possible to detect them by photographing the star over a period of years and charting wobbles in its path through space. Some evidence of this effect has been found for a few stars, though the data and their interpretation remain controversial. When planetary systems are oriented edge-on to our point of view, one can take spectra of the star and look for periodic blueshifts and redshifts in the star's light caused when its planets tug it a little toward or away from Earth. This approach, too, is exacting; one is trying to measure variations in a star's motion of as little as three miles per hour, about the speed of a leisurely walk. But success was reported in 1995, when the astronomers Michel Mayor and Didier Queloz at Geneva Observatory in Switzerland made headlines with news that they had found evidence of a planet orbiting the sunlike star 51 Pegasi. If their calculations are correct, the planet is big—roughly as massive as Jupiter—yet very close to the star, only one-tenth the distance from Earth to the sun. That big a planet, that close to a star, would be startlingly different from anything in the solar system. It could challenge the account of planet formation we have been describing here—unless the planet originated farther out, and was knocked into its present orbit during act four of 51 Pegasi's early evolution.

Further advances in the spectral technique soon followed. In 1996, two astronomers working at Lick Observatory in California and using a method similar to the Swiss observers' identified planets orbiting the sunlike stars 70 Virginis, in Virgo near Arcturus, and 47 Ursae Majoris, near the Big Dipper. Both stars lie within 35 light-years of Earth and are visible to the naked eye. According to San Francisco State astronomer Geoffrey Marcy, who with his research associate Paul Butler discovered the planets while on a fellowship at Berkeley, the planet at 70 Virginis is 8.1 times the mass of Jupiter and lies about as far from its star as does Mercury from the sun.[15] At that distance, it would have a surface temperature of about 185 degrees Fahrenheit. As this is well below the boiling temperature of water, the planet could support oceans and life. The planet of 47 Ursae Majoris is 3.5 times Jupiter's mass, lies at a distance comparable to that of the sun's asteroid belt (between the orbits of Jupiter and Mars), and orbits its star every 1,100 days. It, too, might conceivably have oceans of liquid water. The discovery of these less idiosyncratic planets orbiting nearby stars generated great excitement in the astronomical community. The NASA administrator, Daniel Goldin, promptly announced that the space agency was embarking on a high-priority search for extrasolar planets, and predicted that by the middle of the twenty-first century telescopes would be able to photograph "oceans, clouds, continents and mountain ranges" on Earthlike planets orbiting stars within 100 light-years of the sun.[16] Another astronomer involved in searching for mature planets, William J. Boruchi of the Ames Research Center in Mountain View, California, exclaimed, "We're finding new worlds," and compared the endeavor to "the second coming of Marco Polo or Columbus."[17]

Theoretical studies implicate supernovae in the star-formation process. According to this scenario, density waves sweeping through the galactic disk stack up gas and dust to the point that some clouds become dense enough to collapse under their own weight and congeal into stars. The most massive of the new stars, the supergiants, burn so furiously that they exhaust their core fuel and detonate as supernovae within only about 10 million to 100 million years—early enough so that when they explode they are still embroiled in the star-forming region from which they sprang. Hence the shock waves of one or more supernovae can trigger the formation of many additional stars. Although we tend to think of supernovae as rare, they are frequent enough to rank as central players in galactic ecology: On average, three stars explode every century in an average galaxy, which means that the supernova rate in the observable universe is one per second.

The debris blasted into space by supernovae is rich in radioactive isotopes, some of which are sufficiently long-lived that we could find traces of them locally, were it the case that a supernova touched off the formation of the solar system. Such isotopes have indeed been found, in meteorites—chunks of rock that were either left behind by the formation of planetesimals or knocked off larger bodies in the course of subsequent impacts. Meteorite hunters like to search for them in Antarctica, where they stand out distinctly against the snow. Many come from trails of debris left behind by defunct comets. Some have wandered in from the asteroid belt. A few were blasted loose from the surface of the moon or of Mars.[18] Meteorites are extremely useful for studying the age, chemistry, and evolution of the solar system, since, unlike rocks on the surface of Earth, they have experienced little erosion. Aside from a bit of surface ablation due to the brief heat of their fiery passage through Earth's atmosphere, they have suffered few indignities other than the gentle wafting of the solar wind and "micrometeorite weathering" by interplanetary dust.

Radioactive atoms trapped in meteorites decay into "daughter" atoms at a rate determined by their half-lives. The age of a meteorite can thus be estimated by measuring the proportion of parent to daughter elements. Measurements of parent-daughter relationships, like that of rubidium to strontium and samarium to neodymium, yield an age for the meteorites—and therefore, presumably, for most solid objects in the inner solar system—of 4.56 billion years. This fits comfortably with the age of the sun, about 5 billion years, derived independently by the astrophysicists. Also found in many meteorites are excesses of daughter atoms created by the decay of relatively short-lived radioactive atoms that could only have been forged in a supernova immediately before the solar system formed. Particularly telling is the presence in certain meteorites of magnesium-26, the daughter of aluminum-26. Aluminum-26 has a half-life of only 720,000 years. For it to have been incorporated into the solar system in significant enough amounts for its daughter isotope to show up today, it could not have been traveling through space for more than a few million years. The remarkable conclusion reached through such meteorite research—conducted variously by Donald Clayton, now of Clemson University, Robert Hutchison of the Natural History Museum in London, and Gerald J. Wasserburg of Caltech—is that a supernova less than 100 light-years away caused the birth of the sun and its planets. So we are left to consider that a nearby supernova, which today would sterilize most life on the dry lands of Earth, was responsible for creating the future abode of human life, and that the record of this most titanic of astronomical events has been preserved for us in cold little lumps of rock that rank among the humblest components of the solar system.

Evolution within stars is depicted by astronomers in a fascinating chart called the *Hertzsprung-Russell diagram*. First drawn up early in 1911 by Ejnar Hertzsprung, a Danish engineer who had taught himself astronomy, and independently arrived at two years later by the American astrophysicist Henry Norris Russell, the HR diagram plots the absolute magnitudes of stars against their colors. Blue stars are plotted toward the left and red stars to the right; the brighter a star, the higher it goes. The dominant feature of an HR diagram is the *main sequence*—an S-shaped tree trunk on which most stars reside. Massive, bright, blue-white stars occupy the top of the main sequence; small, dim stars are found near the bottom, and midsized stars like the sun sit in between. Stars spend most of their lives on the main sequence, but once they run out of nuclear fuel they balloon into red giants: This moves them to the right (since they are getting redder) and up (since they're getting brighter). On their way, they pass through several *instability strips*—narrow zones where stars pulsate, sloshing back and forth on the diagram. (The Cepheid variables employed in charting intergalactic distances are massive stars residing on an instability strip.) Eventually the red giants shed their outer atmospheres, leaving behind white dwarf remnants that skid left on the diagram (because they're no longer red) and down (because they're getting dimmer), and plunge into the graveyard of the dwarfs. So an HR diagram for a random selection of stars at all stages of evolution exhibits three main components—the curved tree trunk of the main sequence; a series of tendrils stretching to the right like windblown branches

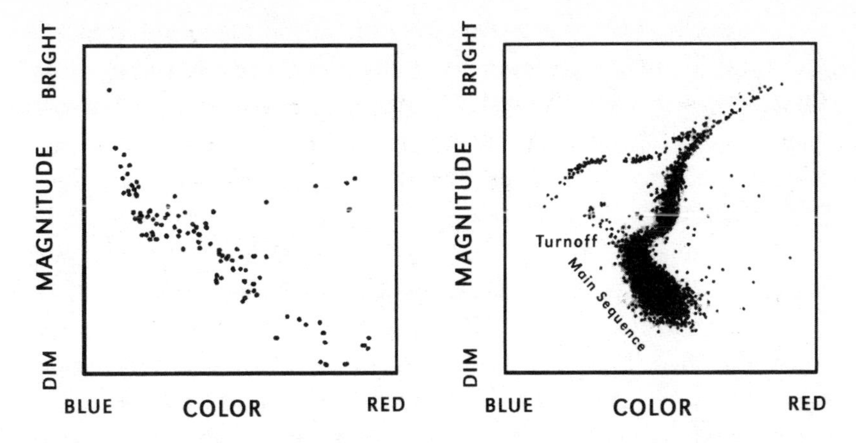

Most of the stars in the younger cluster *(left)* lie on the main sequence, while in an old cluster *(right)* the more massive stars have evolved into red giants. As time passes, this fate befalls ever less massive stars, so the location of the turnoff point indicates the age of the cluster.

and composed of red giants; and a heap of dwarfs scattered at the bottom like fallen leaves.

Astronomers use the HR diagram to age-date star clusters. Virtually all the stars in a young cluster lie on the main sequence—forming what's called a Zero-Age Main Sequence, or ZAMS. As the cluster ages, its most massive stars—the big ones burn up quickest—run out of fuel and exit the main sequence, moving to the right. In this fashion, the top of the main sequence is peeled off and transformed into a branch. As time passes the same fate befalls the lesser stars, in descending order of their mass. Hence the main sequence is pruned from the top down, forming additional branches rooted in the evacuated space where the depleted main sequence used to be. The branch that is attached to the main sequence at the epoch when we observe the cluster is called the *turnoff point.* It clocks the cluster's age: The lower the turnoff point, the older the cluster. Ancient globular clusters that contain the oldest stars in the galaxy have HR diagrams in which the main sequence has been cut down to a stump.

Galaxies, too, evolve. The Milky Way galaxy began, presumably, as a spheroidal cloud of hydrogen and helium gas. Today it contains such a diversity of objects that astronomers, if they could see nothing beyond the Milky Way, would have their hands full studying it for a thousand years to come. To let one modest example stand for many, consider the bubble of relatively empty space through which the solar system happens to be passing at present. The bubble comprises an evacuated area in the plane of our galaxy. Within it, there is less than a tenth as much interstellar gas as is normally found in the galactic plane. It measures at least 300 light-years in radius, enough so that it has probably burst all the way through the galactic disk, creating a hollow tube called a galactic chimney and thus sweeping clear sight lines leading to the depths of the universe. This is fortuitous for terrestrial astronomers—especially those working in the short wavelengths of extreme ultraviolet light, who couldn't see much of anything if they had to peer through the fog of the unswept disk. The bubble almost certainly was swept out by a supernova. In 1992, the astronomers Neil Gehrels and Wan Chen of NASA's Goddard Space Flight Center proposed that a runaway pulsar in Gemini came from the supernova responsible. (The cores of supernovae sometimes collapse asymmetrically, creating a jet that sends the resulting pulsar hurtling through space at velocities in excess of 100 kilometers per second. So we're dealing with a rapidly spinning neutron star, smooth as a marble, only 10 miles in diameter but as massive as the sun, moving a hundred times faster than a rifle bullet.) It is known as the "Geminga" pulsar.[19] Estimates of the pulsar's trajectory and the rate at which its spin is slowing down suggest that the pulsar has traveled as much as 1,000 light-years since originating in a supernova that occurred about three hundred thousand years ago, probably within a few hundred light-years of Earth. That's close enough for our distant ancestors to have beheld the supernova as a "new" star in Orion, bright as the full moon, dominating the sky for about two years. The supernova's shock wave, arriving at the solar system roughly ten thousand years later, would have been deflected by the solar wind before it reached Earth. But it could have left traces farther out, on the pristine surfaces of Saturn's and Neptune's moons. If telltale evidence of the supernova is found there by future space probes, we will have another instance of how cosmic evolution can influence the local environment.

A clue to the long and complicated evolutionary histories of galaxies may be found in the consideration that galaxies seen at great distances (and, therefore, earlier in cosmic time) tend to be bluer in color than those closer to us in space and time. Presumably this means that galaxies used to form stars at a more robust rate than they typically do today: The higher the star-formation rate, the more blue-white giant stars there are, and their light tints the host galaxy blue. Until recently, astronomers could not make clear enough photographs to investigate distant blue galaxies in any detail. But in 1995, a team of English and American scientists announced that by pushing the Hubble telescope to the limits of its sensitivity they had been able to record images of galaxies down to thirtieth magnitude, four billion times fainter than the naked eye can see. There, at distances of some eight billion light-years, they found swarms of galaxies glowing with the exuberant blue light of myriad young stars. Many of the blue galaxies are weirdly shaped, owing perhaps to the powerful shock waves that formed the stars having shaken the galactic disk into tatters. (By 1996, Hubble images of galaxies more than ten billion light-years distant showed that fully a third were too distorted to classify, as compared with only a few percent in the modern universe.) It was not immediately known whether all the blue galaxies survived their ordeals. "Some of them probably self-destructed," said Richard Griffiths of Johns Hopkins University, who headed the survey team.[20]

Since they abound in brilliant young giant stars, distant galaxies tend to be intrinsically more luminous than nearby ones. One extremely young galaxy, discovered in 1996 by astronomers using the Canada-France-Hawaii telescope at Mauna Kea, is estimated to be giving birth to a thousand stars a year—compared with about three a year in a typical spiral galaxy today. It is a bit remarkable, though, to see how steeply the curve of galaxy luminosity rises with distance: Evidently the youthful universe was ablaze with starburst galaxies, compared with which even today's most brilliant galaxies seem pallid. Looking back only a quarter of the way to the big bang, astronomers find galaxies manufacturing stars at twice the modern rate, and the action farther away gets a lot hotter. This is a good thing for populated planets like ours, since galaxies that make a lot of stars also make a lot of supernovae, which if they go off nearby can be lethal. But it also spurs a certain amount of wistfulness among physicists, who, in contemplating the violence of the cosmological past, may be forgiven if they sometimes feel that cosmic history resembles a party at which we have arrived only after the fireworks show has ended, the choice food and drink has been consumed, and most of the guests have faded from view.

The wild youth of galaxies owes a lot to their interactions with other galaxies—an example of what the physicist Heinz Pagels called the "complex interplay between and among all the objects we observe in the heavens," in a process he compared to the ecological network that supports plants and other living things.[21] Galactic interactions inspire bursts of star formation. When two galaxies sail near each other, they shake each other up. They are perturbed by tidal effects produced by the difference in the gravitational force exerted on the side of a galaxy closest to the intruder and the side farthest from it. The difference is appreciable, since gravitational force decreases by the inverse square of distance: The near side of the galaxy can droop deep into a gravitational potential well that the far side barely feels. The tides shock clouds by the thousands into collapse, setting off shellbursts of new stars. The face-on spiral galaxy M51 is a striking specimen of intergalactic tidal meddling. An interloper called NGC 5195 whipped past it only yesterday—that is, within the past few hundred million years—and as a result the spiral arms of M51 have come unsprung and now are salted with millions of hot young stars. Since most major galaxies belong to binary pairs, groups, or clusters, such interactions are commonplace. This suggests that billions of planets—including, plausibly, many on which life has arisen—owe their birth to intergalactic interactions. Along these and many other lines of inquiry, one searches in vain for a point at which one can justifiably declare that from here on down all processes are to be ascribed to local events, while from here on up they are to be regarded as cosmic. In many ways, life on Earth and the geological processes that midwifed it *are* part of cosmic evolution.

Galaxies like M51 and the Milky Way—belonging as they do to small, sparsely populated groups—experience major interactions infrequently and soon recover their composure when they do blunder into one another. Some stars are flung into intergalactic space and a telltale puddle of gas is left behind, but after a few hundred million years things look much as they did before. The more frequent and more intrusive interactions experienced by galaxies in dense clusters, however, can alter them permanently. Edwin Hubble and Milton Humason in the 1930s found that massive galaxies belong to two types—the fertile spirals, where stars continue to form, and the sterile ellipticals, which have stopped making stars and where, consequently, all the stars are old. Astronomers ever since have wondered whether galactic taxonomy harbors an evolutionary message. It now appears that the answer is yes. There have been plenty of interactions in clusters—it is estimated that most cluster galaxies have by now either collided with or passed near to other galaxies—and close encounters can strip the interstellar clouds from the galaxies involved, putting an end to their star-forming careers.

This may be how elliptical galaxies form. Ellipticals are more common today than in the past. In a significant study, Alan Dressler of the Carnegie Observatories in Pasadena, California, used the Hubble Space Telescope to conduct a census of the rich cluster CL 0939+4713, the light from which is about five billion years old. He found that back then, in this one cluster at least, a lot more galaxies were merging than is the case in comparable clusters today, and that spiral galaxies were much more common. "From our images it appears that spiral galaxies were a major constituent of rich clusters only four billion years ago, when they accounted for about 30 percent of all cluster members. In contrast, only 5 percent of the galaxies in nearby rich clusters are spirals," Dressler reported.[22] The implication is that galaxies in rich clusters have collided and often merged, with spirals combining and being stripped of their interstellar gas to end up as ellipticals. Evidence supporting this hypothesis may be found in the hearts of the giant ellipticals commonly found at the centers of rich clusters. They typically have more than one nucleus. The multiple nuclei could be those of galaxies swallowed up by the cannibal ellipticals.

Since quasars evidently are the nuclei of violent young galaxies and are bright enough to be seen at great distances, cosmologists have reason to hope that, by studying quasars, they can learn more about galaxy evolution. Quasars themselves clearly have evolved into something less conspicuous over cosmic time. No quasars are found within two billion light-years; their numbers increase sharply as we look farther back in time; and more distant quasars are more numerous, brighter, and more often variable, flaring up like guttering candles. This is consistent with the idea that quasars form when gas falls into the core of a galaxy. The gas feeds a black hole at the galactic center, heating the accretion disk that surrounds the black hole and firing up its plasma jets, thus kindling the extremely bright galactic core that is called a quasar. Presumably quasars were fed more frequently in the early days of the universe, when there was more intergalactic material around for them to ingest.

Observations of quasars are complicated by indigenous obscuring effects. The clouds that feed the central black hole can also block its light from view, especially if the galaxy is a spiral seen edge-on, and particularly if the quasar has been fired up by the merger of two spirals, which tend to carry a lot of dust. So it may be that there are many more quasars than the observational data would suggest, and that what we detect as quasars are those galaxies that happen to present a nearly face-on view, affording us a look into their fiery cores.

To sum up this brief review, it seems both permissible and helpful—though, admittedly, far from perfectly clear—to regard living creatures, planets, stars, galaxies, and the atoms and molecules of which they are made as products of cosmic evolution. But is this thesis really justified? Evolution certainly shaped life on Earth, but like any analogy it can be pushed too far, and some thinkers are uncomfortable with talk of cosmic evolution.

Critics point out that while a scientific theory should have predictive power, when speaking of cosmic evolution we do not so much predict the future as account for the past. To this we may reply that science does not exclusively predict the outcome of natural phenomena that have not yet occurred; it also predicts what we will learn, *in* the future, about the past. Such "retrodictions" have their place in piecing together cosmic history, of which there has been a great deal and about which we have a great deal to learn. Consider the theory that dinosaurs evolved into birds. The dinosaurs are dead and gone. Either they evolved into birds or they did not. The theory predicts that the paleontological record will disclose further links between dinosaurs and birds. If it does, the theory will survive. If it does not, the theory will not.[23] The theory thus has predictive power, though what it predicts is not the future of dinosaurs and birds but the future course of our knowledge about them. Much the same is true of evolutionary predictions about cosmic history, such as the claim that young rich clusters will, like the one Dressler studied, turn out to contain more spiral galaxies than rich clusters do today.

A more telling criticism springs from the observation that evolution is in some sense *creative.* This point has been made by several thinkers, notably by the philosopher Henri Bergson in his *Creative Evolution,* published in 1907, and, on sounder scientific grounds, by the astrophysicist David Layzer in his 1990 book *Cosmogenesis.*[24] By "creative" these authors mean that evolution is genuinely innovative—that its products cannot be predicted in detail. The superintelligent inhabitants of a starship orbiting Earth 65 million years ago might have foreseen, as they watched the fatal comet hit, that the dinosaurs and most other terrestrial life-forms would not long survive and that new and wildly different ones would appear to take advantage of their misfortune. But they could not, *in principle,* have predicted that this extinction event, unlike those that had preceded it, would in time lead to the rise of intelligent bipeds with opposable thumbs and a talent for science. Therefore, to the extent that we describe natural processes as evolutionary, are we not saying that they are unpredictable? And if so, have we not abandoned the predictive value of science? After all, the proof of Newtonian mechanics is that it predicts the position of a stone two seconds after it is dropped from a tower, and of the moon over Miami at 11:40 P.M. on New Year's Eve in the year 2525. (According to my computer the moon will be at celestial coordinates Right Ascension 7 hours, 46 minutes, 15 seconds, Declination 26 degrees, 12 minutes, 39 seconds, in the constellation Gemini, forming an approximate right triangle with the stars Castor and Pollux.) What's the virtue of introducing "cosmic evolution" into science, if its thrust is to say that lots of things *cannot* be predicted?

Its virtue, I would reply, is that it might be true. Now, to say this is to step on some toes, philosophically speaking. The concept of creative evolution probably cannot be confirmed empirically. To verify it one would have to demonstrate that the emergence of, say, a new species of butterfly on Earth or of monocellular life on a previously sterile planet not only *was* not predicted but *could* not be, and that is probably impossible. To *dis*prove creative evolution would require a theory capable of blueprinting the butterfly or the alien lichen before it appeared, and that seems equally unlikely. So creative evolution apparently does reside beyond the purview of science. It is, in short, a philosophical notion, not a scientific one. As such it is readily dismissed by those who believe that science represents the sole path to truth. But those who prefer to entertain the whole rainbow of human thought, rather than restricting themselves to ideas that bear the tincture of science, may wish to give the doctrine of creative evolution a fair hearing. To find that some phenomena are not predictable would be to encounter a limit to scientific inquiry, and what is so bad about that? The grandiose claims of a few classical scientists that science would eventually be able to predict *everything* were never very pretty; perhaps they were not pretty because they were not true. Pierre-Simon de Laplace, author of the nebular hypothesis, was the author also of the notorious assertion that

> an intelligence that at a given instant was acquainted with all the forces by which nature is animated and with the state of the bodies of which it is composed would—if it were vast enough to submit these data to analysis—embrace in the same formula the movements of the largest bodies in the universe and those of the lightest atoms: Nothing would be uncertain for such an intelligence, and the future like the past would be present to its eyes.[25]

A few scientists may still give weight to this audacious assertion, but the vast majority of humankind does not. The "ordinary" individual laughs aloud at the notion that scientists will one day be able to predict, by consulting a supercomputer, the lyrics of an unwritten song or just when a dancer will trip while rehearsing a ballet as yet unstaged. To argue otherwise seems absurd, and it is. The proximate flaw in de Laplace's argument is that there cannot ever *be* "an intelligence . . . vast enough to submit these data to analysis," unless one invokes theological conceptions of an all-knowing God.

The world is *not* strictly deterministic. This is famously the case on the quantum level, where Heisenberg indeterminacy inhibits our ability to predict the behavior of individual subatomic particles. And it may well pertain wherever evolution may properly be said to operate—precisely because evolution is creative, and creativity is unpredictable. As Layzer puts it, we live in "a world of becoming as well as being, a world in which order emerged from primordial chaos and begot new forms of order. The processes that have created and continue to create order obey universal and unchanging physical laws. Yet because they generate information, their outcomes are not implicit in their initial conditions."[26]

Which leads us to conclude with an examination of cosmic evolution in relation to human thought.

Every evolutionary process has three aspects—one conservative, another innovative, and the third selective.

First there must be a way of retaining older features that "work," meaning that they contribute to the continued survival of the entities involved. If we think of evolution as akin to erecting a building, this *conservative* mechanism resembles the building's foundations and beams. (This is not to say that evolution is progressive, a presumption that, as we have noted, must be handled with long tongs.) Many of these foundations may subsequently be torn down and rebuilt, but they must exist, and at least some of them must *per*sist if evolution is to build more complex and varied entities rather than merely shuffling things around. In biological evolution, the conservative function is accomplished by the stability of the DNA molecule, which preserves inherited genetic information. Strong evidence of biological conservancy may also be found in embryology, where it gave rise to the saying that "ontogeny recapitulates phylogeny."[27] A human embryo grows gills like its fish ancestors, then tears them down and rebuilds them into lungs. Similarly it constructs and then destroys so much brain tissue—a process known as "neural carving"—that most of the brain cell death we experience takes place prior to birth. This is not the most efficient imaginable approach. The engineers at Boeing would not elect to make a 747 aircraft by first building the airframe of a 707 and then removing all but the few parts they could retain for a 747. But it is imposed by the strong conservative streak characteristic of biological evolution, which is obliged to craft the entities of the future using the tools and procedures of the past.[28]

Second, evolution requires *innovation*, a means of modifying characteristics of the entity involved. In biological evolution, the innovative agency is mutation, the altering of the DNA molecule through internal errors of replication—the chemical equivalent of a typographical error—or from outside influences such as radiation, toxic chemicals, or cosmic rays. Mutations arise by chance, not design. They are random. One thinker aptly refers to them as "blind groping."[29]

Conservation alone means stasis; conservation plus innovation amounts but to change. For evolution to take place requires a third component, that of *selection*. Here we come to the kernel of Darwin's great insight into biological evolution. Living species reproduce at a more rapid rate than the environment can support. (Darwin came upon this insight by reading Thomas Malthus, who got it from Ben Franklin.) Most individuals therefore must compete. Some survive long enough to produce offspring; others do not. Random variation produces differing individuals within each species: No two rattlesnakes or rockfish cod are identical. Many of these genetic experiments fail, as every child knows who has mourned the loss of one puppy in a litter or one chick in a nest. But once in a while a variation appears that does better than average, either because it possesses a novel feature useful in an unchanging environment or because the environment has suddenly changed in ways that benefit it.[30] Owing to the reproductive advantage bestowed upon it by genetic innovation, this variety will come to dominate at least the local population and may in some cases lead to the origin of a new species.

Now, if we are to speak meaningfully of "stellar evolution" or "cosmic evolution" it must be possible to apply at least loosely the three characteristics of conservancy, innovation, and selection to nonbiological phenomena. Can this be done?

The answer to this question depends on the scope of the system being considered. If we maintain a reductionist concentration on simple systems and the laws of physics, the world looks deterministic and the concept of evolution superfluous. So one can criticize the term "stellar evolution" by arguing that a star is deterministic. Some astronomers take this position: Tell us the mass and chemical composition of a star, they say, and we can predict every subsequent phase of its career, from protostar to main sequence to red giant to white dwarf. Therefore there is no need to speak of the "evolution" of the star. This point is sound as far as it goes. But if we widen the framework of our inquiry to encompass a whole galaxy of stars, determinism becomes more difficult to maintain. A star is one node among billions in a complex of galactic chemistry. The courses of all these stars depend on a great many considerations, including their interactions with interstellar clouds and with one another, and eventually these variables become too complicated to compute in detail. And if we consider chemical and geological changes in all the stars' planets, the process shades over completely from predictable determinism to unpredictable creativity, and therefore may properly be spoken of as evolutionary.

Does the whole universe evolve? Here again the term seems unnecessary if we restrict ourselves to discussing, say, the physics of the big bang, but justifies itself once we start speculating about the myriad specifics of the historical universe. A deterministic approach might enable a scientist to predict that an extrasolar planet resembling Saturn is likely to have a satellite resembling Dione, but one doubts very much that one could ever draw, in advance, an accurate map of its Dione-like surface. The process that hoisted the universe from the relative uniformity of the big bang to the incredible variety and diversity we see today in the sky seems more properly to be described as evolutionary.

Nevertheless the specific application of evolutionary conservancy, innovation, and selection to cosmic history remains both problematic and invigorating.

The various unified theories imply that at least some of the constants of nature did not always pertain but arose by chance, through symmetry-breaking events in the first moments of cosmic time. The question whether there is an agency of conservation underlying natural laws will depend, ultimately, on whether there turns out to be some sort of macrolaw—a set of precepts (such as the arithmetic principle that one does not equal two, or the topological finding that knots can be tied in three dimensions only) that constitutes the stage on which evolutionary forces could function.

John Wheeler speaks of "law without law."[31] Its outlines at present are hazy at best. Innovation certainly occurred in the universe, but the issue of selection is a lot more problematic. If, say, there are an infinite number of universes, each with differing physics, some of these universes may support life and others not. We might choose to call the uninhabited universes "nonexistent," in which case we could regard the inhabited ones as having been "selected for." We will return to this puzzle at the end of this book.

If there is only one universe, however, it is more difficult to see how the concept of selection might usefully be applied to it. The notion that constants such as the strength of the electromagnetic force have arisen by chance suggests the "nature versus nurture" issue in biological development. Confronted with, say, a child prodigy in chess, one would like to know whether and to what extent his facility springs from his genetic makeup or his having been raised in a family that fostered chess playing. In cosmology, this question takes the form of trying to differentiate between the genuinely fundamental laws of nature—that is, those that could not be otherwise, like the statement that one does not equal two—and "frozen accidents." "Frozen accidents," as the physicist Murray Gell-Mann writes, "are chance events of which the particular outcomes have a multiplicity of long-term consequences, all related by their common ancestry." Gell-Mann illustrates the concept of a frozen accident with "the example of the succession to the English throne of Henry VIII—after the death of his older brother—resulting in the existence of a huge number of references to King Henry on coins and in documents and books. All those regularities stem from a frozen accident."[32] A classic example in physics is antimatter—mirror-image partners of the proton, electron, and other particles but with opposite charge. Antimatter exists in theorists' calculations and in collider experiments but is rare in nature. Presumably the universe began with almost equal amounts of matter and antimatter that mutually annihilated in the fires of the big bang, leaving behind a residue of matter. If so, the fact that matter survived and antimatter did not was an accident, one that has remained frozen into nature ever since.

It is always possible that, as Henry Adams put it, chaos is the law of nature and order but the dream of man.[33] So we must be cautious in ascribing any system to nature, especially one as ambiguous as evolution. The history of ideas is full of attempts to portray nature as organized according to the dictates of one or another orderly scheme. Many a fine mind has preached that the universe is water, or fire, or built on vortexes or swerving atoms, and most such noble attempts, from Aristotle's *élan vital* to Joseph Priestley's doomed attempt to prove that the (nonexistent) substance phlogiston produces fire, were selected against in the fullness of time. Science is distinguished not for asserting that nature is rational, but for constantly testing claims to that or any other affect by observation and experiment. Evolution evidently is a fact—of life, at least, and perhaps also of inorganic nature—but we need to guard against leaping to conclusions about what evolution means. Adams saw the passion for unity and order as a quality of the young: "The older the mind, the older its complexities, and the further it looks, the more it sees, until even the stars resolve themselves into multiples," he wrote—the allusion is to binary stars, a hot topic of astronomical research in Adams's day—"yet the child will always see but one."[34] Given that ours is a young species (and that, as Jacques Monod remarks, "science possesses an ever-youthful eye"[35]), we may be overly eager to fit everything into a unifying scheme in which cosmic evolution is imagined to be both inexorable and invariably progressive. Many thinkers have succumbed to this temptation, among them Herbert Spencer, Teilhard de Chardin, Hegel, Marx, and Engels. A belief that cosmic evolution makes the universe an ever better place was central to the espousers of nineteenth-century scientific progressivism. They drew inspiration from what they took to be the upwardly mobile history of life on Earth, a stirring tale of humble bacteria and mosses rearing up to become pterosaurs, great apes, and builders of telescopes. But the other side of the same coin is stamped with the sobering consideration that terrestrial life remained stuck at the unicellular level for fully 80 percent of its history, and that intelligence—which for simplicity we can define as the ability to fashion tools—did not appear until quite recently. Reduce the age of the earth to one calendar year and you will find that the first vertebrates appeared on November 21, primates on the day after Christmas, and *Homo sapiens* only three and one-half minutes before midnight on New

Year's Eve. If this is progress at work, it's pretty damned halting progress. Democritus said that "everything existing in the universe is the fruit of chance and of necessity."[36] To which do we humans primarily owe our existence? The honest answer is that nobody knows.

Yet there's something enthralling about the undeniable consideration that the universe is not exclusively engaged in shambling down an entropic slag heap but has also, in some times and places, articulated itself into fascinatingly antientropic entities like binary pulsars and mathematicians. We employ the term "evolution" to describe the racheting mechanism by virtue of which this has happened. The rachet evidently runs on random chance, yet we can never be certain that what appears to be chance *is* chance—a point to which we will return—and among its products are thinking beings who search for order in the most disciplined way they can practice, and do seem to find it. Darwin often spun this mystery ball in his mind, and he saw past his reflection in it, but he never found an answer. He wondered at what he called

> the impossibility of conceiving this immense and wonderful universe, including man . . . as the result of blind chance or necessity. When thus reflecting I feel compelled to look to a First Cause having an intelligent mind in some degree analogous to that of man and I deserve to be called a Theist. But then arises the doubt, can the mind of man, which has, as I fully believe, been developed from a mind as low as that possessed by the lowest animal, be trusted when it draws such grand conclusions.[37]

Darwin was a gentle man, who in old age was still hurt by the memory of how as a little boy he had once mistreated a puppy, and he wondered, like the rest of us, how an evolutionary world destined for perfection could produce cruelty and injustice. "With respect to the theological view of the question. This is always painful to me," he wrote. "I am bewildered. I had no intention to write atheistically. But I own that I cannot see as plainly as others, and as I should wish to do, evidence of design and beneficence on all sides of us. There seems to me too much misery in the world. I cannot persuade myself that a beneficent and omnipotent God would have designedly created the Ichneumonidae with the express intention of their feeding within the living bodies of caterpillars, or that a cat should play with mice."[38]

And here we conclude our brief investigation into the concept of evolution, where it shades from the realm of science into that of ethics, and leaves us, as it left Darwin, with the consideration that if *we* are the beneficiaries of an evolutionary process, however widely or narrowly it may cast its net, our obligation is not just to learn but also to love.

8

Symmetry and Imperfection

God, Thou great symmetry,
Who put a biting lust in me
From whence my arrows spring,
For all the frittered days
That I have spent in shapeless ways
Give me one perfect thing.

—ANNA WICKHAM[1]

The task of the physicist is to see through the appearances down to the underlying, very simple, symmetric reality.

—STEVEN WEINBERG[2]

PHYSICS IS WIDELY ESTEEMED—and not just by physicists—as the quintessential science. It casts the widest net: Since every occurrence involves matter and energy, physics in studying atoms and forces lays claim to having something to say about all the sciences, at least at the fundamental level. Much of astronomy today consists of physics, and the same is true of chemistry, optics, materials science, and many other disciplines. Broad at the base and towering in stature, physics stands as a living monument to what many esteem as the paramount intellectual achievement of the twentieth century.

In many ways, its reputation is deserved. Certainly physics has made progress toward fulfilling the glistening promise envisioned by the German physicist Hermann von Helmholtz more than a century ago, when he predicted that through it humans would gain "intellectual mastery over nature," and that material mastery would follow, since "by means of such preliminary knowledge of the laws, we [will be] in a position to let natural forces work for us, after our will and wishes."[3] All of which has pretty much come true, and in a remarkably brief period of time. Physicists today can accurately predict the outcome of every fundamental (meaning simple) process in the known universe, and some are at work on the more complicated ones—although claims that they will soon be able to foretell the weather or the rise and fall of the stock market should be taken with a shakerful of salt, and the time when science can unerringly predict the behavior of an individual human being *who knows the prediction in advance* is probably never. The technological spinoffs Helmholtz foresaw have grown to such grandiose proportions as to raise concern over whether working "our will and wishes" on nature means degrading the livability of the planet.

But the more closely one examines the lofty tower of physics, the less monolithic it looks. This is clearly the case with regard to its methods. Unlike philosophy, physics seldom proceeds by hunting down sweeping answers to big questions. Most often it involves poking around at specific issues which seem to promise some intellectual or practical reward: as Einstein remarked, the scientist looks to an outside observer like an "unscrupulous opportunist."[4]

Similarly, the results of physics research are less finished than popular accounts usually suggest. The standard model of particle physics, though well worth boasting about, remains very much a work in progress. It makes accurate predictions, but to do so requires plugging in a couple of dozen parameters whose values are derived from experiment without benefit of any compelling underlying knowledge of why each should be what it is instead of some other number. Nor do the theories that compose the standard model dovetail into one another with any great logical inevitability. Three of the four fundamental forces of nature (*interactions,* in the jargon) are accounted for by quantum field theories that incorporate the special theory of relativity. But the fourth force, gravitation, remains the province of general relativity, which is excluded from the quantum theories and which, moreover, views the world in a very different way than the quantum theories do. Needed—or, at least, desired—is a single theory that would embrace all four forces, and in doing so lay bare the reasons that the experimentally determined values are what they are. This wished-for account is known variously as a grand unified theory, as quantum gravity, or, most hyperbolically, as the "theory of everything" (though it would remain a theory of fundamental interactions only, not of where a particular blackbird will fly or what bonnet Aunt Ellen will choose to wear next Easter).

Wise hunters stalk the ultimate theory by searching for signs of *symmetry.* The laws of nature are all expressions of symmetry, and all physics is in some sense a search for symmetry.

Most of us first encountered symmetry as a way of classifying how certain aspects of the appearance of objects persevere when the objects are rotated or otherwise moved. A sphere, for instance, casts a circular shadow, and that shadow stays circular regardless of how we rotate the sphere: The sphere is therefore said to be rotationally symmetrical about any axis. That is one reason Plato regarded the sphere as the ideal geometric solid. (The other reason was that a sphere contains the maximum possible volume in a given surface area, which is why some of our more corpulent physicists, among them Ernest Rutherford and Abdus Salam, liked to joke that by gaining weight they were approaching geometric perfection.) There are many other spatial symmetries. Consider the *translational* symmetry of the Xs in this line: XXXXXXXXX. They are symmetrical along the dimension of the printed line for translations approximately equal to their width. (In other words they fit right over one another, which is not the case for a sequence like ABCDEFGHI.) The Xs also have *mirror* symmetry, since they are indistinguishable from their reflections, while a line of Zs—

ZZZZZZZZZ

—if reflected turns into

ƧƧƧƧƧƧƧƧƧƧ

So Xs are mirror-symmetrical and Zs are not.

Symmetries of this kind are limited to our experience in ordinary space; indeed, they form a subset of the many proofs that ordinary space is itself transformally symmetrical. It is owing to the transformal symmetry of local space that we can travel without ill effect and that a beach ball rolling down the sand retains its spherical shape. If we lived in the intensely curved space at the lip of a black hole, which is transformally *a*symmetrical, travel away from the hole would be expensive and travel toward it fatal, and beach balls would be elliptical.

In quantum physics, researchers often invoke *abstract* spaces to solve particular problems. Suppose we want to map the electron and its antimatter partner, the positron. The electron has the same mass and spin as the positron, but opposite electrical charge. (Electrons have negative charge, positrons positive.) So the physicist makes an abstract, three-dimensional "space," the axes of which represent charge, mass, and spin. An electron if transformed into a positron is said to be symmetrical along the axes of mass and spin —since these remain the same—but *a*symmetrical with respect to charge, which inverts when the particle is transformed. This example may not be very edifying insofar as understanding electrons is concerned, but it is meant to broaden our conception of symmetry, by demonstrating that symmetry need not have anything to do with geometrical shapes in ordinary space. In this broader sense we can move beyond appearances, and define symmetry as representing a *quantity that remains unchanged through a transformation.* (Hence the word *symmetry,* from the Greek for "the same measure.") The mathematician Lewis Carroll invoked this sense of symmetry, along with its more familiar geometric properties, in "The Hunting of the Snark":

> *You boil it in sawdust:*
> *You salt it in glue:*
> *You condense it with locusts in tape:*

Still keeping one principal object in view—
To preserve its symmetrical shape.[5]

The word *invariance* is shorthand for "a quantity that remains unchanged." To understand invariance is to see why symmetry is the central theme of physics. The laws of nature express symmetries because symmetries identify invariances. The law of conservation of energy, for instance, describes a quantity ("energy") that remains unchanged (is "conserved") through such transformations as a steam engine's doing work or a star's burning its way toward its red-giant stage. The special theory of relativity exposes several invariances, notably the equivalence of mass and energy ($e = mc^2$). Indeed, Einstein's motive in writing the theory was to show how one could conserve the laws of physics—notably James Clerk Maxwell's equations for electromagnetic fields—for systems in rapid motion relative to each other. Maxwell's equations predicted that radio waves and other electromagnetic fields travel at the velocity of light but did not prescribe from what frame of reference their velocity should be measured. Einstein showed that it didn't matter: The velocity of light is the same for all observers, regardless of their own velocities. To recover laboratory physics in this odd situation, Einstein invoked what is called "Lorentz invariance," which prescribes how time slows down and measuring rods contract aboard a speeding spaceship. This preserved both the inherent symmetries of Maxwell's theory and the wider symmetry by virtue of which the laws of nature remain invariant when experiments are done at various times, places, and velocities. That is why Einstein called his accomplishment "invariance theory," a term that, had it been retained, might have spared the nonscientific world much of the confusion visited upon it by talk of "relativity."

Symmetry was central to the rise of quantum field theory—an opulent story of intellectual attainment that I will here sketch with an unseemly though requisite brevity. It begins with the founding of quantum electrodynamics by the English physicist Paul Dirac.

Dirac was distinguished by his mathematical creativity and his almost complete lack of involvement with anything else. The physicist W. M. Elsasser described him as an individual "of towering magnitude in one field, but with little interest or competence left for other human activities. . . . Everything went into the performance of his great historical mission, the establishment of the new science, quantum mechanics, to which he probably contributed as much as any other man."[6] Dirac was famously taciturn. Interviewed by a reporter for the *Wisconsin State Journal,* on April 31, 1929, he uttered a recorded total of nineteen words. This was part of their exchange:

> REPORTER: Now, doctor, will you give me in a few words the low-down on all your investigations?
> DIRAC: No.
> REPORTER: Will it be all right if I put it this way—"Professor Dirac solves all the problems of mathematical physics, but is unable to find a better way of figuring out Babe Ruth's batting average"?
> DIRAC: Yes.
> REPORTER: What do you like best in America?
> DIRAC: Potatoes.[7]

Dirac's reticence could be irritating—he seldom spoke unless asked a direct question, and even then would remain silent unless certain that he could say something sensible—but his sweet and decent nature won him some good friends, among them the physicist and novelist C. P. Snow, who regarded Dirac as "the greatest living Englishman."[8] No particular psychological sensitivity is required to identify the childhood roots of Dirac's eccentricities. His father, a Swiss émigré who loathed his own parents, was a notorious disciplinarian at the school in Bristol where he taught French, and he brought his work habits home. He permitted the family neither to go out nor to have visitors. Paul ate dinner with his father, in the dining room, while his mother, brother, and sister ate in the kitchen—banished because they did not speak French adequately. "My father made the rule that I should only talk to him in French," Dirac recalled. "He thought it would be good for me to learn French in that way. Since I found that I couldn't express myself in French, it was better for me to stay silent than to talk in English. So I became very silent at that time."[9] The children of this cheerless household were denied domestic contact with music, painting, and poetry—and, for that matter, with the so-called facts of life. "I never saw a woman naked, either in childhood or youth [until] 1927, when I went to Russia with Peter Kapitza," Dirac recalled. "She was a child, an adolescent. I was taken to a girls' swimming-pool, and they bathed without swimming suits. I thought they looked nice."[10] Dirac's brother, who became an engineer, committed suicide. Dirac himself led a life full and narrow as an Antarctic chasm but less cold. He took part in scientific conferences around the world, enjoyed hiking and mountain climbing, and his marriage to Margit Wigner, sister of the physicist Eugene Wigner, produced two daughters.

Of all the physicists, said Niels Bohr, "Dirac has the purest soul."[11] Its purity was the purity of pure reason; Dirac filled the vacuum of his childhood with logic, and sought thereafter to live by logic alone. Yet he was a scientist as well as a mathematician, and therefore habitually empirical. The physicist Jagdish Mehra recalls dining with Dirac at high table at Cambridge. "The weather outside was very bad, and since in England it is always quite respectable to start a conversation with the weather, I said to Dirac, 'It is very windy, Professor.' He said nothing at all, and a few seconds later he got up and left. I was mortified, as I thought that I had somehow offended him. He went to the door, opened it, looked out, came back, sat down, and said 'Yes.' "[12] Dirac had a deep if deeply hidden sense of beauty, one that evinced itself in his famous remark that "it is more important to have beauty in one's equations than to have them fit experiment."[13] He insisted on viewing his research as an aesthetic enterprise; there resided his art, his poetry, his silent dance.

Dirac's signal achievement was his creation of an equation for the electron that incorporated both quantum mechanics and special relativity. The Dirac equation yielded an accurate gyromagnetic ratio describing how electrons couple to an external electromagnetic field. (This is the part that invokes special relativity, itself a theory of electrodynamics.) It also invoked a previously unimagined symmetry, one linking the electron to an unknown, oppositely charged particle that the world now knows as the positron. This was so weird that it disturbed Dirac himself, who was relieved when positrons were discovered experimentally, thus opening the door onto the world of antimatter.[14] Even stranger was a third symmetrical relationship: The Dirac equation revealed that the existence of an electron implies the existence of an accompanying cloud of electron-positron pairs. These are the *virtual* particles now seen to be fundamental to quantum field theory, in which every "real" particle is shrouded in a nebula of virtual particles that constantly bubble up out of the vacuum, interact, and subside.

By midcentury, quantum electrodynamics had been refined into a highly accomplished theory that generated the most accurate predictions of experimental results ever attained by any science. The mathematics of symmetry and invariance was so efficacious in this and related areas of research as to seem almost suspicious. Symmetry not only accounts for the behavior of the fundamental forces of nature but emerges as the reason there *are* forces in the first place. Presented with a particular transformation and wondering how a quantity remains invariant through it, theorists could and did postulate the existence of symmetry-preserving particles that do the job. These particles then proved to be nothing other than carriers of forces—the *gauge bosons.* The gauge boson of electromagnetism, for instance, is the photon. Quantum field theory invokes the photon to preserve a symmetry called local gauge invariance, and predicts that the photon should have zero rest mass. Yet photons are not content to remain an abstract concept useful in theoretical mathematics alone. They actually exist. You can count them, using a photon counter. And the rest mass of the photon is indeed zero.

Symmetry rescued physics from the crisis of proliferating particles that arose once experimenters began building accelerators that could slam protons, electrons, and other particles together to produce tiny but intense explosions, a development that inaugurated the golden age of high-energy physics. At the time it was customary to compare the process to a car crash: Particles collide and scientists inspect what comes out, in a messy business that antireductionists ridiculed as being like trying to find out how Swiss watches work by smashing two watches together. Today, with the big bang model better established, scientists more often think of the process as replicating the big bang. The particles that come out were not exactly in there to begin with. (Heisenberg to Victor Weisskopf, as they sat outside the entrance to a swimming pool on a hot spring day in 1931: "These people go in and out all very nicely dressed. Do you conclude from this that they swim dressed?"[15]) Rather, new particles precipitate out of the energy of the explosion, exhibiting a slice of the physics that predominated when the universe was young. However one looks at it, the result of collider experimentation was to produce a bewildering array of new particle types—hundreds of them. Pessimists warned that physics was getting more complicated, not simpler, and might be approaching a dead end. Optimists urged that matters would become clearer once a way was found to organize all these particles—as Mendeleev, in constructing the periodic table of the elements in the nineteenth century, had set physics on the path to mapping the atom.

The optimists proved right. Enlightening symmetrical relationships were found among the subatomic particles, as well as *broken* symmetries that may bear evidence of events in cosmic history. This work would have proceeded more slowly had the physicists been required to work out all the potentially relevant symmetries before determining which ones were useful to physics. But, happily, they found that mathematicians had already done that job, classifying possible symmetries by organizing them into *groups*. A symmetry group is a set of all the transformations of a given sort that leave something unchanged. Group theory had been developed in the nineteenth century, by mathematicians who, as usual, had no expectation that it would have any practical application (and, if asked, assuredly would have boasted that they didn't care). It provided the physicists with what amounted to a toolbox full of gleaming symmetries. All they had to do—though this was a tall order—was to find which ones might "unite" some of the many particles by revealing family relationships linking them.

A major success along these lines was scored in the early 1960s by Murray Gell-Mann at Caltech and, independently, by Yuval Ne'eman in Israel. Gell-Mann and Ne'eman were investigating the strong nuclear interaction, the force that binds protons and neutrons together in atomic nuclei. From the toolbox of mathematical symmetry they selected the *Lie groups,* named for the Norwegian mathematician Sophus Lie. They found a Lie group known as SU(3)—"SU" for "special unitary"—that exposed a relationship among the hadrons, particles like protons and neutrons that respond to the strong nuclear force, and mesons, which mediate it. The SU(3) group generates octets—sets of eight members each ($3^2 - 1 = 8$) into which the hadrons can be fit, like hanging ornaments dangling from the tips of a mobile, revealing their symmetrical relationship with one another. Specifically, the eight low-mass spin–one-half baryons (of which the proton is one) fit into one octet, the eight pions (pions are a variety of mesons) fit into another, and the spin-one vector mesons fit into a third.

The polymathic Gell-Mann named this approach the Eightfold Way, after the Buddhist precepts taught by Gautama Buddha in his first sermon (right understanding, thought, speech, action, livelihood, effort, mindfulness, and concentration). But the Eightfold Way in physics initially looked wrong, owing to a seemingly fatal flaw that was to become its greatest boon. The group SU(3) generates, in addition to the octets, a complex triplet. This implied that the particles dangling from the octet mobiles were themselves triplets of some kind. In 1964, Gell-Mann, acting on a suggestion made during a casual conversation with colleagues at a Columbia University meeting a year earlier, proposed that protons and neutrons really *are* each composed of three particles. Gell-Mann called these particles *quarks,* a term he'd come across in James Joyce's *Finnegans Wake.* ("I was paging through *Finnegans Wake* as I often do, trying to understand bits and pieces—you know how you read *Finnegans Wake*—and I came across 'Three quarks for Muster Mark,' " Gell-Mann recalled.[16] "I said, 'That's it! Three quarks make a neutron or a proton.' ")

Quarks did not generate much initial enthusiasm. The theory was not all that simple—it called for three kinds of quarks, called *up, down,* and *strange,* to which were subsequently added the *charm, bottom,* and *top* quarks, for a total of six—and the underlying symmetry required that quarks have fractional charge, something that had never been observed in nature. Nor did experiments initially find any clear evidence of quarks. Theoretically oddball and experimentally unsupported, quark theory in the early days was unpopular. Emblematic of its status was the experience of George Zweig of Caltech, who came up with the same idea, at about the same time Gell-Mann did, while on leave at CERN. Zweig lost priority when he withheld his paper in resentment at a CERN requirement that resident researchers publish only in the CERN journal, *Physics Letters.* Then he got the other side of a bad bargain when he was denied an important faculty appointment by a department head who, having learned that he trafficked in quarks, dismissed him as a "charlatan."

But it can be a mistake to handicap scientific theories too hastily, and today quarks are major players in the particle world. All six quarks have now been identified experimentally—in a campaign that culminated in 1995, when the top quark was detected at the Fermilab particle collider in Batavia, Illinois. Quantum chromodynamics, the quark-based theory of the strong force (named after *color,* a quark parameter that has nothing to do with ordinary color) is now nearly as well established as the quantum electrodynamics that preceded it. Protons and neutrons are portrayed by the theory as each made of three quarks, which in turn are bound together by *gluons*—carriers of the color force, which now supplants the "strong nuclear force" picture. The color force does not decrease with distance, as does electromagnetism, but effectively gets stronger as quarks are pulled apart, like a rubber band.[17] As a result it is extremely difficult to pull a quark out of a hadron: If you pump in enough energy to overcome the color force, you just make new quarks. And that explains why it was so difficult to observe quarks in the first place.

With these and other victories in symmetry studies came an enhanced appreciation for *broken* symmetry as a significant aspect of nature. Quantum chromodynamics exposes a broken symmetry: The strange quark has a much larger mass than the up or down quarks, whereas were symmetry preserved, all quarks would be equally massive. In the 1960s, Steven Weinberg, along with Abdus Salam and Sheldon Glashow, found a symmetry linking the massless photon, carrier of the electromagnetic force, with the massive bosons implicated in the weak nuclear force, which mediates radioactive decay. The massive bosons are the negatively charged W^-, the positively charged W^+, and the neutral Z^0. The photon doesn't look at all like the W^+, W^-, and Z^0. The reason, Weinberg realized, is that the symmetry linking them has been broken. As Weinberg later put it, "These particles are siblings of each other, tightly related by a principle of symmetry that says that they're really all the same thing and that the symmetry is broken. The symmetry is there in the underlying equations of the theory, but it's not present in the solutions of the equations. It's not present in the particles themselves. That's why the W and the Z are so much heavier than the photon."[18]

To sum up, particle physics can be regarded as the study of symmetries both realized and broken. The broken symmetries are characteristic of the present, and the unbroken ones may have been manifest when the universe was young. Particles can thus be viewed as evidence of symmetry-breaking events that took place in the big bang. As Weinberg puts it, "The significance of the symmetries being broken is that that's what makes the world the way it is. The reason that electrons are different from quarks, and that up is different from down, and that chalk is different from cheese, all has to do with the breaking of underlying symmetries of equations which we don't yet know—equations that govern everything that goes on in the universe. The task of the physicist is to see through the appearances down to the underlying, very simple, symmetric reality."[19]

Which brings us back to cosmology. In a static, unchanging universe, broken symmetries would be a mystery. But in an evolving universe they emerge as evidence of historical events. Their flaws are informative, just as a broken mirror is more informative than a pristine mirror, since the fact that it is shattered testifies to its having had a more complicated history than a mirror that has remained unchanged since it left the factory. Specifically, the Weinberg-Salam-Glashow theory notes that the symmetry linking the photon to the weak-force bosons would be restored under conditions of high energy. Since the universe once *was* in a high-energy state, during the big bang, the theory indicates that what are today two forces, electromagnetism and the weak nuclear force, split off from what was originally a single, *electroweak* force. "There was a time in the very early universe, when the temperature was above a few hundred times the mass of the proton, when the symmetry hadn't yet been broken and the forces were all the same—not only mathematically the same, at some deep level having to do with the field equations, but *actually* the same," Weinberg says.[20] "A physicist living then, which is hard to imagine, would have seen no real distinction between these four forces produced by the exchange of these four particles—the Ws, the Z, and the photon."

Insights like these point to a new view of cosmic history, one in which the universe is viewed as a kind of paradise lost. In this sense, the cosmos has *de*volved from a state of perfect (or more nearly perfect) symmetry to the rubble heap of broken symmetries we find around us today. This consideration raises two important questions: What agency broke the primordial symmetry, and what would we learn if we had a theory that accurately described the original, symmetrical cosmos?

The first question—what broke the symmetries?—is the more technical of the two and yet, oddly, the more accessible. The favorite candidate is the *Higgs field,* named after Peter Higgs of the University of Edinburgh. The Higgs field can exist in a symmetrical state, but at its lowest energy—which it, like everything else in nature, from water flowing downhill to clocks running down, prefers—it breaks symmetries. In this picture, symmetry-breaking events like the one that divided the electroweak into the electromagnetic and weak forces were mediated by Higgs fields. In the process Higgs fields would have given particles their differing masses. Whether Higgs fields exist can be tested experimentally by looking for their carrier, the Higgs boson. The Higgs boson is calculated to be quite massive, however, so finding one (making one, really) requires generating very large energies. A mission of the Superconducting Super Collider (SSC) was to create particle collisions energetic enough to tease Higgs bosons out of the vacuum. But the SSC was canceled by Congress before its construction could be completed, so that challenge will fall to experimenters at CERN's Large Hadron Collider and Fermilab's Tevatron.

The second question—whether a theory of unbroken symmetries might describe the original universe—opens out onto one of the loftiest paths ever explored by science.

If indeed the history of the universe is a tale of a fall from perfect symmetry, this would have a number of implications. It would mean that much of the way things are today results from chance rather than necessity. The standard example of spontaneous symmetry breaking, first invoked by Werner Heisenberg in 1928, illustrates this nicely. Take a bar magnet. It has a north and a south magnetic pole. Heat it above 770°C, and it loses its magnetic properties. It is now magnetically symmetrical: Its "magnetic moment" is zero in all directions. Let the magnet cool—as the universe cooled, owing to expansion—and when its temperature falls below 770 degrees Centigrade it will regain its bipolar magnetism. *The question of which pole will become magnetic north and which south, however, is decided by chance.* Half the time it will go one way, half the other. If spontaneous symmetry breaking produced the fundamental forces of nature in this manner, we live in a world authored chiefly, and perhaps exclusively, by such flips of the coin.

Another implication central to particle physics today is that a unified theory that laid bare the symmetrical relationships behind today's forces would also identify the law or laws that ruled nature at the outset of cosmic time. The pursuit of this goal has brought about a sea change in the outlook of cosmologists and particle physicists, who now find themselves collaborators in the exploration of cosmic history. Particle accelerators would, like telescopes, be time machines—the telescopes directly observing events of the distant past, the accelerators re-creating events that transpired in the big bang. Physicists searching out symmetries latent in the laws of nature would be digging, like Schliemann at Troy, to bring to light shards of the world as it used to be.

Twenty years ago, the tools that promised to dig deepest went by the rather hyperbolic name of grand unified theories, or GUTs. They were all the rage. Graduate students in astrophysics wore T-shirts emblazoned with the words "Cosmology Takes GUTs." On the back of the shirts was a diagram of three curves—the strength of the electromagnetic, weak, and strong forces plotted against the distance between interacting particles. The lines converge at a distance of 10^{-29} centimeters. The idea was that at this point, equivalent to conditions in the big bang early in the first second of time, there were not three forces but just one. If the right GUT could be written, it would show how this force operated, thus providing a unified account of all the forces other than gravity.

Alas, this pregnant idea, although it generated many insights along the way, hasn't worked out terribly well. The simplest GUT, an SU(5) theory, predicted that protons decay. This was an eye-opening notion. It would mean that *all* atomic nuclei are radioactive—that hydrogen is unstable, as is radium, but on a longer timescale. Protons are found in the nucleus of every atom; if they decay, then all familiar objects, from galaxies to dandelions, are destined to evaporate. The predicted decay rate was slow: In some GUTs, the proton half-life was far too slow to detect, while in others it was roughly 10^{31} years. That's a long time, some 10^{21} (a thousand billion billion) times the current age of the universe, but it falls within reach of experiment. All one had to do was build a big tank—underground, to minimize cosmic-ray interference—fill it with 10,000 tons of water, which amounts to 10^{33} protons, outfit it with detectors to catch the flash of light occasioned by a proton's decay, and wait. If SU(5) was right, one proton decay event would be observed every few days—or, depending on the errors in the calculations, once a month, or once a year . . . or perhaps once every ten years.

I visited one of these detectors in 1982, when GUTs were still riding high. Located deep inside the Fairport Salt Mine, beneath the shores of Lake Erie outside Cleveland, Ohio, it consisted of a cavity the size of a six-story office building, lined with sheets of black plastic, filled with superpurified water, and watched, in total darkness, by 2,048 photomultiplier tubes. Sad-eyed postdocs monitored the computer screens. They saw nothing then, and they have seen nothing since. ("That's what they get for being *experimental* physicists," snorted a Nobel laureate theorist when I expressed sympathy over their plight—though, of course, it was the theorists who had led them there.) These tanks have since come in handy as neutrino detectors, in which role we encountered them when discussing dark matter, but none has recorded the decay of a single proton. So the simplest grand unified theory is all but dead. Unfortunately, this was the only GUT that made an actuarial prediction of the mortality of the proton that the experimenters could test. The Harvard physicist Howard Georgi, who had put a lot of work into GUTs, threw in the towel in 1989. "It does seem likely that the simple version of SU(5) has not been implemented by nature," he conceded.[21] ". . . In the absence of proton decay we cannot tell if some minor modification of SU(5) is required or if the whole idea of grand unification is wrong, or something in between."

Even if the GUTs or something like them were to succeed, we would still not have a fully unified theory, since gravity would be left out of the picture. Without gravity it is impossible to theorize about the Planck epoch, the first 10^{-43} second of cosmic history, when gravitation last played an important role in quantum interactions. Some theorists think that what is needed, if physics is to probe the Planck epoch, is to transcend quantum field theory altogether. Notwithstanding their success in building the standard model, with its accurate predictions of fundamental interactions involving the strong, weak, and electromagnetic forces, the field theories have always had some serious conceptual problems. One difficulty has to do with *renormalization*. Recall that the theories depict every "real" particle as surrounded by a cloud of "virtual" particles that bubble up out of the vacuum and then subside back into it. The production rate of virtual particles is limited by the time each exists: At a given vacuum energy density one can have lots of short-lived virtual particles or fewer long-lived ones. The trouble is that quantum field theories prescribe no minimum distance a virtual particle can travel. The travel distance can be zero, in which case the number of virtual particles becomes infinite. This would mean that every electron, say, is surrounded by an infinite number of virtual electrons and positrons. In that case, electrons would have infinite mass and infinite charge. They don't. So something is wrong. In practice the problem was solved—or avoided—by "renormalizing" the theories, a legerdemain that amounts to introducing new infinities to cancel out the unwanted ones. Renormalization works, but it is neither intellectually nor aesthetically satisfactory. Dirac complained that "even though [renormalization rules] may lead to results in agreement with observations [they] are artificial rules, and I just cannot accept that [they] are correct."[22] The irreverent Richard Feynman, who himself helped renormalize quantum electrodynamics, called renormalization "dippy" and a "shell game."[23] Asked for what he had won the Nobel Prize, he replied, "For sweeping them [the infinities] under the rug."[24]

But as Feynman also liked to say, "If there *is* something the matter, it's interesting!"[25] And what's interesting about the infinities problem is that it might signal an inadequacy in quantum field theory as a whole. Field theory depicts many particles as literally infinitesimal points—in other words, as having zero dimensions, no extension whatever in space. This poses no conceptual difficulties as a matter of mathematics, which is comfortable with such abstractions, but it seems unphysical. Might the ultimate unified theory arise from a different depiction of particles?

The answer is yes, according to the emerging *superstring* theories. Their central conception is that subatomic particles are actually tiny strings made of space. (If the universe began as pure space, strings may be thought of as shards of space that splintered at the outset of cosmic expansion, rather like ice crystals condensing from liquid water as it freezes.) Strings vibrate constantly, in an infinite number of frequencies. Since they are self-contained pieces of space, there is, as Steven Weinberg puts it, "no place for their energy of vibration to go."[26] Strings can interact in various ways, forming loops and crosses. The resulting attributes give rise to the characteristics of all the known particles. Strings are so small that when viewed from a distance—meaning at any wavelength of light or any other form of electromagnetic illumination—they "look like" infinitesimal particles.

To describe string theory we need to consider one further element of quantum physics, and that is *spin*. Spin was invented to explain a doubling in atomic spectral lines that would otherwise have been single. This effect was accounted for by saying that electrons possess an atomic number, designated spin, that is either "up" or "down." The language is arbitrary: Electrons don't have an up or down, and electron spin does not really work as if the electron were spinning on its axis, any more than an electron resembles a billiard ball. (Like other particles, the electron is equally well thought of as a wave.) Spin is not continuous—if it were, electrons would generate all sorts of spectral lines, not just two. Electrons are defined as having a spin of one-half. (More precisely, their spin is represented by the quantum of action, h over 2.) After being discovered experimentally, spin was shown to be central in Dirac's equation for the electron. It has remained central to quantum physics ever since.

Subatomic particles can be divided into two classes: those whose spin is fractional (one-half) and those whose spin is an integer (one). Fractional-spin particles are called *fermions*. Integer-spin particles are *bosons.* The difference is fundamental: Generally speaking, fermions constitute matter while bosons carry force. Physicists were able, by using symmetry considerations, to link fermions with fermions or bosons with bosons, but no theory crossed the great divide. That began to change in the 1970s, with the emergence of *supersymmetry* theories that hinted at symmetries between fermions and bosons. These symmetrical relationships are not realized in the universe today, but they lurk behind it, as ghosts of cosmic history. Attempts to write a fully unified supersymmetric theory incorporating gravitation (the "supergravity" theories) faltered, but physicists had been provided with another indication of where the lamp of symmetry might show the way.

Enter superstring theory. The concept that particles are really tiny strings dates from the 1960s, but it took on wings in 1974, when John Schwarz of Caltech and Joel Scherk of the Ecole Normale Supérieure in Paris (who died young, in 1980) came to terms with what had been an ugly blemish in their calculations. String theory kept predicting the existence of a particle with zero mass and a spin of two. Schwarz and Scherk realized that this unwelcome particle was nothing other than the graviton, the quantum carrier of gravitational force. (Although there is no quantum theory of gravity yet, it is possible to specify some of the characteristics of the quantum particle thought to convey it.) This was liberating: The calculations were saying not only that string theory *might* be the way to a fully unified account of all particles and forces but that one could not write a string theory *without* incorporating gravity. Edward Witten of the Institute for Advanced Study at Princeton recalled that learning this news constituted "the greatest intellectual thrill of my life."[27]

Which was saying something. In the high carrels of theoretical physics, where intelligence is taken for granted, Witten is regarded as preternaturally, almost forbiddingly, smart. A tall, boyish-looking man, he wears the habitual small smile of the theoretician for whom sustained mathematical thinking has something like the emotional qualities that mystics associate with meditation. He speaks in a soft, high-pitched voice, floating short, precise sentences punctuated by witty little silences—the speech pattern of a man who has learned that he thinks too fast to otherwise be understood. Though he is the son of a theoretical physicist, Witten came to science in a roundabout fashion. He graduated from Brandeis College in 1971 as a history major, wrote political journalism for the *Nation* and the *New Republic*, and worked in George McGovern's presidential campaign. Primarily a mathematician, he picked up physics along the way, almost as a hobby. But colleagues who compare him to Einstein have something more specific in mind than his imposing intellect: Like Einstein, Witten is a geometer. "The great ideas in physics," he says, "have geometric foundations."[28] String theory, he believes, provides a geometric basis for particle physics—which means, among other things, a way to make everything out of nothing. He calls string theory "part of the physics of the twenty-first century that fell by chance into the twentieth century."[29] It caught his interest and kept him in physics. He published nineteen papers on strings in the year 1985 alone and has bustled on at a similar pace ever since, laying tracks on which mighty trains can run.

Science in action is a rather haphazard affair, a workshop where tools are fashioned and found inappropriate for the task at hand, only to be picked up later by other workers and tried out on quite different tasks. Modern string theory is no exception. Its origins lie in a series of frustrated efforts, beginning with work by Gabriele Veneziano in the late 1960s and Yoichiro Nambu in the early 1970s, to apply a stringy conception of particles to the strong force.[30] Soon thereafter, Schwarz and Scherk saw how the string conception might be applied to all four forces. At first, few paid much attention. Quantum chromodynamics had come into its own as an account of the strong force, and researchers were crowded over there, in the light, toiling away happily with tools that worked fine. String theory, for its part, produced anomalies—flaws that nearly all physicists quite reasonably regarded as evidence that it was wrong. If something good did turn up in strings, those who noticed tended to borrow the tool, take it away, and apply it to supersymmetry or some other seemingly more promising field.

Undaunted, Schwarz kept plugging away. "String theory is too beautiful a mathematical structure to be completely irrelevant to nature," he insisted.[31] During the summers, he collaborated with Michael Green of the University of London, who had succumbed to the aesthetic appeal of the field and admitted to being "hooked on strings." They had plenty of time—owing largely to their dogged pursuit of such weird ideas, their careers were all but stalled—and they enjoyed the freedom of young innovators left to explore on their own without the burden of elders looking over their shoulders. "I never worked with such intensity in my life," Green recalled. ". . . Not so much physical effort as the concentration of thoughts in my head. I've never been so immersed in a subject."[32] Slowly, string theory took shape. Then, in August 1984, working together at the Aspen Physics Institute in Colorado, Schwarz and Green found themselves on the brink of finding a cure for the anomalies and thus producing a self-consistent string theory that incorporated supersymmetry. Ten years of work had led them to a single hypothesis; technically stated, it was that the anomalies disappeared when one calculated one-loop amplitudes with either of two internal gauge symmetry groups. The question of whether this proposition was true came down to a bit of simple arithmetic: multiplying 31 times 16. If the answer was 496, string theory would be liberated from its long bondage in the thicket of the anomalies. Green chalked 31 and 16 on the blackboard, multiplied, and got 486.

"Oh, dear," he said. "It doesn't work."[33]

"Try it again," Schwarz said. This time Green got 496. So did Schwarz. They had come to the end of the rainbow. When their solution was published later that year, physicists almost instantly turned from supersymmetry, "supergravity," and the other active workbenches to converge on string theory. Witten, the first big-name scientist to have taken up the string banner, was influential in singing its praises. "It's beautiful, wonderful, majestic—and strange, if you like," he said, "but it's not weird."[34] Many other researchers were prompted to investigate strings by reading a 1987 book, *Superstring Theory,* coauthored by Green, Schwarz, and Witten. (In honor of the meanings of their names, it bore a green, black, and white cover.) String became a repository of the highest hopes of the finest minds in physics.

Ten-dimensional string theory (the variety to which I am limiting this discussion; its competitor works in twenty-six dimensions) has many beautiful qualities. As noted, it tends automatically to unite the four forces. It is supersymmetric, so it can account for fermions as well as bosons, drawing all matter into an elegant picture in which particles' attributes are seen as the vibrations of strings, like notes struck on Pythagoras' lyre. It does away with the infinities that dogged quantum field theory, thereby dispensing with the dirty business of renormalization. And, perhaps most appealing of all, it makes everything out of space. Strings are just curved space. The central riddle of genesis—how can the universe have come into being, if, as Shakespeare put it, "Nothing can be made out of nothing"?[35]—is answered thus: Everything *is* nothing, in a sense, for all is made of space, which in this context means pure geometry. String theory can even explain how the universe got from ten dimensions to four. It hypothesizes that six of the original dimensions collapsed by way of a *phase transition*, a mechanism by which forces divided and particles proliferated as the universe expanded and cooled; a new state was achieved, as when liquid water turns to ice. Our one temporal and three spatial dimensions can be thought of as having continued to expand as the universe emerged from the phase transition, while the remaining dimensions expanded no farther. Consequently, six of the original ten dimensions remain minuscule, curled up. They would appear as spacetime foam if we could examine space way down on the minute level of the Planck scale.

But there is, of course, a catch. String theory isn't finished yet, and at its present state of development presents problems. One concerns "compactification"—just how did the six other dimensions curl up? Another has to do with string theory's proliferation of particles. Like supersymmetry, string theory posits many sorts of particles, among them supersymmetric partners of known particles—these go by names like the sneutrino, the squark, and the photino—and many more besides, perhaps an infinite variety of them. It lacks as yet an explanation of why the universe turned out as it did, rather than in one of the tens of thousands of alternative possibilities string theory entertains. (Perhaps, as we will be considering later, there *are* many universes, each sporting its own particle array.) A field theory of strings should derive the masses of the proton and other particles, but no such theory has yet been devised. "The problem," writes the physicist Michio Kaku of City College of the City University of New York, who has banged his head against this last difficulty as much as anyone, "is that *no one is smart enough to solve the field theory of strings.*"[36] One hears variations of this complaint throughout string studies. String theory may be "part of the physics of the twenty-first century," as Witten maintains, but physicists trying to solve it with twentieth-century techniques sometimes feel as the Wright brothers might have if they had been called upon to fly a Stealth bomber.

Ten-dimensional mathematics in itself is dizzying. Working with it calls for a command of *topology,* the science of surfaces. Topology can be strange, as many have discovered upon attempting to work with strings. A sphere, for instance, can be caved in on itself like a deflated basketball, or stretched into the pear shape that Columbus (wondering how he reached Asia so quickly) took to be the shape of the earth. While these changes will alter the definition of something so geometrically straightforward as the description of parallel lines on the surface of the (former) sphere, in topological terms spheres and pears are the same. Cut two holes in the sphere and connect the holes with a tube, however—which is to say, put a handle on it—and the topology has changed. A handle is topologically equivalent to a doughnut. (Hence the campus joke, "A topologist is someone who doesn't know the difference between a coffee cup and a doughnut.") It is both the glory and the curse of string theory that it has lofted physics into the ionospheres of higher mathematics. As Kaku observes:

> One consequence of this formulation is that a physical principle that unites many smaller physical theories must automatically unite many seemingly unrelated branches of mathematics. This is precisely what string theory accomplishes. In fact, of all physical theories, string theory unites by far the largest number of branches of mathematics into a single coherent picture. Perhaps one of the byproducts of the physicist's quest for unification will be the unification of mathematics as well.[37]

Skeptics view string sophistication as a liability, its higher math a castle destined to remain floating in thin air. String theory, they complain, is in many ways more complicated than the old physics it purports to replace—and it cannot yet even specify the old models. For example, string invokes 496 massless gauge bosons, while the standard model is content with 12. Moreover, most of the physics of strings was manifest, if ever, only during the Planck epoch, the first instant of time, at energy levels so staggeringly high as to defy replication by experiment. (To generate Planckian energies using today's technology would require building a particle collider a light-year in length.) "We need revolutionary ideas in accelerator design more than we need theory," says Samuel Ting, an experimental physicist at MIT.[38] Paul Ginsparg and Sheldon Glashow of Harvard sound a similarly cautionary note. "Contemplation of superstrings may evolve into an activity as remote from conventional particle physics as particle physics is from chemistry, to be conducted at schools of divinity by future equivalents of medieval theologians," they write. "For the first time since the Dark Ages, we can see how our noble search may end, with faith replacing science once again."[39]

Witten rejects such sentiments. "Good wrong ideas are extremely scarce, and good wrong ideas that even remotely rival the majesty of string theory have never been seen," he says.[40] He concedes that string phenomenology would bloom only at very high energies, but notes that the theory predicts the existence of some supersymmetric particles detectable near what physicists call the "tree level"—at energies that experimenters expect to muster by the dawn of the next century. "For string theory to turn out to be wrong would be unprecedented in the history of science," Witten argues. ". . . People that predict that something cannot be tested are begging to be proved wrong by some combination of experiment and theory they can't foresee. . . . And the unforeseeable is the most important aspect of that. The preoccupation of physics from about 1920 to about 1970 was field theory. That was fifty years. It might take us that long to work out string theory, which is less than thirty years old."[41] He sees the elevation of science to the supersymmetric perspective of string theory as a step "in which one gets 'out of flatland' to see a wider symmetry from a high-dimensional vantage point."[42]

When (or if) the capstone of string theory is set in place, establishing it as the long-sought ultimate unified theory, it may bear the inscription *extreme black holes.* Unlike ordinary black holes, which are as massive as stars, extreme black holes have the minuscule masses of subatomic particles. Indeed, if string theory is correct, subatomic particles *are* black holes. We encountered extreme black holes earlier, at the conclusion of chapter 3, where we depicted the quest for an ultimate unified theory as akin to building two sides of an arch. One side is rooted in general relativity and builds up through speculative accounts of spacetime geometry, like supergravity and quantum gravity. Here we climb the arch from the side that is based on quantum physics and proceed up through supersymmetry and string theory. Hopefully both routes will converge at the same apex—a theory that combines curved space and the quantum principle, depicting particles as strings and thus as shards of space.

A significant step up the arch was made in 1995 by Andrew Strominger of the University of California, Santa Barbara. He identified a mathematical correspondence between strings and extreme black holes. Then, in collaboration with the physicist Brian Greene of Cornell University and the mathematician David R. Morrison of Duke University, Strominger showed how extreme black holes could take on the appearance of subatomic particles by going through phase transitions. In addition to bringing the particle–black hole arch closer to completion, this work reduced the notoriously high number of ways that string theory had permitted the six extra dimensions of space to collapse. "What we discovered is that the tens of thousands of seemingly different choices are in fact just different descriptions of the same thing under different circumstances—much as water and ice are both descriptions of H_2O," Strominger said.[43]

An elegant man with a diffident attitude, Strominger combines scientific practicality with an artist's urbane fatalism. "I don't believe anyone finds out anything about the universe except how beautiful it is, which we know already. It's just very pretty," he says, adding, "The important thing in my life isn't any particular belief —I just want to know what works."[44] His calculations, which built on work by Witten and Nathan Seiberg of Rutgers University, moved string theory closer to fruition by revealing that extreme black holes and subatomic particles may be two aspects of the same thing—strings.

String theory promises many boons to cosmology. It proffers exotic particles, one or more of which could constitute the mysterious dark matter. It points naturally to *inflation*—an extremely rapid early cosmic expansion that would have rendered the universe much larger than had been thought, flattening local space in the process. More specifically, string theory postulates the existence of the *scalar fields* important to inflationary theorizing. And because string theory is based on only one fundamental parameter, the Planck mass, it would seem able to connect itself and other aspects of the early universe with the "tree level" universe of today, accounting for cosmic evolution without the need for attributing arbitrary initial conditions to the universe.

And so, on to inflation, the balloon that will loft us to a new view of the universe as enormous and, perhaps, as but one among an infinite number of universes.

9
The Speed of Space

"What is the news of the day,
Good neighbor, I pray?"
"They say the balloon
Is gone up to the moon!"

—*Traditional verse*

Inflation is eternal.

—ALAN H. GUTH[1]

LIKE THE ARABIC INVENTION of the zero or Einstein's realization that a falling man feels no force of gravity, inflation is a simple idea with multiple implications. Baldly stated, the inflationary hypothesis doesn't sound like much. It asserts that the infant universe underwent a spasm of *exponential* expansion, a fraction of a second during which it repeatedly doubled its radius over equal intervals of time. (During exponential expansion the radius of the universe increased with each tick of the clock as R = 1,2,4,8,16 . . .) This historic hiccough was over within 10^{-34} second, an interval compared to which the blink of an eye is an eternity. Thereafter the universe settled into the *linear* expansion rate (R = 1,2,3,4,5 . . .) that has characterized it ever since.

Yet inflation may well rank among the most productive ideas to have graced modern cosmology. It provides a physical explanation for cosmic expansion, thus addressing the riddle of how the universe started expanding in the first place. It cures the classic big bang model of its most troubling intrinsic difficulties: the *flatness* problem, having to do with the universe's eerie closeness to critical density, and the *horizon* problem, having to do with its being homogeneous and isotropic over distances too large for information to have traversed since the big bang. In this chapter we revise the big bang theory to incorporate inflation, then trace the history of the inflationary hypothesis and look at some of its implications.

Although the inflationary "epoch" was unimaginably brief, its consequences were literally enormous. During inflation, the radius of the universe would have multiplied many times. At the start of inflation, the universe was smaller than a proton. But if you keep doubling even a minute quantity, it soon gets big. The concept of exponential increase has long been illustrated by the story about the young man who persuades a mathematically challenged monarch to reward him if he succeeds in rescuing the king's lost daughter—the reward to be amassed by placing a penny on the first square of a chessboard, two pennies on the second square, four on the third, and so on, until all the squares on the board are covered. Once his daughter is returned, the king learns to his horror that the reward is more than he can come up with: indeed, it is far more than the total wealth of all the civilizations in the history of the world. Inflation is like putting extra money on the first square of the chessboard. Even if we take a modest inflationary model—one in which the universe is swelled by inflation to only about the size of a medicine ball—the familiar, linear expansion that follows will double *that* radius another ninety times or so over the course of the next ten billion years, yielding a universe that today is much larger than it would have been had inflation never taken place. And if we use the estimates of *chaotic* inflation, the figures become truly Brobdingnagian.

Chaotic inflation is the work of the cosmologist Andrei Linde, now at Stanford University. When Linde was first developing the theory, he explained it to me by jotting down, on a scrap of white paper, its estimate of the radius of the universe. Using a thick, felt-tipped black pen, he first wrote a ten, then an exponent of ten, then raised *that* by another factor of twelve. This was the largest number I'd ever seen. It beggared the term "astronomical." It is equal to a one followed by a trillion zeros. Had Linde written a one followed by a trillion zeros on a strip of tickertape, the paper would have girdled the earth two dozen times. To print it here for the reader's edification, in small type, would require that this book be a million volumes in length—the capacity of a good university library. Gaping at Linde's ten to the tenth to the twelfth, I asked one of the stupidest questions of my life: "In what units is that expressed?" I inquired. "I mean, is it centimeters? Light-years? Hubble radii?"

Linde laughed. "Well," he replied, "when you're dealing with ten to the tenth to the twelfth, it doesn't really *matter* what units you're using."[2] Actually he had in mind the radius of the universe in centimeters, but the point is that the inflationary universe is not just big but, as the variety show host Ed Sullivan liked to say, *"really* big."

Now, while "the universe" comprises, by definition, the totality of all the regions that are potentially observable by us, the *observable* universe is restricted to the region from which light signals have already had time to reach us. In the classical big bang picture this distinction was not terribly important, since in that model the observable universe amounted, depending on the geometric particulars, to something like 75 to 90 percent of the entire universe. But since the universe in the inflationary model is gigantic, the observable part of it is only a tiny fraction of the whole.[3]

One important effect of this radically altered relationship between the size of the observable universe and that of the universe as a whole is to flatten cosmic space. As we noted in chapter 2, the fact that the geometry of cosmic space is anywhere near flat (i.e., that the value of omega is approximately one) was inexplicable in the classical big bang model. To make cosmic geometry come out that way would have required extraordinary fine-tuning of the initial conditions: The primordial value of omega had to be set to one within one part in 10^{60}. God might have been up to it, but why should he have gone to the trouble? Attention was drawn to the flatness problem by James Peebles and Robert Dicke at Princeton years earlier, but within the classical model no solution could be found.

Inflation provides a straightforward explanation of why cosmic space should appear to be flat. The classical model, since it envisioned the observable universe as constituting the lion's share of the entire universe, permitted observers to behold the global geometry of space. Their situation was rather like that of comic strip characters standing on a small, spherical planet, who can tell just by looking around that the surface of the planet is curved, and can easily measure the extent of its curvature. But because the inflationary universe is much bigger, any observer perceives the surface (i.e., the shape of space) as flat. To illustrate this effect, imagine that you are a geographer in ancient Athens and you want to measure the curvature of the earth. You spend a few days down at the docks, estimating how far ships have to sail out to sea before they disappear over the horizon. Soon you have derived a plausible estimate of the curvature of the earth's surface. This represents the classical big bang theory, in which we can see most of the entire universe and so can measure the curvature of cosmic space. Now inflate the earth a zillion times. (Actually many zillions, but enough of that.) The horizon appears to be perfectly flat—you cannot measure any curvature; outgoing ships shrink to dots but *never* disappear—even though the bloated earth remains spherical. That's how inflation solves the flatness problem. Regardless of whether its geometry ultimately is open, closed, or flat, the inflationary universe *appears* to be flat, since we cannot see enough of it to determine its true curvature.[4]

Similarly, inflation resolves a major problem that beset the classical big bang model—the riddle of why the universe should exhibit *homogeneity,* meaning that matter on the large scale is distributed evenly, and *isotropy,* meaning that matter is distributed in comparable ways in all directions. Both difficulties arise from the relationship between the observable universe and the entire universe in the classical model. Since the observable universe is the region within which information can have been exchanged, it also defines the largest area within which cause and effect can have operated. The trouble was that in the standard model, given the velocity of light and the age of the universe, objects on opposite sides of the observable universe could never have been in physical contact or otherwise have exchanged information with one another. If that had been the case, astronomers examining the cosmic microwave background radiation on opposite sides of the sky today would be looking at disconnected and dissimilar regions; indeed, the two regions would be separated from each other by ninety discrete horizons. Not only would they reside outside of each other's observable universe, but eighty-eight observable universes would lie between them. Trying to get a message across the classical big bang universe would have been harder than sending a letter through ninety nations none of which handled the postal traffic of any of the others. In fact it would have been impossible. Yet the cosmic background radiation looks much the same all over the sky. How did photons from one region ever "learn" that they were to emit in a black body spectrum, at exactly the same temperature? How did electrons on one side of the universe "learn" how to be identical with the electrons on the other side? This is the horizon problem.

At first glance, inflation would seem to make the horizon problem worse, since it postulates a more rapid cosmic expansion rate near the outset of time. Instead, inflation eliminates the horizon problem. It does this in two ways. First, since inflation need not have begun at time zero, it permits scenarios in which the primordial material remained concentrated long enough to mix before inflation got under way. Second, inflation delivers a homogeneous universe by getting rid of primordial lumps. Suppose you are observing the Alps. You note that they are lumpy. Now inflate the earth to a zillion times its former radius. The Alps are still there, but the horizontal dimension has increased so radically that you don't notice them. (Even if you were standing on Matterhorn's peak, your surroundings would now look as flat as a tabletop.) No more lumps. So all observers in postinflationary universes find that matter on large scales is distributed homogeneously. As Linde writes, "Inflation remains the only theory which explains why the observable part of the universe is almost homogeneous."[5]

While accounting for the smoothness of the universe, inflation also handily predicts its departure from smoothness—the existence of large-scale structure—on a level, about one part in a hundred thousand, consistent with what astronomers using the COBE satellite have observed in the microwave background. That's because inflation would have amplified random quantum fluctuations in the early universe, producing the inhomogeneities we have previously encountered as the Harrison-Zeldovich spectrum that the COBE data support. Left to themselves these quantum fluctuations would quickly have subsided, to be replaced by others in the perpetually restless shuffle that characterizes the quantum world. But inflation freezes them in, bequeathing them to the postinflationary universe as the seeds of large-scale structure. The mechanism hinges on the relationship between the wavelength of a given field and the length of the Hubble radius. (The Hubble radius is that of the observable universe; it defines the zone of causal contact at a given moment in cosmic history. Today its length is over 10 billion light-years. During inflation it was only a fraction of a light-second.) As the universe inflates exponentially, the fields carrying quantum flux information grow exponentially in wavelength. When the wavelength of a given fluctuation exceeds the Hubble radius, its amplitude freezes, while its wavelength continues to grow. At the end of inflation, such a frozen field emerges looking like a classical field, creating inhomogeneities in the cosmic distribution of mass.[6] So inflation evidently can deliver both the local lumpiness and the global smoothness that characterize the universe we live in.

As a final advertisement, I should mention that inflation does away with the *magnetic monopole* problem. When discussing unified theory we saw that the vacuum of space, in undergoing phase transitions that broke a hypothetical primordial force into the four forces manifest today, would have been fraught with Higgs fields, which are said to have endowed particles with mass. Knots in the Higgs fields should have generated lots of magnetic monopoles. These are massive topological defects that have only one magnetic pole. Never yet observed but reasonably well established in theory, they are alluring objects, little onions whose anachronistic layers retain the primordial physics that ruled the earlier universe. The trouble is that in the classical big bang picture there were far too many of them. The calculated monopole density would have made the universe dense enough to halt cosmic expansion only thirty thousand years after the big bang. Inflation obviates this difficulty by diluting the monopole density, and by sweeping those few that are produced to regions far outside the horizon of the observable universe. Enormous numbers of monopoles could have arisen in the inflationary universe, yet we and all other observers would find them to be observationally far rarer than snowballs in the Sahara. Calculations suggest that inflation would spread them so thin that the average observer could expect to find only a single monopole in the entire observable universe.[7]

So much for inflation's curative powers. What do the inflationary theories actually say, as a matter of physics? Answering this question requires us to take a closer look at the concept of the vacuum.

The word comes from the Latin *vacuus,* for "empty," and idiomatically that's what a vacuum is—nothingness, an absence of everything. Such a vacuum figured importantly in the thinking of the Greek atomists. They regarded objects as made of tiny, indivisible particles. (*Atom* means "indivisible.") To account for motion, the atomists had also to postulate the existence of a void; otherwise the atoms would be crammed together everywhere and motion would be impossible. The void, or vacuum *(kenon),* gave the atoms somewhere to go. Atoms were things; the vacuum nothing. As Democritus put it, "There are only atoms and the void."[8] This picture is superficially pleasing, and indeed is still being taught to schoolchildren today. But it presents a logical difficulty. If the vacuum is, literally, *nothing,* how can it be said to exist? Aristotle pointed this out in critiquing the position of Democritus' teacher Leucippus, who, he writes, "granted that there could not be movement without void, that the void was 'not being,' and nothing of what *is* is not being; for what, strictly speaking, *is,* is completely full."[9] In short, to argue that the void "exists," while also arguing that the void is nothing, is to make the contradictory claim that there exists something which does not exist.[10]

This is no mere quibble. For if we speak of nature as constituting a whole—and that is what *universe* means: "All turned into one"—we must mean that the universe *exists.* But how can this whole exist, if it is shot through with the marbleized veins of a vacuum that we define as *non*existent? At best such a fractured cosmos would constitute not a unity but many separate realms—islands of existence separated by the nonexistent. This satisfied neither Aristotle nor Plato nor others who examined it closely. Many were led to conclude that the vacuum must be full of something.

For many centuries, the concept of the vacuum danced around this axis, with some thinking of it as nothingness and others as a plenum. The plenum held sway in the nineteenth century, when most physicists assumed that space was full of the so-called aether. Then the Michelson-Morley experiment proved that the aether does not exist, and Einstein demonstrated that it was theoretically superfluous as well.

But no sooner had the vacuum been emptied than quantum physics filled it up again. The quantum vacuum is a frothing sea of activity. This is emphatically the case in high-energy environments like the centers of stars, and it remains so, albeit to a lesser degree, in the coldest, emptiest corners of the universe. Owing to what is called wave-particle duality, quantum physics sees nature as if through two eyes, one of which beholds particles and the other waves. Look through the particle eye and we find that for every "real" (meaning long-lived) electron there are countless "virtual" electrons and positrons. Look through the wave eye and we find quantum fields roiling the vacuum like winds across water. We tend to think of fields mostly in terms of energy ("force fields"), but matter, too, may with equal accuracy be depicted, via wave mechanics, as composed of quantum fields. (Asking which is "really" nature—particles or fields—is like asking with which eye you are really seeing when you have both eyes open.)

So the vacuum today is a kind of mix. As the nuclear physicist Hans Christian von Baeyer writes, "The modern vacuum represents a compromise between the opinions of Democritus and those of Aristotle: The former was right to insist that the world consists of atoms and the void, and the latter when he claimed that there is no such thing as true and absolute emptiness. . . . The dynamic vacuum is like a quiet lake on a summer night, its surface rippled in gentle fluctuations, while all around, electron-positron pairs twinkle on and off like fireflies. It is a busier and friendlier place than the forbidding emptiness of Democritus or the glacial aether of Aristotle."[11]

Important where inflation is concerned, the quantum picture of the vacuum urges that all vacuums are not created equal. The fluctuating quantum fields in a vacuum have all possible wavelengths and move in all possible directions. If the values of the fields cancel out when averaged over time, then we have a classical vacuum, which is to say one that resembles old-fashioned empty space. But if the fields do not cancel out, we have what the physicists call a "false" vacuum. A false vacuum contains more energy than a classical vacuum. It is said therefore not to be in its minimum energy state. In the universe today, false vacuums are discerned only in certain circumstances. Quarks, for instance, occupy a false vacuum, its high energy created by the strong nuclear force field that binds the quarks to one another. According to many unified theories, however, during the first moments of cosmic history the ambient energy was so great that the entire universe was in a false vacuum state. The energy of the false vacuum acts as a kind of antigravity, and can cause space to balloon at an exponential rate. This could be the engine that drove inflation. During inflation, the universe was nearly empty, its energy content having been swallowed up into the false vacuum. Once the false vacuum decayed into a classical vacuum, its excess energy precipitated, like raindrops congealing in a mountain thunderstorm, into the myriad hot particles of the big bang.

Just as water runs downhill and clocks run down, natural systems all tend to devolve from high-energy to lower-energy states. So nature abhors a false vacuum, and favors the lower-energy classical vacuum state that we, consequently, regard as normal. This raises the question of why the universe put up with existing in a false vacuum state for as long as 10^{-34} second—an interval that, though abrupt by human reckoning, seems by the standards of early-universe physics as interminable as an indifferent production of *Lohengrin*. One plausible explanation is that the universe, as it inflated, *supercooled*. Supercooling occurs when a substance drops in temperature so rapidly that it falls, without changing its state, below the temperature at which it would normally undergo a phase transition into another state. Owing to supercooling, cosmic space could have remained in a false-vacuum state for longer than we might naively expect before disgorging itself of the energy that drove its inflation. Indeed, the universe may never have completely stopped inflating—a prospect to which I will return a bit later.

The inflationary hypothesis was launched into prominence in 1980, by the MIT physicist Alan Guth, then a research associate at the Stanford Linear Accelerator Center.[12] Guth was looking for a way around the magnetic monopole problem, and for an explanation for the near-critical mass density of the universe. He did the essential calculations on the evening of December 6, 1979, and immediately grasped their importance: When he awoke the next morning, he wrote the words "SPECTACULAR REALIZATION" in capital letters across the top of the page, drew a box around them, and basked in the light of an epiphany that was destined to make his career. But life is seldom simple, and Guth's version, known today as "old inflation," was subsequently much revised. In 1981–1982 it was supplanted by a "new inflationary" model arrived at by Linde in Russia and independently, soon thereafter, by Andreas Albrecht and Paul Steinhardt at the University of Pennsylvania.

The Guth universe inflated just fine, but once inflation ended it crashed out into a froth of bubbles much less homogeneous than is the real universe. For a while, theorists thought the problem might be solved by making the bubbles percolate into one another, but that didn't work; the ongoing cosmic expansion kept the bubbles separate. So "old" inflation could not be fitted to the observed cosmic homogeneity—a conundrum that came to be known as the "graceful exit" problem. "New" inflation solved it, by showing how the vacuum could have made the transition from false to classical more slowly than in Guth's original model. This "slow rollover" got around the bubble difficulty, producing a cosmic structure that answered to the observations.

It did so, however, by fine-tuning some critical parameters. We can illustrate the difference by imagining an abstract quantum space in which the false and normal vacuum states are separated by a barrier. In Guth's original theory the barrier stood high, and the universe made the transition by performing a quantum leap ("quantum tunneling") through it. This was physically plausible, but as noted it made too many bubbles. New inflation lowered the barrier, permitting the vacuum to execute a graceful exit from false to classical. But to make this work was rather like playing one of those Canadian Royal Air Force games in which you manipulate a board that tilts on two axes to maneuver a ball-bearing through a maze without letting it fall through one of the many holes that dot the board. Precisely because the transition out of the false vacuum stage is so easy, one has to insert just the right values to make inflation proceed at an acceptable rate. But there was no satisfactory explanation for why those values should be just so. This prompted the strong suspicion that new inflation, though helpful, was not conclusive. "Most theorists (including both of us) regard such fine tuning as implausible," volunteered Guth and Steinhardt. "The consequences of the scenario are so successful, however, that we are encouraged to go on in the hope that we may discover realistic theories in which such a slow-rollover transition occurs without fine tuning."[13] Theorists continued to tinker with new inflation, and in one development, Steinhardt and Robert Crittenden developed a promising model in which inflation ended before the universe made the transition out of the false vacuum phase. Such work may be expected to continue, with developments in unified theory and improved observations of structure in the cosmic microwave background conspiring to raise the net in the new inflation game.

Meanwhile, in 1983, Linde came up with the theory of *chaotic* inflation. Whereas both Guth's original model and the new inflationary scenario envisioned the very early universe as hot and inflation as analogous to a thermodynamic phase transition, Linde's mechanism did away with the need for heat. "The assumption that the universe was hot before inflation is unnecessary, and typically it is even harmful," Linde asserted.[14] Chaotic inflation is in many ways a more powerful and far-reaching conception than its predecessors. Linde sees it as more "natural," although as the Princeton physicist Ed Turner warned his colleagues, "Your temperature should rise whenever anybody says, 'My model is more natural than your model.' "[15]

To appreciate Linde's approach we should look more closely at a central mechanism of inflationary scenarios, the role played by scalar fields. As we noted earlier, all unified theories (except for string theory) are field theories, and they postulate the existence of Higgs fields that break symmetries and endow particles with mass. Here we take a wider frame of reference and observe that the Higgs is one among many possible scalar fields. Scalar fields are thought to represent the dominant form taken by matter in high-energy conditions such as the big bang, but also to exist (though less evidently) in low-energy regimes. According to quantum physics, scalar fields fill cosmic space but are noticeable only when potential differences between fields occur. Linde compares this characteristic with electricity: "Electrical fields appear only if [their electrostatic] potential is uneven, as it is between the poles of a battery or if the potential changes in time," he writes. "If the entire universe had the same electrostatic potential, say, 110 volts, then nobody would notice it; the potential would seem to be just another vacuum state. Similarly, a constant scalar field looks like a vacuum: We do not see it even if we are surrounded by it."[16]

Scalar fields possess only magnitude—whereas, say, an electromagnetic field, which is a vector field, has both magnitude and direction at each point. The topological maps carried by hikers depict a kind of scalar field; on topo maps, the magnitude is height above sea level, with the "field lines" connecting locations of identical altitude.[17] Also, scalar fields act on all particles impartially—unlike electromagnetic fields, which influence only particles that carry an electrical charge. In this sense scalar fields mimic the behavior of empty space, which treats everything equitably. This makes them handy tools for interpreting the behavior of the vacuum. Finally, and important in this context, scalar fields can generate a repulsive force strong enough to overcome gravity.

Scalar fields can both spur inflation and rein it in. The same general relativity equations that predicted cosmic expansion in the first place show that the inflation rate of the universe is proportional to its mass density. Scalar fields contain energy—which is of course equivalent to mass—so they can enhance the cosmic mass density and thus generate the exponentially rapid expansion that we call inflation. And, having propelled inflation, they can also end it. Inflation is thought to stop once the pertinent scalar fields reach their minimum potential energies. Considerable work has been devoted to understanding this process, with the old and new inflationary models electing different parameters for their scalar fields. The customary way to visualize these is to picture the vacuum as a cowboy hat with an indented crown. The dent at the top of the crown represents the local minimum value of the relevant scalar field. During inflation the field sits in the dent, hung up at its local minimum value. Inflation ceases when the field descends to the brim, which represents its global minimum value. Old and new inflation differ in how they get the field down from the dent in the crown to the brim. In the old model, the field quantum-tunnels through the crown. It the new model, the dimple in the crown is much shallower and the field flops out of it, like a fish escaping a frying pan. Old inflation, as noted, resulted in too inhomogeneous a universe. New inflation had to hold the field in the shallow depression atop the crown long enough for inflation to take place, yet still be able to flop it out so that inflation would stop. Both scenarios were troublesome. Both assumed that the universe began hot.

Linde's chaotic model broadened the framework within which inflation was investigated. Linde showed that the universe could have begun not with a single sort of scalar field with a particular value (an initial condition) but with a seething ocean of all sorts of scalar fields. That's the "chaotic" part. The fields have many different minima, and they also differ as to how far away each is from its minimum value, and in the extent to which each is homogeneous. Our observable universe is said to have evolved from a scalar field that was almost homogeneous and that happened to be far from its minimum value. Such a field "rolled over" slowly, producing the universe we inhabit today. Other fields would have produced very different regions. This is the key to the multiple-universe model we will be investigating in the next chapter. Here the important point is that Linde was able to dispense with most initial conditions. In chaotic inflation, just about the only initial condition is chaos. The universe need not have been hot at the outset. The blossoming of particles out of the vacuum at the end of the inflationary spasm, often called "reheating," may, Linde noted, have been the *first* heating of a previously cold universe. Nor did the universe necessarily evolve from a single scalar field with just the right values. All that's necessary is that *our* universe came from such a field. To sweeten the deal, Linde also provided a mechanism for (re)heating, one in which the required energy was generated by oscillations in the scalar field produced when it fell clanging into its minimal-energy state. What we call the big bang might, in Linde's view, have been the first violent eruption produced when the field danced up out of the braking cosmic vacuum.

To sum up, inflation seems a good bet to become part of

the standard big bang cosmology—or, more properly, to paint a broader picture within which the big bang would itself be subsumed. It is original; the Berkeley astrophysicist Joseph Silk exaggerates less than one might think when he asserts that "inflation is the only new idea in cosmology since Einstein."[18] It is generous in scope, interfacing adeptly between the classical physics of general relativity and the quantum physics of subatomic particles and fields. It resonates well with string theory, since both would build the foundations of the cosmos out of space itself. It offers adept explanations of why the universe is isotropic and homogeneous (the stretching of a wrinkled balloon) and how galaxies nevertheless came into being (they were seeded by density perturbations that originated in quantum flux). It has already made two testable predictions—that the cosmic matter density should be critical (as it is confirmed to be, within a factor of 10) and that the spectrum of primordial density fluctuations should be scale invariant (the Harrison-Zeldovich spectrum, subsequently detected in the cosmic microwave background by the COBE satellite).[19] And always important in science as in art, the inflationary hypothesis is fecund; it raises new and intriguing avenues of inquiry. Linde speaks grandly but not groundlessly of inflation as "a new cosmological paradigm . . . much more interesting and complicated than what we have expected."[20]

Some of the implications of inflation are intellectually, as well as spatially, enormous, and will occupy much of the remainder of this book. Here we look at just one narrower and more specific prospect—the possibility that inflation is still going on today, or might accidentally be repeated.

As noted, inflation, once instigated, tends to persist. The implication is that inflation may never have completely stopped. This could have happened for various reasons. If, for instance, more than one field drove inflation, one of the fields could have "rolled over" into a postinflationary state while others remained in a false vacuum state. The false vacuum that drove inflation certainly contained a much higher energy than does the cosmic vacuum today, but that doesn't mean that the modern vacuum necessarily has descended to its minimum energy. Conceivably, the universe could still be hung up in a false vacuum state, one lower in energy than during inflation but more energetic than its true ground state. If that is the case, a gentler sort of inflation might persist even today, manifesting itself as a mild form of cosmic antigravity.

Which brings us back to the cosmological constant—the antigravity force that Einstein introduced into general relativity and later repudiated as the worst blunder of his career. The universe during inflation had antigravity to burn. If mild inflation continues today, the modern value of the cosmological constant is relatively small but not zero. This prospect was discussed in a 1995 paper titled "The Cosmological Constant Is Back," by Lawrence Krauss of Case Western Reserve University and Michael Turner of the University of Chicago. "A diverse set of observations now compellingly suggest that the universe possesses a nonzero cosmological constant," they write, noting that the constant "corresponds to the energy density associated with the vacuum and no known principle demands that it vanish."[21] Their argument rests on those observations that suggest a high value for the Hubble constant—results that, if verified, would require that the universe be younger than are the oldest stars, *unless* a force of cosmic repulsion is acting to speed the expansion rate. (If cosmic antigravity is operating today, observations of the local expansion rate would have led astronomers to underestimate the age of the universe, just as a speed-trap reading on the velocity of a drag racer crossing the finish line will, if extrapolated backward linearly, lead the radar-gun operator to think that the car cleared the quarter-mile in much less time than was actually the case.) Opinions differ as to just how compelling these data really are, but further observations should resolve the issue. If the Hubble constant is about 50, the need for a cosmological constant will vanish. If it is more like 70 or 80, the cosmological constant will emerge as a hot issue. It should be possible to test for a nonzero cosmological constant: In such a universe, for instance, galaxies will be farther away than their redshift would indicate under the standard model.

There is a scary side to all this. If, as Krauss and Turner suggest, "we are currently in the midst of a phase transition where the Universe is hung up in the false vacuum (a mild period of inflation)," then it is always possible that something might yet trigger the collapse of that false vacuum into a lower energy state.

Concentration of too much energy at one point in spacetime could set off the collapse. Years ago, a few physicists worried aloud about whether a collider experiment might inadvertently pump enough energy into the vacuum to do the trick.[22] To call this event the worst possible laboratory accident is to understate the case. The false vacuum would tear open, creating a bubble of lower-energy vacuum that instantly propagated in all directions at the velocity of light, destroying everything it touched. We may, however, take comfort in a continuity argument. The part of the vacuum located within our observable universe has remained stable for over 10 billion years now, and so perhaps its reliability in the past may constitute a warranty on the future.

Looming behind all such considerations is the intriguing possibility that what we are really talking about are not just alternate histories of our universe, only one of which actually was realized, but the actual histories of many extant universes. The longer one thinks about Linde's chaotic tangle of scalar fields—and he has thought about it longer and more deeply than anyone else—the more one begins to appreciate that inflation need not have produced our universe alone. Instead it seems more "natural" (to use that loaded word again) that many different domains gave birth to many inflationary bubbles, each a universe unto itself. This may be idle speculation—or it may represent the Kitty Hawk of human thought about the origin of the universe.

10
The Origin of the Universe(s)

Omnibus ex nihil ducendis sufficit unum.
(For producing everything out of nothing one principle is enough.)

—LEIBNIZ[1]

Let your soul stand cool and composed before a million universes.

—WALT WHITMAN[2]

THE STUDY OF GENESIS—called *cosmogony,* from the Greek for "world-begetting"—is beset by paradox. Science as we know it is built on cause and effect, space and time. How can it comprehend an uncaused effect that, by definition, could not have occurred within a preexisting framework of space and time? Many scientists think it can't. "Ultimately, the origin of the universe is, and always will be, a mystery," writes the astronomer Stuart Bowyer.[3] Says the physicist Charles H. Townes, "I do not understand how the scientific approach alone, as separated from a religious approach, can explain an origin of all things. It is true that physicists hope to look behind the 'big bang,' and possibly to explain the origin of our universe as, for example, a type of fluctuation. But then, of what is it a fluctuation and how did this in turn begin to exist? In my view, the question of origin seems always left unanswered if we explore from a scientific view alone."[4]

Yet paradoxes don't usually denote genuine limits of inquiry. More often, by signaling where accepted concepts break down, they indicate where new concepts are called for. The history of science supports Søren Kierkegaard's view that paradoxes are "grandiose thoughts in embryo," Oscar Wilde's declaration that "the way of paradoxes is the way of truth," and Leibniz's remark that "there is hardly a paradox without utility."[5] The paradox Einstein encountered when attempting to imagine what an oscillating electromagnetic field would look like to an observer riding a light beam led him to special relativity, and Niels Bohr was urged, by paradoxes in his original model of the atom, toward the quantization of electron orbits, itself part of the wider (and still rather enigmatic) conception of subatomic objects as both particles and waves.

Cosmogony raises at least three paradoxes—the paradoxes of a first cause, of getting something from nothing, and of infinite regress. Examining their thrown bones suggests that each has the potential of being resolved by shifting from a classical to a quantum paradigm. Attaining a quantum perspective is difficult. Living in a macroscopic world where quantum phenomena are rarely manifest, we humans came upon classical physics first, and tend to think of quantum physics as a special case. Nevertheless it's beginning to look like the universe is fundamentally a quantum system. This view has unsettling implications, which we will examine in this and the following chapters. We begin by sketching its solvent effect on the three cosmogonic paradoxes.

The first paradox may be stated: *There can be no effect without a cause. Whatever events transpired near the outset of time, each must have been caused by some prior event. So we can never attain an account of the very beginning.*

This is a noble and venerable argument. It was, for instance, the basis of Thomas Aquinas's "cosmological proof" of the existence of God.[6] But we today understand the doctrine of causation to be rather more problematical than was appreciated in the thirteenth century. As the modern philosopher John William Miller points out, the notion of cause when used broadly is too vague to be very helpful: "Suppose one substitutes the word 'God' for 'cause' in order to understand some actual event. It may then be true that what has occurred is God's deed, his purported decision; but it is also true that no understanding of the given event can be so obtained."[7] And if instead we restrict its application, we are left wondering, "If cause makes no sense apart from the restricted, what sense does it make when applied to the whole universe?"[8]

The doctrine of causation erodes considerably when applied to the subatomic realm of quantum physics, and therefore seems a dubious tool for understanding the early universe, when virtually all structures were subatomic. (To construct even an atom in the big bang would have been like building a house of cards in a firestorm.) Logically speaking, strict causation is equivalent to the statement, "If A, then B." But in the probabilistic world of quantum physics one is frequently in the position of saying, "There is a fifty percent chance of A and therefore a fifty percent chance of B." The probabilities are said to be inherent in nature and not merely a reflection of our limited knowledge—a consideration that profoundly alters one's sense of how the world works. Strictly speaking there seems to be, for instance, no such thing as a "cause" of the radioactive decay of a radium atom. The radioactive isotope radium-224 has a half-life of 3.64 days. So if we study an atom of radium-224 for 3.64 days we will have an even chance of witnessing its decay. But we cannot know just when it will decay—this particular atom might wait for years—nor can we, in principle or in practice, assign a *cause* to its decay. All we can know are the probabilities. Similarly, there is in quantum mechanics no such thing as a strict cause of a particular vacuum fluctuation, such as the fluctuation that some versions of inflation theory postulate as the agency of creation: Rather, fluctuations arise statistically. So strict causation may break down both in quantum physics and in considering the origin of the universe. Possibly this is not a coincidence, but a clue that the quantum principle holds the key to understanding genesis.

The second paradox: *You can't get something from—or for—nothing. The "origin" of the universe, if that concept is to have any meaning, must create the universe out of nothing. Therefore there can be no logical explanation of genesis.*

Here the major premise restates the law of conservation of energy—that a zero-energy system to which no energy is added must remain in a zero-energy state. But it may be that the energy content of the universe *is* zero. As the physicist Edward Tryon, then at Columbia University, proposed in the 1970s, gravitation is a purely attractive force and so should be entered on the negative side of the cosmic energy ledger. Sum it against all the matter and energy in the universe, and the result, remarkably, is zero.[9] If this analysis is correct—admittedly, a big if—genesis isn't a matter of getting something from nothing but of getting one zero-energy system from another zero-energy system.

Tunneling is a quantum mechanism that could have generated the universe as a zero-energy balloon from a preexisting space. We encountered quantum tunneling earlier, in Andrei Linde's versions of inflation. Alexander Vilenkin of Tufts University has created models of this sort. He finds that quantum chance permits the creation via tunneling of many sorts of zero-energy ballooning universes, in some of which life as we know it is possible. So the something-from-nothing paradox, too, might be resolved by employing quantum concepts.

The third, and most telling, cosmogonic paradox holds that: *Regardless of its net energy, the universe must have originated from another system, and that system must in turn have had an origin of some sort. And so we are caught in infinite regress.*

The regress envisioned here may be of two varieties, temporal and logical. If temporal, it is based on the assumption that time is infinitely divisible. In that case scientists might contrive an infinite number of theories that successively approach the moment of genesis without ever reaching it. Cosmogonic research would then resemble Zeno's paradox in which Achilles never overtakes the tortoise. But men *do* pass tortoises—which was Zeno's point in the first place—so perhaps time is *not* infinitely divisible. Quantum physics suggests that this is the case. If the very early universe—during the Planck epoch—comprised a quantum spacetime foam, then time as well as space was fragmented. In these conditions there was no "arrow of time," so it would be meaningless to talk of something as having come "before." Hence no problem of infinite temporal regress arises.

The problem of *logical* regress is more robust. Certainly it is very difficult to imagine a theory in which the universe originates out of absolutely nothing. Science lacks even an agreed-upon conception of what "nothing" might be. As we have seen, some versions of inflation theory postulate cosmic expansion as having arisen from a chaotic array of scalar fields, but one may always ask where the chaotic fields came from. String theory makes everything from ten-dimensional space—but what selected that particular geometry for primordial space? Leaving aside genesis for a moment, consider the fundamental constants of nature. Either they are inevitable, like the statement that $1 + 1 = 2$, or they are the random result of phase transitions or other chance occurrences. If inevitable, they must be based on some other system (e.g., the logic of basic arithmetic), which must in turn be based on another system, and so forth. If seemingly random, there will always remain the possibility that a deeper order lies concealed behind the seeming randomness. (One cannot *prove* that anything is genuinely random. We'll return to this point in the Afterword.) So there's potential regress either way.

Ultimately, logical regress may turn out to be a bedrock paradox, but it's just too soon to say. In any event, progress would be made if science were to demonstrate that the universe emerged from some quite different state, regardless of whether future generations were able to explain that state's origin. The big bang is itself a theory of that kind. It depicts our quasi-classical universe as having emerged from a quantum state, the particulars of which may be investigated experimentally. It remains to be seen whether such investigations will lead to a new scientific paradigm capable of resolving the paradox of logical regress.

To sum up, the cosmogonic paradoxes urge upon us a quantum approach. We have at present only two kinds of physics to choose from, classical and quantum; and classical physics, as Alex Vilenkin notes, "fails to describe the beginning of the universe."[10] Its breakdown is clearly signaled by the fact that general relativity invokes a singularity at time zero, which is to say that its equations yield infinities and can produce no meaningful results. Roger Penrose and a youthful Stephen Hawking proved this in 1970, in theorems demonstrating that if gravitation is always attractive and if the universe has anything like the matter density we observe it to have, then there must have been a singularity at the outset of time. So we are left with *quantum cosmology*—the attempt to apply quantum precepts, previously employed in studying subatomic particles and fields, to the universe as a whole.

Since it deals in probabilities, quantum physics proffers a rather fuzzy picture of nature. The assertion that it does so not because our *knowledge* of subatomic nature is fuzzy but because nature on that scale really *is* fuzzy has sparked a lot of philosophical anguish, but for cosmologists, hard-core quantum fuzziness may be a blessing in disguise. If, for instance, the geometry of cosmic space and the behavior of time could be shown to have originated in the chaos of a robustly fuzzy early universe, then the cosmogonic paradoxes could, as we've noted, dissolve in that primordial stew. Once we enlarged our frame of reference, the classical universe of seamless space and time would be seen to have emerged from the foamy quantum universe in something like the way that chaotically swarming bees emerge as a swarm with a distinct flight path.

To turn such hints and glimpses into quantitative theories will require overcoming formidable obstacles. Yet the potential of quantum cosmology is commensurately grand. "We aim . . . to provide a theory of the initial condition of the universe that will predict testable correlations among observations today," writes the University of California cosmologist James Hartle.[11] "We demand of physics some understanding of existence itself," says John Wheeler.[12] So let's see how this brash new game is played.

Quantum cosmology can be described as the attempt to find the *wave function* of the universe. A wave function is a mathematical description of a quantum system. Equations that express the wave function are commonly called Schrödinger equations, after Erwin Schrödinger, the Austrian physicist who derived the basic nonrelativistic wave function for atomic systems. The attempt to apply this tool to the entire universe dates from the 1960s, when John Wheeler at Princeton and Bryce DeWitt at the University of Texas at Austin formulated a cosmological version of the Schrödinger equation. The Wheeler-DeWitt equation treats the radius of the universe as analogous to the position of a subatomic particle, and its rate of expansion as analogous to the particle's momentum. Another helpful tool, also contributed by Wheeler, is *superspace,* an abstract plenum that contains all possible three-dimensional geometries of the universe. Wheeler, who was born in 1911 and remembers when automobiles had fenders, likes to compare superspace to a junkyard containing one specimen of every conceivable fender, each stacked next to the ones that most closely resemble it in shape. The wave function of the universe could, if correctly formulated, select the actual cosmic geometry out of all the fenderlike spaces.

A step toward turning these aspirations into practice came in 1982, when Hartle and Hawking postulated a wave function of the universe. Hawking unveiled the theory at a 1983 general relativity conference in Padua, Italy. Speaking in the vaulted, wood-roofed hall where Galileo delivered incendiary lectures on behalf of the Copernican cosmology to crowds of thrilled students and indignant professors, Hawking spoke from his wheelchair in a voice so badly slurred by his paralysis that a student had to translate his every word.[13]

"At the present time, the universe is accurately described by classical general relativity," Hawking began. "However, classical relativity predicts that there will be a singularity in the past. Near that singularity, the curvature will be very high, and classical relativity will break down, because quantum effects will have to be taken into account. In order to understand the initial conditions of the universe, we have to turn to quantum mechanics, and the quantum state of the universe will determine the initial conditions for the classical universe. So today I want to make a proposal for the quantum state of the universe. . . . This proposal incorporates the idea that the universe is completely self-contained, and that there is nothing outside the universe. In a way, you could say that the boundary conditions of the universe are that there is no boundary."[14]

The "no boundary" aspect of the Hartle-Hawking wave function arises from its authors' having employed a set of geometries that place time and space on equal footing. The elegant result is that the "arrow of time"—time moving only forward, as it does in the classical universe we occupy—emerges internally, from the geometry, rather than being imposed from without.[15] By doing away with any initial time, this method also dispenses with the initial singularity. (Singularities, remember, are made of spacetime, not just space.) In the Hartle-Hawking model, there is thus no moment of creation. Rather, the existence of "moments" is a consequence of the spatial geometry. As Hawking describes it, in this model "there would be no boundary to spacetime and so there would be no need to specify the behavior at the boundary. There would be no singularities at which the laws of science broke down and no edge of spacetime at which one would have to appeal to God or some new law to set the boundary conditions for spacetime. One could say: 'The boundary condition of the universe is that it has no boundary.' The universe would be completely self-contained and not affected by anything outside itself. It would neither be created nor destroyed. It would just BE."[16]

Behind Hawking's user-friendly rhetoric lies a point of central significance to quantum cosmology—that the fuzziness of nature on the quantum scale provides ways of avoiding singularities, and in doing so renews hope of composing a coherent scientific account of genesis. As Penrose and Hawking demonstrated in their singularity theorems, any two world lines constructed in general relativity will lead back to a singularity, where the (nonquantum) laws of nature break down. On a graph in which time moves upward and space expands sideways, such world lines can be depicted like this:

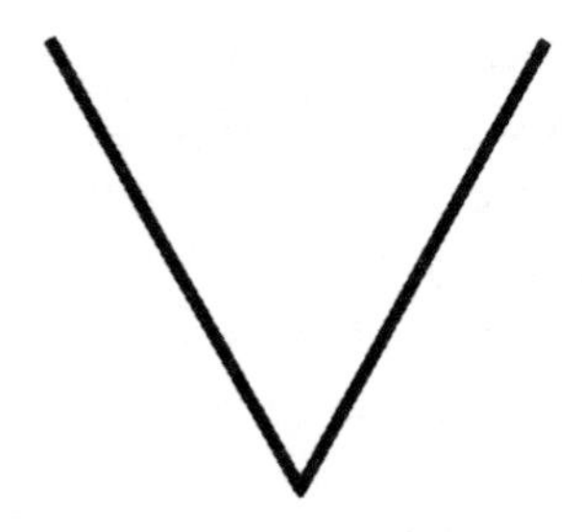

By incorporating quantum physics one can select geometries that round off near time zero without reaching a singularity point. That's what Hartle and Hawking did. As Hawking describes it: "When quantum mechanics is taken into account, there is the possibility that the singularity may be smeared out and that space and time together may form a closed four-dimensional surface without a boundary or edge, like the surface of the earth but with two extra dimensions. This would mean that the universe was completely self-contained and did not require any boundary conditions [and] there would not be any singularities at which the laws of physics would break down."[17] World lines in these geometries look like this:

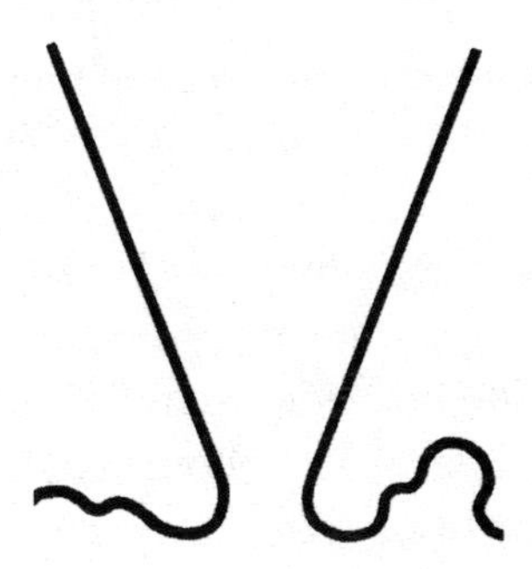

Such no-singularity models are closely related to inflationary scenarios like the ones proposed by Vilenkin and Linde, in which the universe begins as a runaway bubble emerging from spacetime foam.

It is perhaps unnecessary to caution that the Hartle-Hawking wave function does not explain the origin of the universe. Nor will I be giving up much of the potential suspense of my tale if I concede that none of the other theories discussed in this chapter does, either. All are actually rather limited. They omit quantum gravity, for which there is as yet no realized theory, and they consider only closed cosmic geometries—that is, the spherical ones, those in which omega is equal to or greater than one. (This comes about for various technical reasons—mainly that nobody knows how to calculate quantum particle densities for open universes.) Their elected geometries and spacetime foam are not quite the "nothing" from which a genuine cosmogony would fashion a universe. Still they represent a start, providing tools for research and examples of how genesis might be calculated.

Hartle was in the audience for the Padua talk and afterward the two chatted, Hartle's tall form curving over Hawking's wheelchair like a crescent moon bowing to the evening star. "I think the response [to the talk] resembles the response when I suggested the radiation from black holes," Hawking said. "Some people didn't believe black hole radiation at all. Most didn't understand it. It's probably about the same today."

Hartle, working on his own and in collaboration with Hawking, Murray Gell-Mann, Jonathan J. Halliwell, and others, has tried to fashion the foundations on which mature theories of quantum cosmology could be erected.[18] To appreciate his work we need to touch on a topic that will occupy much of the next chapter —the thorny issue of quantum *observership,* also known as quantum *measurement.*

Central to quantum physics is Heisenberg's indeterminacy principle. We recall that owing to Heisenberg indeterminacy, certain information about subatomic systems can be obtained only at the cost of remaining ignorant about other information. If, for instance, we ascertain the exact position of an electron, we lose information about its momentum, and vice versa. Indeterminacy mandates that quantum calculations incorporate probabilities. The probabilities, in turn, produce the characteristic fuzziness of the quantum realm. There are compelling reasons to accept the ex *cathedra* view that this fuzziness is not just an expression of incomplete knowledge but an essential aspect of nature on the subatomic scale. Without indeterminacy, the sun would not shine, since the indeterminacy of particle position is what makes quantum leaping possible, and quantum leaps, across the Coulomb barrier posed by their like charges, permit protons to fuse with other protons in quantities sufficient to sustain fusion reactions at the center of the sun. Nor would the retina of the human eye, or the pixels on the CCD chip, work as they do in response to the photons they gather. So Heisenberg indeterminacy evidently is characteristic of the universe. If so, it must also characterize any quantum theory of the origin and early evolution of the universe.

Heisenberg indeterminacy is not a problem in itself. All scientific theories have limits, and nobody thinks we could learn *everything* about cosmic history, even if nature were strictly deterministic and quantum indeterminacy didn't exist. The trouble is that indeterminacy creates situations in which a given quantum system can take on one or the other of two mutually contradictory appearances, depending on how it is observed, and this leads to puzzlements when viewed in terms of commonsense concepts arrived at on the classical scale. A conspicuous example concerns whether subatomic particles—electrons, say—are particles or waves. One can write equations that accurately predict the behavior of electrons by using either wave equations or particle equations. The two approaches are mathematically equivalent, and both will yield reliable results. But the electron cannot *be* both particle and wave, since particles and waves have mutually exclusive characteristics. If, for instance, you drop two stones into a pond, making waves, the waves where they intersect will produce an interference pattern. (Interference patterns occur where the two sets of waves impinge on each other: Wave peaks amplify other peaks where they coincide, forming bright lines, and valleys amplify conjugate valleys to form dark lines.) But if you fire two shotgun blasts of buckshot through each other, most of the individual bits of shot will pass unscathed and a few may collide, but there will be no interference pattern. So what are electrons "really"—particles or waves? The indeterminacy principle prohibits us from answering that question, not just in practice but in principle. All we know is that if you run electrons or other particles through an apparatus designed to detect waves, you see waves, and if you run them through an apparatus designed to detect particles, you see particles. This is the *wave-particle duality*, and it threatens to make a hash of the belief that there is an objective reality out there. It is as if the phase of the moon obligingly switched from crescent to full every time somebody *asked* if it was full.

The consensus way of making sense of the wave-particle duality and related quandaries is the *Copenhagen interpretation*, named after the institute of theoretical physics where Niels Bohr and his colleagues did pioneering work in quantum theory from 1921. It asserts that quantum systems cannot be said to *have* either state (e.g., wave or particle) *until they are observed*. Put in slightly more technical language, the Copenhagen claim is that electrons and other quantum-scale objects exist in a *superposed* state, until an act of observation "collapses the wave function," resolving the system into one or the other of its potential, and mutually contradictory, aspects.

The Copenhagen interpretation is troubling enough in the here and now—it makes one wonder what, exactly, an observation *is,* and what is so magical about it—and it becomes positively disastrous when applied to the big bang, in the fires of which there could have been no observers. As Hartle writes: "The 'Copenhagen' frameworks for quantum mechanics, as they were formulated in the 1930s and 1940s and as they exist in most textbooks today, are inadequate for quantum cosmology on at least two counts. First, these formulations characteristically assumed a possible division of the world into 'observer' and 'observed,' assuming that 'measurements' are the primary focus of scientific statements and, in effect, posited the existence of an external 'classical domain.' However, in a theory of the whole thing there can be no fundamental division into observer and observed. Measurements and observers cannot be fundamental notions in a theory that seeks to describe the early universe when neither existed."[19]

So Hartle, Gell-Mann, and others have formulated an alternative interpretation of quantum physics. Hartle calls it "post-Everett," by which he means that it is based on the "many worlds" interpretation advanced in the 1957 doctoral thesis of Hugh Everett III of Princeton University. We will investigate Everett's original idea in the next chapter. Here what matters is Hartle's version. Sometimes called the "many histories" approach, it incorporates Richard Feynman's "sum over histories" method of applying probabilities to retrodiction. It views the cosmic past as not one but many pasts—a vast tree-branching of events. Some events occurred, some did not, and for some—for many—it is impossible, owing to indeterminacy or other limitations, to know whether they occurred or not. The task of the (quantum) cosmologist is to reconstruct as much of cosmic history as is accessible to him and is pertinent to his inquiries. The goal is to trace the branches back to the origin of the universe. In doing so, Hartle emphasizes, the cosmologist like any historian is obliged to work with incomplete and imperfect records. "In classical physics reconstructing the past history of the universe, or any subsystem of it, is most honestly viewed as the process of assigning probabilities to alternatives in the past given present records," he writes. "We assign the date 55 B.C. to the Roman conquest of Britain on the basis of present textual records. . . . History becomes predictive and testable when we predict that *further* present records will be consistent with those already found.

Texts yet to be discovered are predicted to be consistent with the story of Caesar. . . . In such ways history becomes a predictive science."[20]

We cannot know the specifics of every particle interaction in cosmic history, so the cosmic historian works with the aspects of history that are disentangled from quantum indeterminacy. Gell-Mann and Hartle call these "coarse-grained." "As observers of the universe," Hartle writes, "we deal with coarse grainings that are appropriate to our limited sensory perceptions, extended by instruments, communication, and records, but in the end characterized by a great amount of ignorance."[21] Coarse-grained histories can be *decoherent.* A "coherent" system is like our electron prior to its being observed experimentally—before it has exhibited, say, either wavelike or particlelike behavior. Because its states are superposed, its "real" state remains speculative, and cannot be assigned a branch in cosmic history. *De*coherent systems have, so to speak, come clean: They've declared their state—particle or wave, located *here* rather than over *there*—and so have emerged from the Heisenberg fog.

In the Copenhagen outlook, decoherence occurs when a quantum system is observed. In the post-Everett interpretation, what matters is not observation, which would require the presence of observers, but decoherence. A decoherent system is in a state that is at least potentially observable. If, for instance, light has shone on a molecule in such a way that an observer *could* learn the molecule's exact position and thus resolve the quantum indeterminacy of its position, then the molecule has decohered, regardless of whether there was actually an observer there to see it. The key to observability is whether the system has enough inertia so that its state can be extracted from the random noise of its surroundings. Hence decoherent systems tend to be fairly big. But they don't have to be all *that* big. The history of a single grain of dust adrift in deep space, recorded when photons from the cosmic microwave background scatter off it, can qualify as coarse-grained by the time the dust particle has moved a distance of only one millimeter.

Only coarse-grained histories constitute branches on the tree of cosmic history. Free from the fuzziness of Heisenberg indeterminacy, they can be assigned genuine probabilities of their having transpired. The question of which branches actually were realized, however, is a matter to be learned through observation. So observers remain important in the post-Everett approach, though in a history-writing sense and not the more grandiose, reality-creating sense of the Copenhagen interpretation.

By including human beings *in* the universe, quantum cosmology shatters the conceptual pane of glass that separated the observer from the observed in classical physics. As Hartle puts it, "The most fundamental, assumption free way of 'including the observer in the universe' is to see it [the observer] as a system that has evolved within the universe."[22] Indeed, the entire classical world of human experience is viewed by quantum cosmology as an evolutionary product of what is essentially a quantum universe. In much the same way that the big bang theory reveals the element abundances of the earth and the sun to have resulted from the chemical evolution of the galaxy and the universe, quantum cosmology depicts the classical side of nature as an emergent property of a universe that began in a quantum state. Since humans and other observers require a predominantly classical environment in which to evolve—since it is on classical scales that nature behaves deterministically enough to provide the stability that living creatures require—quantum cosmology places our existence, our science, and our very curiosity about the universe squarely in a cosmological context. The universe of quantum cosmology is an *involving* universe.

Which brings us to Andrei Linde's recent work on quantum genesis. All cosmogonic theories are vulnerable to the criticism that they involve a certain amount of special pleading, since they posit some sort of initial conditions that typically serve to rule out one or another circumstance, such as an initial singularity, that scientists consider repugnant but that nature, or God, may not have. As the Oxford philosopher John Lucas observes, such theories are guilty of "explaining a particular in terms of the general—why this particular universe in terms of its best satisfying some rational *desiderata.*"[23] How can this fault be minimized? Linde's answer is to begin with an absolute minimum of initial conditions—the state we call chaos.

In the previous chapter, we described Linde's work on "chaotic inflation," which suggests that "the" universe began as a bubble that ballooned out of the spacetime of a preexisting universe. The maternal spacetime is chaotic in that it contains scalar fields of all possible parameters; one scalar field—an unlikely but possible one—emerged as the driving force of the inflationary event that launched the expansion of our universe. Linde's approach benefits from a refreshing straightforwardness. "There is no need for quantum gravity effects, phase transitions, supercooling or even the standard assumption that the universe was hot," Linde writes. "One just considers all possible kinds and values of scalar fields in the early universe and then checks to see if any of them leads to inflation. Those places where inflation does not occur remain small. Those domains where inflation takes place become exponentially large and dominate the total volume of the universe."[24] Reductive though it may be, chaotic inflation has two startling implications. First, it mandates the existence of a universe precedent to our own. Second, less evidently but just as forcefully, it implies the existence of countless other universes. Linde's is a theory of a multiverse.

Like Robert Wilson, the sculptor who returned to physics to build the Fermilab particle accelerator on the Illinois plains and adorned it with the looming steel stabile *Broken Symmetry* and others of his own works, Linde originally studied art, and he continues to be almost as much an artist as a scientist. ("Some said I would be a painter, not a physicist, but I decided otherwise."[25]) Like Einstein, he came to physics through philosophy. "I started studying physics as a way of answering my questions in philosophy," he recalled in a 1993 interview. "When I was in high school, I developed a nice theory of how extrasensory perception might work. But then I learned that it violated the special theory of relativity. I realized that unless I learn physics I may come up with all sorts of ideas that will sound nice, but always I will be talking nonsense."[26] His early career, in Russia, was punctuated by bouts of illness—he came up with the "new inflationary" theory shortly after spending two months in the hospital—and he displays a creative artist's manic-depressive combination of wicked wit and surging self-doubt. "Sometimes I wake up in the middle of night and wonder, What if it is just a great myth you are creating?" he said recently. "Maybe God is fooling me. Maybe he tells me something which looks like fun, which looks like I am rewarded, for what I don't know, so that I've got this nice understanding of how the universe was built, but in the end I will discover that I was fooling myself and I was fooling other people by saying that I know how the universe was produced. Maybe I am too proud right now. Maybe actually the solution is absolutely quite aside from what we're thinking about. It is a very dangerous feeling, this feeling that you're not totally secure in what you're doing. But it makes life so exciting."[27] One sometimes gets the sense that Linde is as taken by the boldness of his ideas as by their plausibility. "We are touching on a very dangerous subject," he told the audience at a 1995 scientific symposium. "And the more dangerous a subject, the more *interesting it is.*"[28] His theory of eternal inflation requires, as does a vigorous new work of art, not only that we examine it by existing standards but that we alter and expand our frame of reference, our way of regarding the world, to make room for it. In making this effort, I will begin with our universe, the one we live in, then look back to the universes that, by Linde's lights, "preceded" it.

In the wide frame of Linde's cosmological picture, the inflationary event that ballooned the universe to enormous size was but one among many. We noted in the previous chapter that the scalar field that drove inflation would have frozen out when the Hubble radius of the universe grew to equal the field's wavelength. But that was not the whole story, according to Linde. The freezing of the field that brought inflation to an end in one domain amplified it in other domains where inflation therefore persisted. "Such regions are extremely rare, but still they do exist," Linde writes. "And they can be extremely important."[29] They are important in that they drive further evolution of fields, producing a new array of inflating bubbles, some of which make more bubbles . . . and so on, *ad infinitum.* "The total volume of all these domains will grow without end," Linde writes. "In essence, one inflationary universe sprouts other inflationary bubbles, which in turn produce other inflationary bubbles."[30] The implication is that the universe is not only much larger than had been imagined in the classical big bang model, but incredibly more vast than even the gigantic inflationary bubble on which our observable universe is perched. The multiverse contains innumerable bubbles like the one in which we find ourselves, and other regions even larger, and still other regions in which inflation is going on right now. (That's why Linde calls the theory *eternal* inflation.)

Working with his son Dmitri, who at age fifteen wrote a word-processing program since popular on the Internet and later went off to study at Caltech, Linde worked up a graphical representation of his theory. They used a powerful graphics computer loaned them by its manufacturer for one week. Writing the code took six days, and on the seventh day Andrei and Dmitri had a depiction of the many-universe theory—an Oz-like set of multicolored spires, separated by interposed valleys. Its vertical axis indexes cosmic energy density. The peaks represent regions of the multiverse that are continuing to inflate or have just stopped inflating and are passing through their thermalizing, "big bang" phase. These have yet to crash through the phase transitions that will determine the values of the constants of nature, so their physics is as yet unsettled. The valleys represent regions of the multiverse that have settled down into a low-energy state. There the laws of physics are graven in stone. We live in a valley.

"When I see pictures of my beautiful universe in color on the big screen it was so lovely I almost am crying," Linde recalled in 1993, speaking a burred but literal English that encourages in the Anglophone listener the pleasant delusion that he or she is comprehending Russian. "I know what I was after, but I didn't expect it to be so beautiful."[31] His rapture was short-lived. "We made the universe and it was good, but on the eighth day the computer's gigabyte hard disk crashed, taking our beautiful universe with it." Newer and more powerful computers were set up in the living room of the house on the Stanford campus that Linde and his wife, the quantum gravity theorist Renata Kallosh, share with their two sons. On their screens more universes blossomed as the scalar field equations were altered and explored. One of them, a fractal model based on a related theory, produced such wildly original patterns that the Lindes nicknamed it "a Kandinsky universe," after the Russian abstractionist.

In some of Linde's models, the bubbles remain out of contact with one another forever, separated by gigantic inflationary wastelands. In others, some bubbles eventually collide, producing fiery domain walls where differing sets of physical laws contend. This makes Linde the author of the most remotely futuristic predictions ever offered as observational proof of a serious scientific theory: If bubbles collide, another part of the multiverse could swim over our observational horizon, a thousand billion years hence. "It might be going away or it might be coming towards us," Linde commented laconically. "Better be going away. . . . If you go there you encounter a boundary. You cross the boundary, you die. So you don't want to go there."[32]

If universes collide—or if they remain connected by the intercosmic umbilicals called wormholes, as the "baby universe" school of quantum cosmologies has it—might communication between them be possible? Probably not, for here "cosmic amnesia" rears its head. A note in a bottle tossed down a wormhole might be shredded by tidal effects or could perturb the wormhole, slamming it shut.[33] And if the message got through and landed in an inflating universe, it might get too big to read. Linde speculates that "if you have written anything on the surface of the universe, you can't read it for a long time, because it's all stretched out beyond the horizon. Your great-grandchildren will still be living in a tiny corner of one letter."[34] What if, instead of sending a message reading, "Here are the laws of physics on our side—what are yours?" you dispatched an engine capable of altering *their* laws of physics. This machine would generate what the physicist Heinz Pagels called a "cosmic code"—a set of laws the eventual decipherment of which by alien physicists would tell them something about us, or our world. "Maybe you could make a law that would define the state of the vacuum," Linde suggests. "That state could be a message. String theory has so many [permissible] vacuum states that people have stopped trying to count them."[35]

Let's assume for the sake of argument that Linde is right—that we live in a low-energy bubble in an incredibly huge and complicated universe, parts of which are bubbles something like ours, while other parts are only now passing through their big bang baptisms, and still more are continuing to balloon in ghostly vacuum states at velocities far greater than light. Where did it all come from? Linde may be the world's greatest scalar-field virtuoso;

"However," as K. A. Bronnikov and V. N. Melnikov of the Center for Surface and Vacuum Research in Moscow note, "there is always a problem of the origin of this field."[36]

To this question Linde has an answer that peers back past time zero and beholds there a wonderland even more fantastic than the one on this side. If, as Linde has it, the bubble of which our observable universe constitutes a small part is but one among many bubbles that continue to nucleate more bubbles as they go, then it is pointless to inquire which was the "original" bubble. Each bubble owes its birth to another bubble, which came from another bubble, and so on. As the lady in the apocryphal story says to Bertrand Russell, who has sought to deflect her assertion that the universe sits on a giant turtle's shell by asking her what the turtle stands on ("Another, larger turtle") and what *that* turtle stands on: "It's no use, sonny: It's turtles all the way down!" In Linde's view, asking whether there was "a" universe before "the" big bang is to be quaintly parochial. Our bubble emerged, he says, from "not *the* big bang but a *pretty* big bang."[37] There were—are—innumerable "pretty big bangs," with countless more to come. The history of the cosmos is darker than the depths of the sea, and its myriad futures richer and less predictable than all the unpainted paintings and uncomposed songs yet to emerge from the minds of all the humans to be born from now till the sun goes red and dies.

As Linde sees it, "The classical big bang theory is dead. . . . The big bang remains a very interesting theory that we must study, but the original big bang is somewhere in the distant past."[38] He likens cosmic history to an apple tree. We live in one apple. There are many other apples on our branch. Eventually, tracing the branches back, one reaches a trunk; and at the root of the trunk sits the original big bang, if indeed there was one. "The evolution of the universe as a whole has no end, and it may have had no beginning," he says.[39] His calculations of scalar-field evolution indicate that the longest branches produce the most apples. This implies that even if an average branch lies within a finite temporal distance from the putative roots, an average *apple* will find itself so far from the ground that the time elapsed since genesis—if there was a genesis—is for all practical purposes infinite. "The long branches are untypical, but they produce more apples than the typical branches do," Linde says. "So most apples grow infinitely far away from the trunk. The notion of 'typical' depends whether you're interested in roots or in fruits."[40]

So Linde's magnificent multiverse removes the question of cosmic origins to the extremities of the ponderable, and perhaps beyond. It's a matter of taste whether the prize is worth the price —whether Linde's "eternally self-reproducing universe" is a garden of Eden or just a cosmogonic rewording of the ancient warning "Abandon hope, all ye who enter here." In the long run (locally speaking) Linde's contribution may have less to do with whether he proves to be right or wrong—and it is unclear how, insofar as proof or disproof is concerned, his models can be distinguished from simple inflation, with its similar predictions that omega equals one and that the cosmic microwave background should evince a chaotic energy-density spectrum—than with his artistic contribution in tearing down the Globe Theater that was quantum cosmology and erecting a gilded opera house in its place. Since Linde, the quest for a theory of genesis looks less like a search for the best route to Eiger's summit than like a mountain pass that opens onto limitless vistas.

But if human beings are to explore those distant and wished-for lands, we must first come to grips with some of the perplexing conceptual issues that have dogged quantum physics since its inception. These riddles dance round the enigma of quantum observership. Its contemplation brings us back from the realm of the multiverse to the intimate confines of our own skin, where we ask what it means to say that "we" "observe" "nature."

11
Quantum Weirdness

"What is the answer?"
(Silence)
"In that case, what is the question?"

—GERTRUDE STEIN, *last words*[1]

The quantum is the greatest mystery we've got. Never in my life was I more up a tree than today.

—JOHN ARCHIBALD WHEELER[2]

GERTRUDE STEIN said of modern art, "A picture may seem extraordinarily strange to you and after some time not only does it not seem strange but it is impossible to find what there was in it that was strange."[3]

Quantum physics isn't like that. The longer you look at *it,* the stranger it gets. The colloquial term is *quantum weirdness,* and it's not just a matter of getting used to the Alice-in-Wonderland oddities of a world in which particles are also waves and can leap from one place to another without traversing the intervening space.[4] Quantum weirdness goes deeper: It implies that the logical foundations of classical science are violated in the quantum realm, and it opens up a glimpse of an unfamiliar and perhaps older aspect of nature that some call the *implicate* universe.

There's no crisis within quantum physics itself. The standard model of quantum mechanics is internally consistent, and its equations accurately predict the behavior of all natural phenomena to which they have been applied. (Indeed, they have produced some of the most precisely verified predictions in science.) The trouble is border trouble. It arises along the quantum-classical frontier, when we try to reconcile quantum mechanics with the characteristics of the macroscopic world—to conform quantum phenomena to a more general philosophy that would satisfy what Vladimir Nabokov called the "ominous and ludicrous luxury . . . of human consciousness."[5] Limited though they may be, these border skirmishes raise questions sufficiently baffling as to constitute the scientific equivalent of a Zen koan. Quantum weirdness is so counterintuitive that to comprehend it is to become not enlightened but confused. As Niels Bohr liked to say, "If someone says that he can think about quantum physics without becoming dizzy, that shows only that he has not understood anything whatever about it."[6]

The subject is a large one, and its subtleties have filled the pages of many capable books.[7] Here we will restrict ourselves to sketching its essentials, then reviewing how this mystery has been analyzed by the thinkers who have looked into it the most deeply.

Quantum weirdness arises when a quantum system is enlarged to a macroscopic scale and then measured in a way that would violate the indeterminacy principle if all the measurements were fruitful. In a typical experiment of this sort, we start with a beam of, say, light, and run it through a beam splitter that divides it in two. (A garden-variety beam splitter consists of a pane of glass half of which has been silvered, so that it's half mirror and half transparent glass. If we think of the photons as waves, the beam splitter divides each into two waves. If we think of them as particles, then the division occurs because each photon has an even chance of hitting the mirror rather than passing through the transparent glass.) The two beams are allowed to travel apart for some macroscopically significant distance, typically several meters. Then they are bounced off mirrors and reconverged at the input of a detector. The apparatus in its initial state looks something like this:

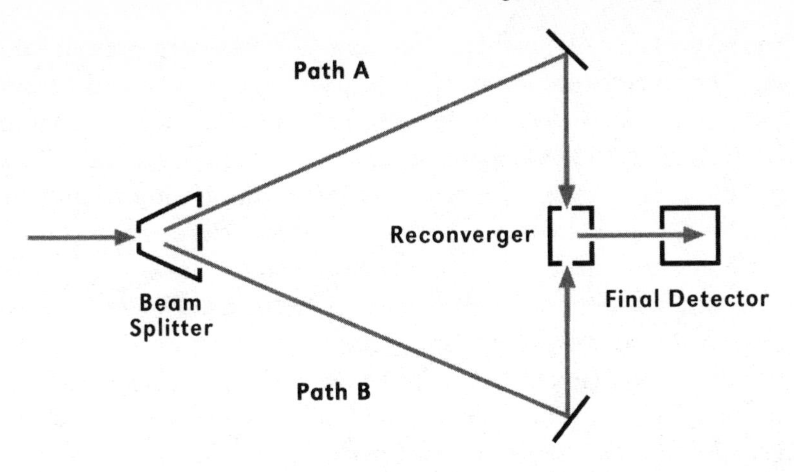

Schematic diagram of a beam-splitter apparatus.

The two beams can be regarded as parts of a single, quantum system. We can verify the validity of this analysis by using a wave-checking device as our final detector and firing a single photon through the system. The photon (having, as it does, wavelike as well as particlelike properties) will display an interference pattern in the detector. The photon is interfering with itself, confirming that even though it's just one photon, it's been deployed across meters of space.

Now that we've dragged the photon out onto a macroscopic stage, what if we try to make one measurement of it on path A, and simultaneously try to make another measurement on path B—such that the two measurements, together, would yield information forbidden us by quantum indeterminacy? The answer is weird: The system *denies us the forbidden information on path B, instantly, as soon as we make a measurement on path A.* Fiddling with the system *here* results in an instantaneous change way over *there.* It does so *even if a signal would have to travel at a velocity faster than light in order to convey news of our fiddling from A to B.*

Let's look at the situation more closely. This time, just in case we're worried that there's something peculiar about photons, we'll use electrons.[8] To minimize technical jargon, let's suppose that electrons have two sets of states, a complete knowledge of both of which is prohibited by the indeterminacy principle. We'll call one set sweet/sour and the other hard/soft. The words don't matter:

The important point is that, according to Heisenberg, we can learn whether an electron is sour or sweet, or whether it is hard or soft, but not both. To test this assumption, we employ two kinds of measuring devices. One box separates sour electrons from sweet ones, spitting the sour ones out of one output window and the sweet ones out the other. The other box does the same, according to whether the electrons are hard or soft.

In an effort to circumvent the indeterminacy principle, we put a sweet/sour box upstream of the beam splitter, near the origin of our electron stream, and admit only sweet electrons to the rest of the apparatus. This works well: Another sour/sweet box, employed as our final detector, confirms that the system now contains sweet electrons only. We're ready to make an assault on indeterminacy. Into each path we insert a hard/soft box. The boxes divert all the hard electrons, permitting only soft ones to continue through the apparatus. At this point, according to classical concepts, we have set things up so that all the electrons arriving at the final detector are both soft and sweet. We know they're all sweet, because we only allowed sweet electrons into the apparatus to start with, and we know they're all soft, because we thereafter discarded all the hard ones. Yet Heisenberg says we cannot know both these things about any one electron. So we've got around the indeterminacy principle, right?

Wrong. When this is done, the final detector ceases to report that all the electrons are sweet. Instead, it starts spitting out electrons in equal numbers from both the sweet and sour windows, even though we admitted none but sweet electrons in the first place! So Heisenberg was right: We can know about sour/sweet or hard/soft, but not both. Indeed, all we've done so far is to verify the validity of quantum indeterminacy. This much of the problem is often explained by saying that the act of making a measurement "interferes" with a particle in such a way as to alter its state—that is, that making sour/sweet measurements randomizes the hard/soft characteristics of electrons, and vice versa. But does that really get to the heart of the issue?

To find out, we change the setup, while continuing to admit only "sweet" particles to the apparatus. This time, we remove the hard-soft detector from path B, and we also divert the output of

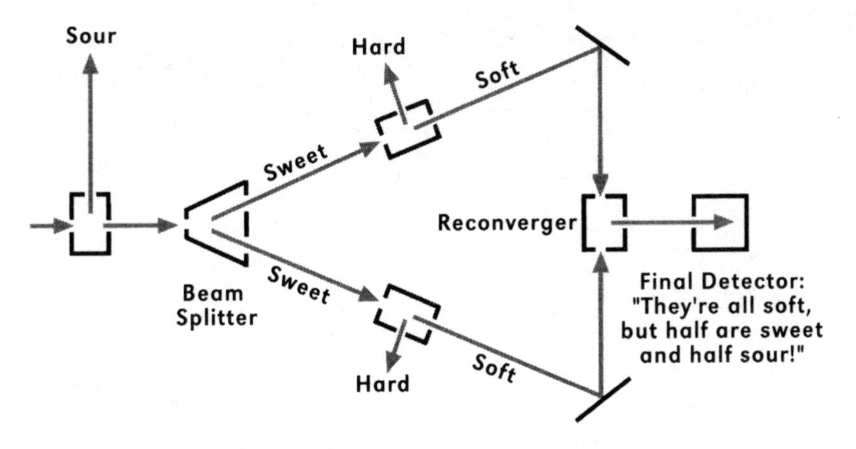

Measuring "hardness" randomizes "the sweet/sour" results.

the hard/soft box on path A so that none of its output gets to the final detector. The result is *really* weird. The final detector again reports that the particles are half sweet and half sour. Yet it is receiving only path B particles—and we didn't interfere with those particles! So how did path B "learn" to randomize its flavor output? How did it "know" that we had made a forbidden measurement way over on path A?

This is quantum weirdness: Interfering with one part of a quantum system alters the results observed in another part, even when the system has been enlarged to enormous dimensions. The result is the same even if only a single particle is admitted into the apparatus at a time. It's the same if we wait until the particle has cleared the beam splitter before making a random decision whether to insert a detector into path A. It would be the same if the two paths were diverted to opposite sides of the galaxy. In every case, the system reacts instantaneously. It is as if the quantum world had never heard of space—as if, in some strange way, it thinks of itself as still being in one place at one time. Such behavior is called *nonlocal*. Classical physics assumes *locality*—that is, it assumes that changes in systems are caused by direct physical contact, comparable to the push-and-pull interactions characteristic of internal combustion engines and other machines (which is why the science of dynamic systems is called "mechanics"). Since measuring one part of a quantum system instantly alters the other parts of the system,

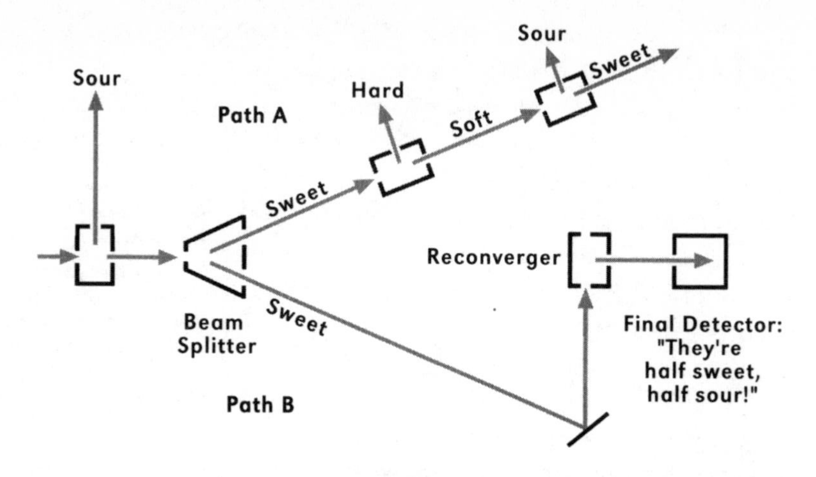

Making measurements on one side of the apparatus *instantly* alters results obtained on the other side.

even if the two parts are too far apart for a message to traverse the intervening distance by any identified agency, quantum systems are said to exhibit *nonlocality:* They act like an intimately connected whole, regardless of whether their parts are far removed from each other.[9]

Whatever we elect to call it—nonlocality, "quantum observership," or the "quantum measurement" problem—weirdness is as knotty a conundrum as the physical world has ever presented to the human mind. Three explanations for it, called "interpretations," have emerged. The first, the *Copenhagen interpretation,* asserts that we should simply accept that we cannot know the state of a quantum system until it is measured, and so should stop worrying about it. The second, or *many worlds* interpretation, begins with the astounding premise that the entire universe splits, with each act of measurement, into two universes, in one of which the particle has the qualities that we measure and in the other of which it resolves itself into the other potential state. This doctrine was advanced in the 1957 doctoral thesis of Hugh Everett III of Princeton University. (We encountered a latter-day version of it in the previous chapter, as the "many histories" approach.) The third interpretation preserves locality: It portrays quantum systems as mechanically linked, so that the particles on either side of our beam-splitter experiment do have a definite state throughout, and actually do alter that state when part of the system is interfered with. They are said to accomplish this by means of a "guiding wave" that has not yet been observed, and perhaps never can be. Consequently, this view is also known as the *hidden-variables* interpretation. Originally advanced by the French theorist Louis de Broglie, it was worked out more fully by the American theorist David Bohm.

Before examining the three interpretations more closely, we should consider another position, popular among working scientists disinclined toward philosophy. They simply shrug their shoulders at quantum weirdness and ask, "So what?" As Isidor Rabi advised Gerald Edelman, "Quantum mechanics is just an algorithm. Use it. It works, don't worry."[10] Richard Feynman told a seminar audience, "The theory of quantum electrodynamics describes nature as absurd from the point of view of common sense. And it agrees fully with experiment. So I hope you can accept nature as she is—absurd."[11] Their point is that quantum physics is successful within its own domain, and can account for all of classical physics, too, so why fret over whether it "makes sense" in classical terms?

This minimalist position is perfectly satisfactory as a matter of pure science. One can simply say that we live in a quantum world, of which classical physics is a subset, and that quantum phenomena are not obliged to make sense in classical terms. But there is more to life than just science, and we all, scientists included, live in a world that we are accustomed to making sense of. Scientists aren't really content just to wield equations; they expect them to relate, not just to one another, but to the "real" world of experience. And they are, like the rest of us, accustomed to thinking of reality in terms of images—metaphors, really—drawn from experience. As the Dutch physicist Peter Debye put it, "I can only think in pictures."[12] Similarly Lord Kelvin: "I never satisfy myself until I can make a mechanical model of a thing. If I can make a mechanical model I can understand it. As long as I cannot make a mechanical model all the way through I cannot understand."[13] And Einstein: "Physical theories try to form a picture of reality and to establish its connection with the wide world of sense impressions. Thus the only justification for our mental structures is whether and in what way our theories form such a link."[14]

As with pictures, so with words. Scientists put a lot of stock in their ability to explain their theories in ordinary language. The indeterminacy principle can be expressed in a few lines of mathematics—in terms, say, of the noncommutative matrix algebra that Heisenberg originally employed for this purpose. Yet scientists don't just leave it at that. They also go out of their way to tell stories and construct explanations of indeterminacy in words, and these tales and models form a nimbus surrounding the hard-core scientific literature—a nimbus that is very much a part of the scientific culture. Scientists know that they belong to a wider society and find it appropriate to relate their work to outsiders, for much the same reasons that architects and athletes do. As Erwin Schrödinger said, "If you cannot—in the long run—tell everyone what you have been doing, your doing has been worthless."[15] The scientists' insistence on resorting to nontechnical language also serves a utilitarian function, that of promoting objectivity and clear thinking while discouraging the subjective and obscurantist tendencies that can beguile even the most caustic mind. Niels Bohr was a lifelong champion of the view that theoretical physics is no place for fancy talk. "Our task must be to account for experience in a manner independent of individual subjective judgment and therefore objective in the sense that it can be unambiguously communicated in the common human language," he wrote.[16] Ernest Rutherford used to advise his students to distrust any concept (or their command of any concept) that they could not explain to a barmaid. Leon Lederman said, "If the basic idea is too complicated to fit on a T-shirt, it's probably wrong."[17] Einstein objected to theories that can "be judged only on the basis of [their] mathematical-formal qualities, but not from the point of view of 'truth.' "[18] Admittedly, the effort to translate physics into common sense becomes more difficult as physics becomes more sophisticated. But the tradition endures, and as long as quantum weirdness remains baffling, there will be physicists and philosophers who keep trying to make sense of it.

How, then, do the three leading interpretations seek to reconcile quantum weirdness with the common sense pictures and language of the macroscopic world?

The *Copenhagen* interpretation was the first, and for decades has remained the foremost, method of keeping peace along the quantum-classical borderline. It declares that the wave function describing a particle constitutes a *complete* description of that particle. Since the uncertainties expressed by the wave function are not resolved until the particle is observed, the particle cannot be said to *have* any definite state until it is observed. Its potential states (such as whether it is a particle or a wave, or has a certain position or momentum, or possesses, in our schematic illustration, the qualities of being hard or soft and sweet or sour) are said to be "superposed." The act of measurement turns potentiality into actuality, resolving the question of what the particle actually "is" through a combination of the particle's inherent potentials and the manner in which it is observed. So the Copenhagen interpretation implicates the observer in what he or she observes. Observers cannot arbitrarily alter reality—cannot violate the laws of nature, any more than a painter can paint a square that is both all white and all black —but they can make of a photon either particle or wave.

Heisenberg discovered quantum indeterminacy while working under Bohr, who was quick to appreciate its implications. Bohr was a wide-reaching thinker—Heisenberg regarded him as "primarily a philosopher, not a physicist"—and it was due chiefly to his influence that the world soon came to regard quantum weirdness as a significant philosophical problem.[19] Although many capable theorists are like composers who play only the piano, Bohr and Einstein were both universalist thinkers, akin to those composers who can play every instrument in the orchestra. The world knows Einstein; perhaps we may take a moment to meet Bohr.

He was one of the *physical* physicists, blessed with a lifelong appetite for fresh air and exercise. He saw life as a whole and was immune to the scholarly delusion that brain power is superior to muscle power. Heisenberg tells a story that illustrates Bohr's integrated view of thought, action, and mystical philosophy: "Once, when on a lonely road I threw a stone at a distant telegraph post, and contrary to all expectations the stone hit, he said, 'To aim at such a distant object and hit it is of course impossible. But if one has the impudence to throw in that direction without aiming, and in addition to imagine something so absurd as that one might hit it, yes, then perhaps it can happen. The idea that something perhaps could happen can be stronger than practice and will.' "[20] Bohr's younger brother Harald was a soccer star—a member of the Danish team that won a silver medal in the 1908 London Olympics—and Niels might have matched him athletically had he not been so preoccupied. Playing goalie against a German club, he busied himself tracing equations with his index finger on the goalpost, nearly letting an errant ball roll slowly into the goal. Like Einstein, Bohr was a sailor, but while Einstein liked to trace broad reaches on lakes, Bohr preferred blue water. (The greatest tragedy of his life came when his eldest son, Christian, was swept to his death from the deck of Bohr's cutter, the *Chita,* in a summer storm in 1934. Only the restraining grip of friends on deck prevented Bohr from leaping into the sea after him.) Bohr viewed ignorance as an integral part of the learning process and regarded confusion and paradox as signposts on the road of inquiry. He complained on his deathbed that the philosophers too often "have not that instinct that it is important to learn something, and that we must be prepared to learn."[21].

Blunt and tenacious to a fault, Bohr was too serious to be pompous and too honest to be facile. If his way of speaking was often confusing, that was because he was himself frankly confused and liked to think out loud, and held that one should, as he put it, "never express yourself more clearly than you think."[22] (When Carl Friedrich von Weizsäcker wrote in his diary on meeting Bohr, "I have seen a physicist for the first time. He suffers as he thinks," he meant that Bohr suffered *out loud.*[23]) His habit of being both frank and frankly uncertain could get Bohr in trouble. Winston Churchill, having been urged by Bohr to reveal nuclear secrets to the Soviets since they were bound to learn them anyway, responded in an outraged note to his science adviser, Lord Cherwell, who had arranged the meeting, "It seems to me Bohr ought to be confined or at any rate made to see that he is very near the edge of mortal crimes. . . . I did not like the man when you showed him to me, with his hair all over his head, at Downing Street. . . . I do not like it at all."[24] Bohr fared little better with the American secretary of state, Dean Acheson, with whom he met in the spring of 1950 to discuss a planned open letter to the United Nations. "The meeting began at, say, two o'clock, Bohr doing all the talking. At about two thirty Acheson spoke to Bohr about as follows. Professor Bohr, there are three things I must tell you at this time. First, whether I like it or not, I shall have to leave you at three for my next appointment. Secondly, I am deeply interested in your ideas. Thirdly, up till now I have not understood one word you have said."[25]

Bohr's explications of the Copenhagen outlook can sound as oracular as if he had uttered them from atop a tripod while chewing laurel leaves, but he was earnestly trying to bring as much clarity to quantum weirdness as he could, and his position is not all that difficult to understand. Briefly put, it is that since, owing to quantum indeterminacy, neither we nor any other observers anywhere in the universe can know everything about a given microscopic particle or system, it is pointless to speculate about whether the missing information "exists." Physics is not the pursuit of imaginary ideals, and physicists need not waste time speculating about quantities (such as whether a photon is "really" particle or wave) that are known to be unascertainable: "It is wrong to think that the task of physics is to find out how nature *is,*" Bohr wrote. "Physics concerns what we can *say* about nature. . . . Our task is not to penetrate into the essence of things, the meaning of which we don't know anyway, but rather to develop concepts which allow us to talk in a productive way about phenomena in nature."[26] The Copenhagen interpretation asserts, to paraphrase John Wheeler (who was paraphrasing Bohr), that no elementary phenomenon *is* a phenomenon until it is an *observed* phenomenon.

To clarify this ontology, Bohr spoke of what he called *complementarity.* The wavelike or particlelike potential states of an undisturbed photon (or its polarization states, or the hard/soft and sweet/sour states of the particles in our schematic experiment) complement each other, like the black and white sides of the yin-yang diagram that Bohr incorporated into his family coat of arms. Bohr saw complementarity as a kind of chiaroscuro, an essential embracing by nature of opposites and contradictions that had been revealed to us by Heisenberg indeterminacy but that has wider implications. The more closely one looks at one side of the issue (e.g., studies the photon as a wave), the more paradoxical the other side (but it's a particle!) becomes.

Every interpretation of quantum weirdness amounts to sweeping the weirdness under one or another carpet, and a magic carpet at that. The magic carpet of the Copenhagen interpretation is the act of observation. It is by making an observation—a *measurement*—that one "collapses the wave function," thus resolving the superposed system into one or the other of its states. But what, exactly, is an observation? From this question have sprung the most enduring thought experiments to have probed the dark realms of quantum weirdness.

The best known of them is "Schrödinger's cat." It consists of a system with two potential states, A and B. This could be a piece of radium with a 50 percent chance of decaying within one hour, or a sweet/sour box into which is introduced a single particle that has a 50 percent chance of emerging from the sweet output window —any probabilistic quantum setup. The important point is that, according to Bohr, the system *has* no definite state—neither decayed nor undecayed, neither sweet nor sour—until it is observed. Instead it exists in a *superposed* state, one fully designated by the probabilities of its wave function. The radium or other quantum object is set up to trigger one of two devices located inside an opaque box that also contains a cat. If the system goes one way (if, say, the radium atom decays) it opens a canister of cyanide gas inside a sealed box, killing the cat. If it goes the other way (no decay), the cat survives. We set up the apparatus, then wait one hour before opening the box. Question: Right before we open the box, is the cat dead or alive? The Copenhagen interpretation answers that until we open the box and observe it, the cat is neither dead *nor* alive but exists in a superposed state of dead/alive. This seems implausible, and that is the point of the thought experiment: Schrödinger's cat critiques the Copenhagen interpretation by reducing it to absurdity. Its object is to deny the plausibility of a bifurcated, quantum-classical universe by demonstrating that such segregation yields nonsensical results. (Minimalists comfortable with a bifurcated physics can and do shrug it off. Stephen Hawking, paraphrasing Hermann Goering, says, "When I hear of Schrödinger's cat, I reach for my gun."[27])

The issue can be illuminated by considering our frame of reference. Suppose that the cat experiment is conducted in a locked laboratory, at night, with only one scientist keeping watch. At the end of the hour, he opens the box and sees . . . what? Until the scientist picks up the phone and announces the result, or runs into the street shouting "Eureka!" we don't know the outcome.[28] The wave function was collapsed in that scientist's frame of reference, but not in ours. That this is problematical is not terribly surprising: In science as in art, the choice of frame counts for a lot. (G. K. Chesterton: "Art is limitation; the essence of every picture is the frame."[29]) It amounts to saying that the Copenhagians are vague when it comes to defining just what, exactly, is meant by "measuring" or "observing" a phenomenon or "collapsing the wave function"—all of which mean the same vague thing.

Another thought experiment, more subtle than the cat but no less telling, was composed in 1935 by Einstein and two of his young associates at the Institute for Advanced Study in Princeton, Boris Podolsky and Nathan Rosen. It is known as the Einstein-Podolsky-Rosen ("EPR") "paradox," and works rather like our beam-splitter experiment. We start with a particle that decays into two other particles, X and Y, that must have a total spin equal to zero. So if one particle has a spin of +1, the spin of the other must be −1. We let the particles fly far apart—this is the now-familiar amplification part of the experiment—and when they are separated by, say, one light-year, a physicist measures one of them, particle X, and finds that its spin is −1. He then *knows* that particle Y, a light-year away, must have a spin of +1, as can be verified by a second physicist, off yonder where particle Y is. That would be perfectly sensible for a macroscopic system—if, say, the particles were replaced by a pair of one-ton gyroscopes that had been spinning in opposite directions all the way out. But according to the Copenhagen interpretation, remember, the particles were in *neither* spin state until their spin was observed. It seemed to Einstein—and has seemed to like-minded thinkers since—that if in fact a particle's spin is indeterminate, then the only way for Y to "know" that X had suddenly resolved itself into a spin −1 state would be if some sort of signal propagated *instantaneously* across a light-year of space, bringing the news from X to Y. And that, of course, would violate both special relativity and common sense. Einstein called it "spooky action at a distance." "No reasonable definition of reality could be expected to permit this," wrote Einstein, Podolsky, and Rosen.[30]

Much of the subsequent discussion of the Copenhagen interpretation—and such critiques of it as Schrödinger's dead-and-alive cat and the EPR "paradox"—has been infected with confusion. It helps in dispelling the mists to keep in mind that Bohr did not exactly maintain that a quantum system *has* no state prior to its being observed. Rather, he said that its state, prior to observation, cannot in principle be determined, and that attempts to define it are therefore meaningless. Bohr was an agnostic on the issue of what might be going on in nature beneath the threshold of its theoretical observability. Einstein used to poke fun at the Copenhagen interpretation by asking colleagues whether they really believed that the moon existed only when they looked at it. Bohr's answer was not that the moon does *not* exist when unobserved, but that we cannot *know* whether it, or some thoroughly unobserved moon of a remote and uninhabited planet, exists, until it *is* observed. His position sports a certain tough-minded bluntness: It confronts quantum weirdness and refuses to blink. But in doing so, it amounts, in the words of David Z. Albert, a physicist who holds a chair in philosophy at Columbia University, to a "radical undermining . . . of the very idea of an objective physical reality"[31]—which, I would add, has long been regarded as the whole point of science.[32] So it is understandable that at least a few philosophically minded scientists kept searching for a more accommodating way to draw quantum weirdness into the embrace of macroscopic logic.

Of these, some came to favor Hugh Everett's *many worlds* interpretation. Everett arrived at Princeton in 1955, the year of Einstein's death, and did his graduate work there under Wheeler, who took the problem of quantum weirdness seriously and never succumbed to the scientific conceit of dismissing it as a philosophical superfluity.[33] Like Einstein, Everett was troubled by the fact that to accept the Copenhagen interpretation is to entertain a worldview in which the probabilities expressed in a particle's wave function are said to exhaust our potential knowledge of that particle. If, for instance, an electron that had a 10 percent likelihood of turning up in a detector field X is actually observed to land at X, we are asked to accept that prior to its being observed the electron really *was* 10 percent here, at X, and 90 percent at other locations. This seems nonsensical. It's like saying that a woman is 10 percent pregnant, or a cat 50 percent dead, rather than making the more sensible statement that these are the odds produced by our limited knowledge of the system in question. Einstein regarded it as unworthy of the Old One, as he called the universal Logos without which science would in his view be reduced to the status of a casino game. (It was to this aspect of Copenhagenism that Einstein objected with his famous declaration "God does not play dice with the universe."[34]) Everett's formulation paints nature in the old-fashioned, classical way, as operating according to strict rules of cause and effect, uncomplicated by concerns about who constitutes an observer or how a measurement of a system is made. In the many worlds picture, the photon in our experiment *is* a particle, or a wave, and we simply record its existence, as we might that of a planet or a trumpet blast.

The interpretation attains this simplicity, however, at the price of making a genuinely flabbergasting supposition: It states that the universe is constantly splitting apart, making copies of itself that are identical except for the outcome of each particular observation. Every time a physicist checks to see whether a photon is a particle or a wave, the universe divides, creating two laboratories containing two physicists, one of whom sees a particle and the other a wave. Every time the position of an electron is observed, an infinity of other universes are born, each containing an electron at each of its other possible locations.

This notion is certainly sufficiently bold to satisfy Bohr's demand that new ideas be "crazy enough" to contribute to quantum theory. But it is also vulnerable to straightforward criticisms of the sort memorialized by Samuel Johnson, who, on being asked about Bishop Berkeley's belief that nothing can be shown to exist except ideas, kicked a stone and said, "I refute it thus." Such critiques have been abundant, their tone ironically understated. "The idea of 10^{100+} slightly imperfect copies of oneself all constantly splitting into further copies, which ultimately become unrecognizable, is not easy to reconcile with common sense," writes Bryce DeWitt.[35] The theorist Philip Pearle archly calls it "uneconomical."[36] David Lindley remarks that "when you think about how many of these parallel universes you have to provide"—to account, for instance, for the universe having split every time a photon bumps off a proton in the course of its long climb out of the sun—"the whole idea begins to seem cumbersome, to say the least."[37]

Nevertheless, a derivation of the many worlds interpretation has become the most widely employed approach to quantum cosmology today, in the form of the "many histories" formulation that we encountered in the previous chapter. There are several reasons that so radical an idea has managed to evolve into something approaching a working set of scientific tools. For one, Everett was among the first theorists to take seriously the central idea of quantum cosmology—that one can apply quantum mechanics to the universe as a whole—and so his treatment lends itself rather well to ongoing efforts to accomplish that goal today. Specifically, it seems to make sense when combined with Richard Feynman's "sum over histories" method—the approach that equates the probabilities in the wave function with various alternative developments that might have occurred in cosmic history but didn't (or didn't, at least, in the part of the universe that we observe). From this perspective, a cosmologist can make quantum calculations without concerning himself overmuch with the vexing question of whether the outcomes that we don't observe actually exist, in some of the infinite number of alternative universes. So science marches on, even if its philosophical implications here seem at least as preposterous as under the Copenhagen interpretation.

That leaves the *hidden-variables* interpretation of David Bohm. Bohm was a young physicist whose Marxist convictions encouraged him in the belief that nature is fully deterministic—in which case, any theory that restricts itself to probabilities cannot be complete. He studied the Copenhagen interpretation and even wrote a book defending it, but a subsequent conversation with Einstein left him dissatisfied with the limitations the Copenhagians placed on the scope of scientific knowledge. ("He talked me out of it," Bohm told Murray Gell-Mann. "I'm back where I was before I wrote the book."[38]) In the Copenhagen approach, Bohm complained, "All that counts in physical theory is supposed to be the development of mathematical equations that permit us to predict and control the behavior of large statistical aggregates of particles. . . . This sort of presupposition is indeed in accord with the general spirit of our age, but . . . we cannot thus simply dispense with an overall world view. . . . Indeed, one finds that physicists are not actually able just to engage in calculations aimed at prediction and control: They do find it necessary to use images based on *some* kind of general notions concerning the nature of reality, such as 'the particles that are the building blocks of the universe'; but these images are now highly confused (e.g., these particles move discontinuously and are also waves)."[39]

Bohm's search for a simpler and more complete interpretation led to his formulation of a new, deterministic account of quantum theory, which he published in 1952. By then, however, his career had been shipwrecked in the political typhoons of the times. Held in contempt of Congress for refusing to testify before the House Un-American Activities Committee, Bohm was fired from his post as assistant professor at Princeton and banned by a pliant university administration from visiting the campus in any capacity. He spent the rest of his life in a species of exile, teaching in Brazil, Israel, and thereafter at Birkbeck College in England. His insistence on examining quantum weirdness in a broad context further separated him from most of his fellow scientists, among whom arose the common judgment that he was a talented physicist who had squandered his potential by mucking about in philosophy.

But Bohm was onto something—a view of nature so revolutionary that he himself could not see it clearly at first. His interpretation has at least two levels, one relatively straightforward and the other, which followed clarifying research by John Stewart Bell, as startling and new as anything to have come from quantum mechanics and relativity. Let's consider each level in turn.

Bohm started with the deterministic premise that subatomic particles really are in one state or another—that quantum uncertainty is a statement of human ignorance and not a state of nature. Schrödinger's cat is dead *or* alive, and there is no need to imagine that it or any other system abides in a "superposed" state. This much is happily commonsensical, but it is paid for in two heavy coins.

First, Bohm was obliged to invent an agency—a guiding wave—to manipulate the particles. He called this guiding wave the "quantum potential." He envisioned it as a gently acting field with the unique property that its strength does not decrease with distance. To modify an analogy of Bohm's, consider a B1 bomber flying on autopilot in its ground-hugging mode. The B1's flight is powered by its mammoth twin jet engines (which here stand for conventional quantum force fields) but its guidance comes from the much weaker pulses emitted by its radar equipment, which reads the ground and adjusts the flight controls accordingly. (The guidance system represents the quantum potential.) Appealing as this picture may be, there is no experimental evidence to indicate that Bohm's quantum potential exists. Nor is it clear how such evidence can ever be found, since Bohm's equations produce exactly the same predictions as those of conventional quantum mechanics. (That's why the variables are "hidden.")

The other problem confronting Bohm's interpretation is that the quantum potential would seem to violate special relativity. In order for it to control the behavior of far-flung particles (in, e.g., an EPR experiment), it must act *simultaneously* on them. From the perspective of contemporary physics, this would mean sending signals that travel at faster-than-light speed. This is a lot to swallow, especially for the likes of Albert Einstein—who, on the day that he talked Bohm out of belief in the Copenhagen interpretation, was motivated by a distaste for just the "spooky action at a distance" that Bohm was to resurrect.

Nevertheless, Bohm's interpretation clarifies aspects of quantum weirdness and continues to gain advocates. David Z. Albert recently has been championing a Bohmian interpretation on the ground that, its philosophical penumbra aside, it is simpler than the Copenhagen approach. "What's so cool about this theory," writes Albert, going on to sound a bit like Gertrude Stein, is that

> this is the kind of theory whereby you can tell an absolutely low-brow story about the world, the kind of story (that is) that's about *the motions of material bodies*, the kind of story that contains nothing cryptic and nothing metaphysically novel and nothing ambiguous and nothing inexplicit and nothing evasive and nothing unintelligible and nothing inexact and nothing subtle and in which no questions ever fail to make sense and in which no questions ever fail to have answers and in which no two physical properties of anything are ever "incompatible" with one another and in which the whole universe

> always evolves ***deterministically*** and which recounts the unfolding of a perverse and gigantic conspiracy to make the world ***appear*** to be ***quantum-mechanical***.[40]

That it does so by invoking faster-than-light effects doesn't ruffle Albert's hair. If a relativistic version of Bohmian quantum mechanics can be written, he claims, its predictions will be in accordance with special relativity "even though the underlying [i.e., Bohmian] theory won't be; and so taking Bohm's theory *seriously* will entail being *instrumentalist* about special relativity."[41] But there is as yet no Bohmian relativistic quantum field theory, and there may never be; nor is it clear that other theorists will be as blithe about demoting the status of special relativity.

Yet what Bohm's interpretation lacks as a scientific theory it gains as an admittedly clouded but evocative glimpse into the mists of a possible future science. Bohm was unable to describe this Ultima Thule with any great clarity, but he insisted on its existence and predicted that its elucidation would bring about not just a new theory but a new "order," a revolution comparable to the world-shaking changes we associate with such names as Copernicus and Einstein. Bohm was a modest man, but he insisted on this one great claim. "We have . . . yet to perceive a new *order,*" he wrote. "We are in a position which is in certain ways similar to where Galileo stood when he began his inquiries."[42] In his view, quantum weirdness is a keyhole through which we have caught a first glimpse of another side of nature, one in which the universe is neither deployed across vast reaches of space and time nor harbors many things. Rather it is one, interwoven thing, which incorporates space and time but in some sense subordinates them—perhaps by treating them as important but nonfundamental aspects of the interface between the universe and the observer who investigates it.

The quantum universe may be thought of as the other side of the coin from the spatiotemporal, relativistic universe that has to date dominated cosmological thought. We humans, having come along when the universe was already billions of years old and being rather *big* creatures, able to see stars in the sky but not atoms in an apple, naturally got into cosmology from the large-scale side of things—by observing galaxies and developing theories, such as relativity, to interpret their behavior. But the universe was not always big and classical. Once it was small and quantum, and possibly it has not lost the memory of those times. It may well turn out that over there—or, more properly, inside and underfoot, marbled through the very fabric of the space that is in turn marbled through every material object—the universe remains as it was in the beginning, when all places were one place, all times one time, and all things the same thing.

To investigate that side of the coin we need to consider one final technical development, and that is *Bell's inequality*. John Stewart Bell was an Irish physicist who concerned himself with the hidden-variables interpretation and worked out a way of testing it experimentally. Without going into specifics, Bell's proposed experiment was a variation on the EPR apparatus—a setup in which two particles that start out together are dispatched across a macroscopic distance before one is observed in a fashion that instantly defines the state of the other. Bell's contribution was to outline how an EPR-like experiment could be employed to test the classical assumption that nature works in a "local"—that is, mechanistic—way. The results were to reveal that the classical assumption is wrong—that nature is in some sense *nonlocal*. From this odd finding sprang considerations so astonishing as to render plausible the physicist Henry Stapp's opinion that Bell's theorem constitutes "the most profound discovery in science."[43]

We encountered the concept of *locality* earlier in this chapter, as the supposition that one system can change another only if there is some sort of mechanical interaction between the two. According to relativity, no such interaction can occur at faster-than-light speed, and what bothers physicists about the hidden-variables interpretation is that it seems to mandate such superluminal interactions.[44] To say that fiddling with one particle over here can *instantly* influence its sister particle over there is to assert that subatomic particles behave in a *nonlocal* way. This would overthrow the time-honored assumption of locality, and that is what Einstein found so repugnant about the situation, and why he constructed the EPR thought experiment to highlight its apparent irrationality.

Bell—a red-bearded experimentalist who spoke with a soft, Northern Irish burr, and whose unassuming wit concealed an exceptional tenacity of mind—pondered this matter for years, focusing on its essential question of whether natural processes obey locality, as had traditionally been thought, or are in some way nonlocal on the quantum level. In a paper published in 1964, he proposed an experiment that could finally settle the matter. Years passed before technology had advanced to the point that it could be implemented. Then, in the 1970s, John Clauser and Stuart J. Freedman at Berkeley, and later Alain Aspect and colleagues at the University of Paris's Institute of Theoretical and Applied Optics, in Orsay, conducted Bell experiments. The specifics need not detain us: They involved testing the polarization of large numbers of photons. Their significance was that they would produce different results if the particles behaved in a local way, as Einstein insisted, or in a nonlocal way, as the quantum mechanics equations mandate. This distinction is, after all, what all the bother over quantum weirdness is about. In both cases, and in all experiments conducted since, the verdict is clear: Bohr was right (nonlocal effects do occur in quantum systems) and Einstein wrong (there are no hidden variables to explain nonlocality). Nature—on the subatomic scale at least—really is nonlocal. Fiddling with one particle really does mean that its sister particle is altered, instantly, even if it is far away, and neither hidden variables nor any other mechanistic scheme can rescue Einstein's belief in locality. As the physicist F. David Peat puts it, "The choice before us is either to abandon any hope of knowing the nature of quantum reality or to accept a nonlocal universe."[45]

Some are comfortable with the first of Peat's alternatives. They believe that we cannot reconcile common sense with quantum reality, and so shouldn't try. But history has dealt harshly with many previous efforts to declare absolute limits to human inquiry, and had that option been popular in this case, there would not have been seventy years of debate about quantum weirdness. So let's look at the alternative—"to accept," in Peat's words, "a nonlocal universe."

What might that mean? It *might* mean that the universe is interconnected in some deep and as yet only dimly perceived way, on a level where time and space don't count. Bohm, who lived long enough to absorb the experimental results confirming that quantum effects are nonlocal, wrestled with this remarkable idea in his book *Wholeness and the Implicate Order*, published in 1980. A capable etymologist, Bohm used the word "implicate" in its sense of "enfolded." He suggested that nonlocal effects are woven through the universe in something like the way that a chef folds a cream into a sauce. For Bohm, classical physics dealt with an *explicate* order, the mechanical world of Newton's gravity and Einstein's relativity, while quantum mechanics was the first science to examine the implicate world of nonlocalities. A scientific clue to this new vision may be found in the odd consideration that photons do not "experience" time. We understand from special relativity that time slows down for space travelers as they approach the velocity of light. At light speed, the speed that photons move in a vacuum, there is no time at all. So a photon "traveling" from point A to point B does so, from its point of view, in zero time—meaning that, in some sense, the two points aren't separate! Another clue comes from the work of John Wheeler and others on the hypothesis that space is interconnected by multitudes of wormholes, little tunnels linking localities that to us seem far apart. A similar outlook has been investigated by Roger Penrose, who sees spacetime as jumbled and dynamic on the quantum scale. Penrose compares space to a photographic plate, one that develops into a "normal," macroscopic picture only when "fixed" by measurement.

Bohm and others have likened the implicate universe to a *hologram* (from the Greek, "to write the whole"). One makes a hologram by illuminating the subject with a beam of laser light that has been run through a beam splitter, creating two beams—a process akin to the dual-slit experiments central to thinking about quantum weirdness—and exposing a sensitized glass plate to the light reflected from the subject. The plate contains no visible image, but when illuminated by a similar pair of coordinated beams of light it produces a three-dimensional replica of the hologrammed subject that seems to hover in space. This image is intriguing in itself—there is no theoretical lower limit to its resolution except that imposed by the wavelength of the light used to make it—but of particular interest, in terms of a cosmological metaphor, is the way information is recorded on the plate: Shatter a hologram, put one of its fragments in the laser beam, and what you see is not a piece of the original image but *all* of it. The image is dimmer and a bit "noisier," but spatially the whole thing is there, in this and every other fragment.

What if the universe is like that? I don't know how to frame such a concept in contemporary scientific terms, so I won't try. Such difficulties may, of course, be a signal that there *is* no "implicate" side to the universe—that this line of thought is just hot air. But they also might mean that, as Bohm believed, we are indeed dealing with a new "order," which must therefore evolve its own concepts and language and cannot properly be analyzed, in Bohm's words, "to make it fit well-defined and preconceived notions as to what this order should be able to achieve."[46] So let me describe the concept more generally, as a kind of fable.

Suppose that, as string theory implies, the universe began as a hyperdimensional bubble of space, all but four of the dimensions of which compacted to form what we today call subatomic particles. Those particles look to us like zillions of individual things, but that is merely their appearance in the four dimensions of spacetime. In hyperspace they could very well still be *one* thing—could, therefore, be not only connected but identical. (Wheeler to Richard Feynman: "Feynman, I know why all electrons have the same charge and the same mass." "Why?" "Because they are all the same electron!"[47]) In that case, we live in a universe that presents two complementary aspects. One obeys locality and is large, old, expanding, and in some sense mechanical. The other is nonlocal, is built on forms of space and time unfamiliar to us, and is everywhere interconnected. We peer through the keyhole of quantum weirdness and see a little of this ancient, original side of the cosmos.

To assert that the universe is deeply interconnected is to echo what mystics have been saying for thousands of years. This can be a liability in the scientific community, which has heard more than enough of complacent, shallow-draft assertions to the effect that science amounts to little more than proving what Lao Tzu and Chief Seattle were saying all along. Yet some of the most important scientific and philosophical thinking in history has been impelled by mystical motives. (Einstein: "The most beautiful emotion we can experience is the mystical. It is the source of all true art and science. He to whom this emotion is a stranger, who can no longer wonder and stand rapt in awe, is as good as dead."[48]) We remember the paradoxes of Zeno of Elea, the philosopher and mathematician who sought to demonstrate that motion is impossible because, for example, a flying arrow must keep traversing half the distance to the target in finite intervals of time, and the number of times the distance can be halved is infinite. But we less often recall *why* Zeno constructed his paradoxes. He did so to support the assertion of a fellow Eleatic, Parmenides, that all is one, and to disprove, by reduction to absurdity, the contrary philosophy espoused by Pythagoras, that nature is made not of one but of many things. Science up to the present is descended mainly from Pythagoras. Zeno's point is that this Pythagorean, many-things view of the universe is incomplete, that from a deeper outlook we would realize that the many-things, spacetime universe reflects but one side of creation. As Dante's Virgil says:

> *Yes, many things there are, which seem to be*
> *Perplexing, though quite falsely so, because*
> *They have good reasons which we cannot see. . . .*[49]

If the many threads of history are revealed to be cuts through one knot Parmenides' sun will rise anew, and Zeno's insight will emerge as an intimation of the nonlocal universe.

That would mean that the role of the observer has scarcely begun, and the poking of one's head through a starry local sphere, as illustrated in the famous woodcut depicting the Copernican revolution, is to be recapitulated on a grander scale. One of the first to see this was Wheeler—who, like Gertrude Stein, has sometimes been underestimated as a thinker, owing to his taste (like hers) for examining questions without pretending to have answers to them. Wheeler wonders aloud whether quantum observership represents an alternative vision of genesis. "How did the universe come into being?" he asks.[50]

> Is that some strange, far-off process, beyond hope of analysis? Or is the mechanism that came into play one which all the time shows itself?
>
> Of the signs that testify to "quantum phenomenon" as being the elementary act of creation, none is more striking than its untouchability. In the delayed-choice version of the split-beam experiment, for example, we have no right to say

> what the photon is doing in all its long course from point of entry to point of detection. Until the act of detection the phenomenon-to-be is not yet a phenomenon. We could have intervened at some point along the way with a different measuring device; but then regardless whether it is the new registering device or the previous one that happens to be triggered we have a new phenomenon. We have come no closer than before to penetrating to the untouchable interior of the phenomenon. For a process of creation that can and does operate anywhere, that reveals itself and yet hides itself, what could one have dreamed up out of pure imagination more magic—and more fitting—than this?

All interpretations of quantum weirdness sweep it under a carpet, but carpets have patterns that cannot be perceived by isolating their parts. If we have to date perceived only the spacetime threads in the carpet, that may be because perception is itself the discernment of parts. If we could take in the whole, would we see that the *here and now* is the same as the *then and there?* Can mere observers behold both the disparate, explicit universe and the implicate universe that wove it? What does it mean, anyway, to be an observer, to be alert, alive? With that question, we descend from the icy cliffs of genesis and quantum weirdness to wander in grassy valleys, where, amid luxuriant life, our wondering and discontented species asks what might be its proper place in the universe.

12
A Place for Us

God said to Abraham, "But for me, you would not be here." "I know that, Lord," Abraham answered, "but were I not here there would be no one to think about you."

—*Traditional Jewish tale*[1]

The optimist proclaims that we live in the best of all possible worlds; and the pessimist fears this is true.

—JAMES BRANCH CABELL[2]

HUMAN BEINGS must be just about the *loneliest* species in the universe. We've only recently begun to learn about the universe, don't yet know quite what to make of it, and haven't anyone else with whom to discuss it. So we talk among ourselves, in musings that necessarily are limited to our necessarily human perspective. In this sense, our dialogues are monologues, hobbled by what might be called the agony of uniqueness. We are conversant with only one kind of intelligence, our own, only one kind of life, for all earthly life is kin, and only one observed universe. How, then, are we to work the calculus of chance and necessity in order to understand which if any of the laws and constants of nature are inevitable and which are accidents, and to judge whether life and intelligence are central or peripheral to the cosmic scheme of things?

Prior to the Renaissance, we lived in a cozy little hut of a cosmos, and our place within it seemed secure.[3] The prescientific universe was *about* us, in two senses of the word: It consisted of our immediate surroundings (and not much more) and we belonged to it—were appropriate to it. The Copernican revolution changed all that, of course, but its shock was not so much that it "dethroned" us from a privileged central position, as the textbooks say, but rather that the universe no longer seemed to be about us. The vast reaches of the Copernican cosmos, if uninhabited, seemed useless and pointless. (This consideration was specifically raised, by the philosopher Giovanni Agucchi in a letter to Galileo, as an argument against Copernicanism.[4]) If inhabited, they obliged us to consider that we share our cosmic home with other beings, some of whom might be smarter or better than us, more worthy of God's concern.

Thus began an age of apprehension, in which it became fashionable among popularizers of science to wield the nasty vastness of the universe like a club. To be scientifically hip was to parade one's unblinking acceptance of the unflattering proposition that we are but slime, clinging to a speck of dirt in a galactic outback, hurtling ignorantly through a lethal vacuum dotted with uncaring stars. Sir James Jeans read this mood perfectly, and wrote a bestselling astronomy book that so stressed how inhumanly big and small and hot and cold everything was that one critic asked whether Sir James's intention was to educate his readers or scare them to death.

Lately the pendulum has begun to swing the other way, with scientists and philosophers reconsidering whether our existence is really all that incidental to the wider world. The Copernican revolution bred the *cosmological principle,* which rules out theories that place humankind in the center of the universe or any other special location. It would, for instance, violate the cosmological principle to argue that the big bang took place at a specific location in a preexisting space, and that we just happen to sit at the site of the explosion, so that the galaxies are all rushing away from *us.* The cosmological principle is fine so far as it goes, but some have begun wondering whether it's too one-sided. Perhaps cosmology would benefit from the addition of a second principle—one that took our existence into account without postulating that there was otherwise anything special about us. The *anthropic principle* attempts to do this. It starts with human existence as a given and scours the universe for clues as to which of its characteristics are essential to the existence of life. Its thrust is to portray the cosmos less as an impersonal machine and more as, in John Wheeler's words, a "home to Man."

To investigate this outlook, we pose two questions relevant to the relationship between humankind and the universe it has so recently come to behold:

Are we alone? Is life cosmically commonplace or rare? Is human intelligence a fluke or a spark of universal fire?

Does our existence tell us anything about the universe? The anthropic approach assumes that it does, and seeks to learn about the universe by taking the existence of life as a starting point.

The question of whether we are alone in the universe is ancient. What's new is that we are coming to possess tools that could give us a shot at answering it.[5] Existing radio telescopes are capable of detecting signals transmitted by an alien civilization of comparable capacity anywhere in our quarter of the Milky Way galaxy. The next generation of optical space telescopes may be able to discern telltale signs of life—specifically, carbon dioxide, which indicates the presence of an atmosphere; water, suggesting oceans; and ozone, a form of oxygen—in the spectra of starlight reflected off extrasolar planets. And, conceivably, some future space probe might yet sniff out life-forms or fossils right here in the solar system —hidden beneath the opaque clouds of Saturn's satellite Titan, perhaps, or lurking in the frozen tundras of Mars. (Interest in the prospect of Martian life was lent fresh impetus in 1996, when NASA scientists found what conceivably is evidence of microscopic fossils in a rock that was knocked off the red planet 16 million years ago—probably in an asteroid impact—fell to Earth thirteen thousand years ago, and was found in an Antarctic ice field in 1984.) Until a search turns up something, or an overwhelming amount of nothing, discussions of extraterrestrial life will remain largely speculative. The quality of such speculations has improved a bit, though, as science has gained a better comprehension of how life works and how life on this planet began.[6]

Such theorizing has produced two sharply different estimations of whether there is life out there, and if so how much of it, and whether any of it is "intelligent" (which for these purposes we define, pragmatically, as possessing the ability and willingness to communicate with humans).[7] One camp, composed mainly of astronomers and physicists, argues that extraterrestrial life is abundant. "I'm sure they're out there," declares the physicist Paul Horowitz, of Harvard, who runs a SETI (Search for Extraterrestrial Intelligence) search from a modest, 84-foot radio telescope equipped with receivers and analyzers which Horowitz and his students built largely by hand.[8] The other camp, made up mostly of life scientists, maintains that while life may exist on other planets, the odds of there being extraterrestrial *intelligence* are so small that we are almost certainly alone in our galaxy, and perhaps in the entire observable universe. The biologist Ernst Mayr, also of Harvard, holds this position. He declares that "SETI is a deplorable waste of taxpayers' money."[9]

Both sides are reasoning from the same basic data. That they reach opposite conclusions demonstrates how difficult it is to calculate the odds of a phenomenon (in this case, life) of which we have but a single example. Let's examine their reasoning.

The optimistic (meaning pro-SETI) argument, stripped to its essentials, goes like this: There are so many stars in our galaxy that even if only one percent of them are orbited by an Earth-like planet, that would still mean that there were more than a billion Earths in the galaxy. Life began promptly in the history of our planet—the earliest fossil cells date from within a few hundred million years of the formation of the earth's crust. This suggests that life arises readily—at least, on terrestrial planets blessed with liquid water—and therefore does not depend upon some sort of extraordinary luck. Once established, life is robust: Terrestrial life has survived numerous catastrophes that decimated living species, yet evolution proceeded apace. Over the course of billions of years of evolution, intelligence will emerge, sooner or later, because it confers survival advantages on the species that possesses it, which is what evolution is all about. Where there is intelligence, soon there will be technology. Humans went from dugout canoes to spaceships in a scant fourteen thousand years. So it makes sense to use radio telescopes (part of the time) to listen for signals from other technological civilizations, as there are likely to be thousands of them in our galaxy.

The pessimists, to simplify *their* argument, reply as follows: First of all, Earth-like planets are probably much rarer than the optimists assume. The earth is unique in the solar system in that it's at just the right distance from the sun so that water, which we all agree is essential to life as we know it, exists here in all three states, as liquid, ice, and vapor. Were Earth's orbit slightly larger or smaller, this would not be the case, and life might very well not exist here. Even if we accept the hypothesis that there's lots of life in the galaxy, the optimists' reasoning collapses when it comes to the advent of intelligence. The optimists claim that intelligence confers a survival advantage on the species blessed with it, and therefore is selected for in the course of biological evolution. But if that is the case, why did intelligence not appear earlier in Earth's long history? The optimists stand accused of inconsistency: They say that life is in the cards because *it* originated early in our planet's history but they also say that intelligence is in the cards, and it originated late. The error (if it is an error) springs from the discredited assumption that evolution is a stairstep progression, a slow-grinding machine aimed at producing, eventually, human beings. It's not. Evolution is pointless and largely random, and the sequence of events that led to *Homo sapiens sapiens* is so long and tangled, so replete with chance events that might well have gone otherwise, that intelligence almost certainly has never appeared anywhere else in the universe. Such extraterrestrial intelligence as may exist might well be akin to that of whales, spiders, insects, and the millions of other species that have lived on Earth, gifted with quite enough acumen to conduct their own affairs but uninterested in building radio telescopes. Finding such life will be difficult, communication with it harder than with owls and earthworms here at home. So, the pessimists conclude, we're alone, and we might as well get used to it.

My purpose in bringing up this ongoing debate is not to resolve it. (As a practical matter, the resolution is simple: We should continue to operate SETI projects because receipt of a signal *would* solve the riddle, while if we don't listen we shall have relinquished most hope of solving it.) Rather, the debate is interesting for what it reveals about our estrangement from a cosmos we have done so much to try to comprehend. It's quite true, as the pessimists assert, that we don't understand the origins of human intelligence. Scientists have produced useful ideas here, such as the hypothesis that the encephalization of the human brain (its rapid growth relative to body weight) was racheted upward (or "pumped," in the jargon) by the selection pressure of our ancestors' having been obliged to come down out of the trees and start hunting on the savannah —this due to climatic changes exaggerated by the ice ages. The neurobiologist William Calvin has stressed the possible role of hunters' throwing stones and spears in selecting the human brain for a command of dynamics.[10] But we don't really know. As the anthropologist Loren Eiseley wrote, "Man, the self-fabricator, is so by reason of gifts he had no part in devising."[11] We fail to understand not only how or why the gift was given us but, more to the point, why it keeps on giving—why our brains equip us not only for domination of our planet but also for a comprehension of the atoms' stuttering and the galaxies' glide. Ignorant of our origins, we are orphans in our own world.

So we *claim* that we are somehow central to things, that science is so ennobling and rewarding an activity—a way to know, and surely therefore something divine—that *it* must reside somewhere near the central fire even if we as its paltry torchbearers do not. But wishing doesn't make it so. The evolutionary record can be interpreted as testifying that intelligence is a fluke, and human history can be read as demonstrating that the invention of science was, too. The ancient Greeks were superb philosophers, who in some sense set us on the road to science, but they produced virtually no technology and almost no real science of their own. From China and India came thousands of volumes of brilliant thought and many technological innovations as well, but few scholars who have studied the matter believe that if science had not arisen in the West a few centuries ago it would by now have emerged in Asia.[12] We're not as much in the dark about the origin of science as about the origin of intelligence, but neither are we in a position to say with confidence that intelligence leads inevitably to science. And the jury is still out as to the alleged Darwinian benefits of having a human-style mind. Science and technology have served us well so far, delivering liberty and prosperity to more people than any other development in history. But they have also raised the specter of global catastrophe through overpopulation, pollution, resource depletion, nuclear disaster, and other agencies set in motion by science and technology themselves.

The proponents of SETI like to calculate the odds of success by way of the "Drake equation." It was fashioned by the astronomer Frank Drake, who in 1960 conducted the world's first SETI observations, training an 85-foot radio telescope at Green Bank, West Virginia, on two nearby sunlike stars and listening at a single frequency. The equation begins with the number of stars in the Milky Way galaxy and then multiplies them by a series of fractions representing the estimated number of sunlike stars, Earth-like planets, and so forth. Some of these values are rather well established and others are much more speculative, but when one plugs in the consensus values, something intriguing emerges: The number of communicating societies in the galaxy turns out to be roughly equal to their average lifetime in years. If a typical technological civilization stays on the air for only about a century, then there are only about one hundred of them in the galaxy, in which case our chances of finding one are rather slim (one in a billion, per star observed). If they last ten thousand years, there are roughly ten thousand of them, and SETI is a better bet. So to listen for a signal is, in a sense, an expression of faith in science and technology. It evinces the belief that "intelligent" creatures—here defined, again, as those with big radio sets—generally manage to survive, rather than fouling their nests or blowing themselves up. The dark ocean in which the putative civilizations float is made mostly not of space but of time.[13]

The cosmological point here is statistical in character. Two scientists, both using the same facts and both innocent of logical error, can reach wildly different estimates of the abundance of intelligent life in the universe. Why? Because it's exceedingly difficult to make reliable calculations of probability based on a single example. If you draw the king of hearts from a magician's deck of cards, how are you to calculate the odds of your having drawn that particular card if you don't know the contents of the rest of the deck? You can't. You need to see more of the cards. Estimating the likelihood of extraterrestrial life is rather like that—and so, interestingly, is the business of trying to understand why the constants of nature have the values they do. Suppose that the card you draw bears not the king of hearts but the equation $G = 6.67259 \times 10^{-11}\ m^3kg^{-1}s^{-2}$. That's useful information: It expresses the gravitational constant, and thus tells you the strength of the gravitational force. But if you're a cosmologist, you'd like to know whether this value is accidental—and how are you going to calculate the odds that the gravitational strength would have *this* value and not some other? We have access to only one universe, and it has just this one gravitational constant, so there's no basis for calculating probabilities.

Casting around for some reference point, we might compare gravity's strength with that of the three other fundamental forces. Doing so, we find that gravitation is remarkably weak. The weak nuclear force is 10^{28}—ten billion billion billion—times stronger than gravity. Electromagnetism is one hundred billion times stronger than that, and the strong nuclear force is a hundred times stronger than electromagnetism. This seems mighty asymmetrical. If the forces were a toy poodle whose shortest leg, the one representing gravity, were one inch in length, the leg representing the strong nuclear force would be far longer than the radius of the observable universe. Does this mean that it's highly unlikely for gravitation to be so weak, or are we calculating the odds in an inappropriate manner? Just how many possible strengths are there for gravity, anyway?

To find out, let's turn to our second question—whether the fact of our existence can tell us anything about the universe. Imagine what would happen were gravity a little stronger. The consequences, it turns out, would be dire. Cosmic expansion would have halted and the universe would have collapsed long before life could have evolved anywhere. Even if expansion somehow continued, the stars would burn out too rapidly to incubate intelligent life on anything like a terrestrial timescale. The sun, for instance, would have lasted only about a billion years.[14] Planets might not even exist: A planet represents a balance between the gravitational force that seeks to collapse it and the electromagnetic force that props up its molecules. Were gravity stronger, planets would light up and become stars, or further collapse to become white dwarfs, neutron stars, or black holes. So life probably could not exist in a strong-gravity universe. If, on the other hand, we decrease the strength of gravity, we find that the primordial material of the big bang simply dissipates, like hot air from a blown tire, before the gravitational fields can gather it into planets, stars, and galaxies. Life seems unlikely in that universe, too. So we've learned something interesting about gravity—that if it didn't have just about exactly the strength it does, we wouldn't be here to inquire into the matter.

Similar arguments can be applied to many other aspects of nature. Why is the universe so old? Because living creatures need carbon (the basis of terrestrial life) as well as iron and other metals (which is why a good multivitamin pill contains minerals), and for a planet to have abundant carbon and iron it had to have formed from material that had been processed through precedent stars, all of which takes billions of years. Why are neutrons slightly more massive than protons? Because if protons were just one percent heavier they would spontaneously decay into neutrons, in which case hydrogen atoms could not exist, nor stars shine: No stars, no life as we know it. Why does space have three dimensions rather than two or four? Because the knots and tangles of genetic material in living cells and the walls of organs can exist only in three dimensions.

To reason in this fashion is to invoke the anthropic cosmological principle—*anthropic* meaning "of humankind," and *cosmological* in that the principle attempts to constrain facts about the universe by taking into account our presence here. To "constrain" means, in this context, to improve our ability to calculate the odds of nature's being the way it is, by reducing its potential states from an infinite number to the much smaller set of states in which it is possible for life to exist. This approach allows us to calculate that the odds of our drawing the gravitational constant card that we did are pretty good: The deck can hold only cards with values quite close to the one we drew. Otherwise there would be nobody around to draw the card.

The anthropic principle has complicated historical roots in the various "design" arguments, which see order in nature as requiring an intelligent agency, and the teleological philosophies, which see nature as having a purpose. But these need not detain us, and we can pick up the story in 1974, when the term *anthropic principle* was coined by the British cosmologist Brandon Carter.[15] Carter was out to put limits on the cosmological principle. The cosmological principle is useful, as we've noted, but it raises two difficulties. First, it requires that we set the constants of nature and other facts about the universe against an infinite field of all other possible values—and this makes it next to impossible to calculate the odds of things having come out as they are. The other problem is more specific. It arose when Fred Hoyle and his colleagues composed the steady state theory. They supported the theory with what they rather grandly called the "perfect" cosmological principle. Why limit the cosmological principle to space? they asked. Why not also deny us a privileged position in *time*? The perfect cosmological principle asserts that the universe not only *is* much as we observe it, everywhere, but always was so. Therefore no big bang. This struck Brandon Carter as amounting to an abuse of the cosmological principle, one that the anthropic principle would prevent. The anthropic principle limits acceptable cosmological theories to those that take human existence into account. "What we can expect to observe must be restricted by the conditions necessary for our presence as observers," Carter said.[16]

Nowadays the anthropic principle comes in three flavors—weak, strong, and "participatory." The weak anthropic principle (WAP) simply states that (as we have been saying) scientists, in considering how nature might be otherwise, need calculate the odds not against an infinity of all other possible values but only against those that permit the emergence of life. The strong version (SAP) goes further: It declares that the universe must be constrained so as to allow for life. As Carter put it (for purposes of definition, not as an expression of his own beliefs), "The universe must be such as to admit the creation of observers within it."[17] In other words, no observers, no universe. The participatory version (PAP) is due principally to John Wheeler. It emphasizes the role of quantum observership in resolving potentiality into actuality and attempts to construct a new conception of the universe as observer-dependent, in the sense that (as we heard Wheeler say earlier) phenomena *are* phenomena only when they are *observed* phenomena—meaning that things cannot be said to exist until they are observed.[18]

Of the three, only the WAP enjoys much currency in scientific circles, where it has provided some illuminating insights into just how life "as we know it" depends on a wide range of cosmic conditions. Yet even this mild potion remains potentially toxic, and arguably the WAP continues to confuse as many scientists as it illuminates. Some of the controversy stems from the fact that the WAP is less scientific than philosophical, and philosophizing is about as popular among working scientists as is bird-watching among professional golfers. But much of it arises from a level-headed sense that the WAP represents philosophizing of a particularly dangerous kind.

Trouble brews, for instance, whenever scientists confuse *constraining* a phenomenon with *explaining* it. If they think they have explained it by showing it to be necessary to life, they may be discouraged from seeking a deeper and more productive explanation. This has already happened. The phenomenon in question was the isotropy of the universe—the fact that it looks the same in all directions. As we have seen, the universe is isotropic to a remarkable degree. In 1973, Stephen Hawking and Barry Collins of Cambridge University invoked the anthropic principle to "explain" cosmic isotropy by noting that if the universe were *not* highly isotropic it would have been difficult for stars and planets to form and, therefore, for life to exist. Fortunately, this argument was not accepted as final, and soon thereafter, isotropy was accounted for in a much more natural and elegant fashion, as resulting from inflation. So we need to be careful that anthropic constraints don't blind us to deeper explanations. If, for instance, the value of the gravitational constant is not an accident but the inevitable consequence of a deep natural structure to be revealed by superstring or some other theory, the anthropic principle will at best have been a red herring when it comes to understanding gravity.

Another danger is that because the anthropic principle is *post hoc*—the universe having been here before we were—it is susceptible to the *post-hoc* fallacy. This fallacy consists of assuming that because B followed A in time, A must therefore have caused B or have been an essential condition for its existence. Because terrestrial life arose from the universe as it is, we may, if we are not careful, lapse into an unduly rigid interpretation of nature as having to be just as it is in order to accommodate life of any kind. But life may be much more various than we, who have witnessed no other form of life, can as yet imagine. A biosphere can be broadly defined as a system, itself highly ordered, that maintains or increases order. To perform this feat, which locally reverses the law of entropy, requires two things—an energy differential (e.g., a hotter region close to a colder one) and enough local stability to permit, but not stifle, evolution.[19] These conditions may well be satisfied in a far wider range of circumstances than our experience indicates. Perhaps it is sheer science fiction to think that our belief that life depends, say, on carbon and oxygen will one day be disproved by the discovery of alcohol-guzzling jellyfish adrift in giant molecular clouds or non-carbonous crabs creeping across the surfaces of neutron stars. But enough of the predictions of the science fiction writers have come true to warn us against underestimating the power of the human imagination, and the anthropic principle can degenerate into stupefaction if it pretends to constrain the unknown simply by identifying antecedents of the known. One may stroll across the Brooklyn Bridge and buy a flounder at the Fulton Fish Market, but that does not mean that the flounder's existence required that of the Brooklyn Bridge, or that there are no other fish in the sea.

Still, ideas are like explosives: The fact that they are dangerous does not mean they ought not to be employed. And the anthropic principle looks a lot more sensible if we entertain the hypothesis that there are many universes, each with its own set of physical laws. Some of these universes are stillborn and wink out of existence in moments. Many have large cosmological constants (a condition the particle physicists regard, by the way, as a much more natural vacuum state than the vanishing small value found in our universe). They remain pure space, forever ballooning, incredibly big, eternally empty. Others contain life wildly different from ours. Others are vast and full of things, but never give rise to life. If they are unobserved, may they be said to exist? If a sterile universe eventually swims over the horizon of an inhabited universe, does it *then* exist? Understandably, some working cosmologists are impatient with these ideas; they find it difficult to see what purpose is served by postulating the existence of universes whose existence we cannot confirm or deny. But perhaps such theories will prove valuable as stimulants to the reasoned imagination, as scientists begin seriously to envision what nature may be like beyond this universe, out where the rules are different, and to sift out which rules *can* be different. A thinker engaged in panuniversal speculations may never know whether he or she has at last broken through to a realm where the imagination has surpassed the inventiveness of the real. But if the answer is yes—if there is and ever has been but one universe, so that all such speculations are hollow—it will be the first time that nature proved less clever and less resourceful than we are.

The cosmologist is like William Blake's builder of a ladder to the stars. "I want! I want!" he cries, but there's nothing to hold up the top of the ladder, and he's not too sure about who's holding the bottom, either. Who are we, and what do we want? Cosmology like every other human endeavor comes back to us in the end, but it's not just *about* us. That's the beauty of it—that we return from the voyage altered. Galaxies, like ocean coral, work a sea change, and make of us something rich and strange.[20] T. S. Eliot wrote:

We shall not cease from exploration
And the end of all our exploring
Will be to arrive where we started
And know the place for the first time.[21]

If this poem ended with the third line, it would rank among the dreariest of modern times. Science is too much trouble if its point is to bring us back to where we started. But the fourth line is cosmology's credo. For to find *our* place, we must know *the* place, cellar to ceiling, from the taproots to the stars, the whole shebang.

Contrarian Theological Afterword

I had only prayer, prayer
and science.

—Gillian Conoley[1]

In a Jewish theological seminar there was an hours-long discussion about proofs of the existence of God. After some hours, one rabbi got up and said, "God is so great, he does not even need to exist."

—Victor Weisskopf[2]

What about God?

The deity has been implicated in cosmology since the dawn of human history. Every monotheistic religion credits God with having created the universe. Plato, Aristotle, and scores of other philosophers have declared God responsible for the natural order revealed in the regular motions of the planets and stars. Theologians assert that it is owing to God's grace that human reason can comprehend the laws of nature. God has even been invoked as a solution to the observership problem posed by the Copenhagen interpretation of quantum physics.

So it seems reasonable to ask what cosmology, now that it is a science, can tell us about God.

Sadly, but in all earnestness, I must report that the answer as I see it is: Nothing. Cosmology presents us neither the face of God, nor the handwriting of God, nor such thoughts as may occupy the mind of God. This does not mean that God does not exist, or that he did not create the universe, or universes. It means that cosmology offers no resolution to such questions.

Many thinkers, living and dead, would disagree. They throw the cosmological bones and divine signs that there is a God, or that there is not. Let us briefly critique their points of view.

Historically, three intellectual proofs of God's existence loom large. They are the *argument from design, the cosmological proof,* and the *ontological proof.*

The argument from design says, in effect, that God is in the details. When we examine the wonderful efficiency and appropriateness of things, it seems impossible that they got to be that way other than through the work of a divine intellect. As the English clergyman William Paley famously pictured it, if one were to come across a pocket watch in the woods, one would conclude, on examining its intricate structure, that it had been designed by an intelligence for a purpose. The design argument is deistic, meaning that it addresses God as a creator and not as an intervenor who works miracles in the universe he has created. (And this Afterword restricts itself to deism, as cosmology in its notorious impersonality has little to say about the existence of a personal God.) It has been treated with respect by philosophers and others who are repelled by the notion of a personal God who answers prayers by influencing the outcome of battles and football games, but believe that the marvelous architecture of nature requires a supernatural explanation.

The anthropic principle is the design argument in scientific costume. Its appeal is demonstrated in Sir Fred Hoyle's evaluation of his own research into the "resonance states" of carbon atoms. Carbon is the fourth most abundant cosmic element, after hydrogen, helium, and oxygen. It is also the basis of terrestrial life. (That's why the study of carbon compounds is known as *organic* chemistry.) Carbon atoms are made inside stars. To make one takes three helium nuclei. The trick is to get two helium nuclei to stick together until they are struck by a third. It turns out that this feat depends critically on the internal resonances of carbon and oxygen nuclei. Were the carbon resonance level only 4 percent lower, carbon atoms wouldn't form in the first place. Were the oxygen resonance level only half a percent higher, virtually all the carbon would be "scoured out," meaning that it would have combined with helium to form oxygen. No carbon, no us, so our existence depends in some sense on the fine-tuning of these two nuclear resonances. Hoyle says that his atheism—and atheism is, let's face it, a faith like any other—was shaken by this discovery. "If you wanted to produce carbon and oxygen in roughly equal quantities by stellar nucleosynthesis, these are just the two levels you have to fix, if your fixing would have to be just about where these levels are actually found to be," Hoyle told a Caltech gathering in 1981. "Is that another put-up, artificial job? . . . I am inclined to think so. A common sense interpretation of the facts suggests that a superintellect has monkeyed with physics, as well as with chemistry and biology, and that there are no blind forces worth speaking about in nature. The numbers one calculates from the facts seem to me so overwhelming as to put this conclusion almost beyond question."[3]

But despite its having been entertained by thinkers from Paley to Hoyle, the design argument suffers from at least two serious defects.

First, it was always woefully anthropocentric. Design implies a purpose, and God's purpose in designing the universe was assumed to be either to make us, or to make things nice for us, or both. The French science writer Bernard de Fontenelle lampooned this position in his 1686 book *A Plurality of Worlds.* "We are all naturally like that madman at Athens, who fancied that all the ships were his that came into the Port of Pyraeus," he wrote. "Nor is our folly less extravagant. We believe all things in nature designed for our use, and do but ask a philosopher, to what purpose there is that prodigious company of fixed stars, when a far less number would perform the service they do us. He answers coldly, they were made to please our sight."[4] The larger the universe looms, the sillier it becomes to maintain that it was all put together for us. To posit a human-centered purpose to the heavens smacks of a lamentable humorlessness about the human condition, as Bertrand Russell was quick to point out. "The believers in Cosmic Purpose make much of our supposed intelligence but their writings make one doubt it," Russell wrote. "If I were granted omnipotence, and millions of years to experiment in, I should not think Man much to boast of as the final result of all my efforts."[5]

More damaging was the historical fact that believers in the design argument habitually drew their evidence from the biological world, citing as evidence of God's handiwork the marvelous adaptations of rattlesnakes and bower birds. This proved to have been an unfortunate choice of fields, once Darwin demonstrated that biological systems evolve by chance and not design.[6] (Darwin himself, though respectful of religion and loath to enter into theological disputation, nevertheless admitted, "I can see no evidence of beneficent design, or indeed design of any kind, in the details."[7]) Driven from the biological arena, the design argument has since sought refuge in physics and cosmology. Some thinkers expect that it will fare better there. I doubt it. A unified theory that showed the constants of nature to have resulted from phase transitions or other chance events would erode the design argument. So would the admittedly more speculative hypothesis that there are many universes with many different sets of laws, in only some of which life may be expected to arise. Darwinism does not dispel the mystery of life. Rather, it equates the mystery of life with the mystery of existence, of being. But the fact that something seems mysterious does not mean that God did it.[8]

Flawed though it may be, the argument from design is more robust than the cosmological and ontological proofs.

The cosmological proof goes back to Aristotle, who held that the existence of motion requires an ultimate source of dynamics, an "unmoved mover"—that is, God. It claims that any hierarchy of existence requires some overarching state of existence, that of an extant God. Descartes, similarly, interpreted his moment-to-moment existence as depending on the existence of a being beyond himself. The cosmological proof has enjoyed a long reign, due in part to the sentiments of thinkers who regard the origin of the universe as a problem inaccessible to science. But it has also encountered serious objections. Why, for instance, must we think of existence as a slippery slope, such that divine intervention is constantly required to prevent things from sliding down into the despond of nonexistence? And is causation really so deep a precept of nature as to render God requisite? Another problem, much discussed in theological circles, turns on the question of whether God had free will when he created the universe. If so, he was free to make the universe in a random, haphazard way. But if the universe is random, what need have we to postulate the existence of God? And if it is not—if, say, God could have made the universe only the most reasonable way, or in a way that promoted human existence —then God cannot be all-powerful. As the philosopher Keith Ward puts it, "The old dilemma—either God's acts are necessary and therefore not free (could not be otherwise), or they are free and therefore arbitrary (nothing determines what they shall be)—has been sufficient to impale the vast majority of Christian philosophers down the ages."[9]

The ontological proof (*ontology* is the study of the nature of being) dates from the eleventh century, when Saint Anselm, the archbishop of Canterbury, made the following argument: We conceive of God as "something than which nothing more perfect can be conceived." From the fact that we have this concept, it follows logically that such a being must exist. Why? Because if he did not, we would be able to conceive of something still more perfect—namely, a perfect being that *does* exist—and it is an absurdity to conceive of something more perfect than the most perfect conceivable being. Just as it is better to have ten real dollars than ten imaginary dollars, it is more perfect to be perfect *and exist* than to be perfect but nonexistent. So the concept of a most perfect being requires that such a being exist. The ontological proof is rather more subtle and persuasive than it looks at first blush, but so logically slippery that it aroused indignation even in the Middle Ages. (Gaunilo of Marmoutier inveighed against it while Anselm was still alive.) Its most telling refutation came from Immanuel Kant.

In *The Critique of Pure Reason,* Kant effectively demolished both the cosmological and the ontological proofs. The ontological argument, Kant pointed out, conflates two quite distinct realms of thought—that of pure reason (e.g., mathematics), in which premises internally dictate conclusions, and that of things, in which we reach judgments based on experience.[10] As Kant writes, "Having formed an *a priori* conception of a thing, the content of which was made to embrace existence, we believed ourselves safe in concluding that, because existence belongs necessarily to the object of the conception (that is, under the condition of my positing this thing as given), the existence of the thing is also posited necessarily, and that it is therefore absolutely necessary—merely because its existence has been cogitated in the conception."[11] In other words, having postulated that things exist, the purveyors of the ontological proof argue that existence is an attribute of things. But this is circular reasoning. And it is false, further, to think of "existence" as a property of things on a par with, say, their inertia or electrical charge. I might reasonably announce that I have ten dollars in my pocket, but not that I have in my pocket five existing and five nonexisting dollars. And, as Kant noted, the cosmological proof recapitulates the same error. It pastes the tag of "existing" on things, then asserts that the existence of any being requires the existence of an ultimate being. Since Kant, the ontological and cosmological proofs have continued to sail the philosophical seas, but they are ghost ships, and we cannot expect their tattered sails to carry us very far.

There remains one further argument, in which the participatory anthropic principle is employed to establish God's existence by way of the riddle of quantum observership. As we've seen, the Copenhagen interpretation of quantum mechanics treats as real only observed phenomena, raising the riddle of how the early universe could have evolved in the absence of observers. The riddle may be "solved" by invoking God as the supreme observer, who by scrutinizing all particles converts their quantum potentials into actual states. The same thesis has long been used by believers to resolve one of the oldest (and most tiresome) ontological questions —the one about whether trees exist when nobody observes them, or make a sound when they fall and there's nobody around to hear it. This position is summarized in a hoary bit of doggerel:

There once was a man who said, "God
Must think it exceedingly odd
If he finds that this tree
Continues to be
When there is no one about in the quad."

"Dear sir, your astonishment's odd
I am always about in the quad
And that's why the tree
Will continue to be
Since observed by,
Yours faithfully, God."[12]

But to go through such gyrations just to salvage the Copenhagen interpretation is to make a very small tail wag a very big dog, or God. And that's the trouble, really, with all cosmological invocations of the deity. Belief in God explains everything about the material universe; therefore it explains nothing. George Bernard Shaw made this point nicely in a toast to Einstein at a black-tie banquet in 1930. "Religion is always right," Shaw said. "Religion solves every problem and thereby abolishes problems from the universe. . . . Science is the very opposite. Science is always wrong. It never solves a problem without raising ten more problems."[13]

Atheists, meanwhile, draw sustenance from cosmological findings indicating that the universe emerged from chaos. Evidence in their support is mounting. The Harrison-Zeldovich spectrum of density fluctuations in the cosmic microwave background, confirmed within observational limits by the COBE satellite, suggests that random flux originated the stars and galaxies. Andrei Linde's chaotic inflationary theories presuppose a random distribution of primordial scalar fields. And, as we've been saying, Darwinian evolution depends on chance genetic mutations. If the world emerged from chaos and works by chance, what role can there be for an omniscient creator?

Because atheists have in many times and places been accused of intellectual arrogance it seems appropriate to examine this charge, if only to dismiss it. The accusation of arrogance stems from the claim that atheists pretend to know everything, since one would have to "walk the whole expanse of infinity," in Thomas Chalmers's words, to prove that God exists nowhere in the universe.[14] This position is simply insupportable. A scientist need not examine every proton in the universe in order to establish that protons originated in the bonding of quarks in the big bang. Nor is theism stoutly served by maintaining that even if there is no evidence of God's existence here, there might be on some other planet. On a logical level, a believer who argues that atheism is overweening, in that atheists cannot disprove every possible definition of God, is putting too much water in his wine: Whatever can it mean to say that "God" exists if you are unprepared to defend some particular conception of God? As the nineteenth-century English essayist Charles Bradlaugh wrote, in his "Plea for Atheism," an atheist is certainly justified in saying, "The Bible God I deny; the

Christian God I disbelieve in; but I am not rash enough to say there is no God as long as you tell me you are unprepared to define God to me."[15]

Ad hominem slurs aside, however, when the cosmological arguments propounded by atheists are subjected to reasoned criticism, they fare no better than the comparable arguments of believers. To find evidence of randomness in nature does not prove that there is no God. This is evident from two considerations.

First, it is impossible to prove conclusively that what appears to be random really *is* random. The sequence of numbers 4159265358 . . . looks pretty random, until we notice that it is the second through the eleventh decimal places of pi. (From another perspective—that, say, of a police detective—it might be significant that it's also a San Francisco telephone number.) It's actually quite difficult to obtain numbers that will pass for random, as cryptographers at the CIA understand: They do things like bounce radar signals off the ionosphere and use digitized strings of the radar echoes for encoding, and they're *still* not certain that the results are truly random. Seeming chaos can conceal design.

And even if the universe did arise from chaos, a believer could reasonably argue that God elected chaos as best suited for that purpose. What better way, for instance, to create the infinite and diverse worlds envisioned in the many-universe models? There is a "joke" about this issue, in which an atheist, asked by a believer where the universe came from, says, "It came from chaos," to which the believer responds, "Ah, but who made the chaos?" This isn't really a joke at all, but rather a succinct statement of two ways of thinking about the problem of creation, one of which is satisfied with chaos as an ultimate explanation and the other which sees chaos as just another system. From these and other considerations we may conclude that atheism is no more soundly footed in cosmological science than is theism.

So we are left with—what? In my view, a situation in which we would clearly be better off if we left God out of cosmology altogether. The origin of the universe and of the constants of nature is a mystery, and may forever remain so. But to assign to God the job of doing everything we don't (yet) understand is to abuse the concept of God. Such thinking posits, in the lingo, a "god of the gaps"—a deity who is hard at work making nature do those things that we fail to understand. I am unaware that God ever proffered such a job description for himself. Anyway, it seems unworthy of him: Feeble indeed is a machine that requires the constant intervention of its designer to keep it running. Nor is it clear why, if God's dominion consists of tending to the unexplained aspects of the phenomenal world, he would have conferred upon humans their aptitude for science, which in constantly closing the gaps has steadily diminished the putative realm of the god of the gaps. Nor does it seem satisfactory for God to have created the universe with a specific purpose in mind, the realization of which required billions of years to attain. As the German philosopher Friedrich Schelling asked, "Has creation a final purpose at all, and if so why is it not attained immediately, why does perfection not exist from the very beginning?"[16] Equally paltry is the notion that God set in motion a deterministic universe, one that runs perfectly but produces only phenomena that he knew in advance would occur. That's just too dull.

More appropriate, I should think, is the view that God created the universe out of an interest in spontaneous creativity—that he wanted nature to produce surprises, phenomena that he himself could not have foreseen. What would such a *creative* universe be like? Well, it would for one thing be impossible to predict in detail. And this seems to be the case with the universe we inhabit. The information theorists find that even if the entire universe were a computer, or could be converted into a computer of the maximum theoretically possible capacity, that computer would be incapable of predicting all future phenomena. Further, a creative universe should give rise to agencies that are themselves creative, which is to say unpredictable. There is in our universe such an agency, spectacularly successful at reversing the dreary slide of entropy and making surprising things happen. We call it life. It would be suitable if this agency were to inquire into the workings of the universe, winnowing out the predictable from the unpredictable and inventing theories to account for the difference. And that is what intelligence does. Better still if thinking creatures were to perceive that they are all in the same boat—"Poor, benighted members of the same ship's company," in Adlai Stevenson's phrase—and hence treat one another kindly and assert that God is love.[17] And so we do, though not often enough.

Finally, in a creative universe God would betray no trace of his presence, since to do so would be to rob the creative forces of their independence, to turn them from the active pursuit of answers to mere supplication of God. And so it is: God's language is silence. The Old Testament suggests that God fell silent in response to the request of the terrified believers who said to Moses, "Speak thou with us, and we will hear: but let not God speak with us, lest we die." Whatever the reason, God ceases speaking with the book of Job, and soon stops intervening in human affairs generally, leading Gideon to ask, "If the Lord be with us, why then . . . where be all his miracles which our fathers told us of?"[18] The author of the Twenty-second Psalm cries ruefully, "My God, my God, why hast thou forsaken me?"

Whether he left or was ever here I do not know, and don't believe we ever shall know. But one can learn to live with ambiguity —that much is requisite to the seeking spirit—and with the silence of the stars. All who genuinely seek to learn, whether atheist or believer, scientist or mystic, are united in having not *a* faith but faith itself. Its token is reverence, its habit to respect the eloquence of silence. For God's hand may be a human hand, if you reach out in loving kindness, and God's voice your voice, if you but speak the truth.

Notes

Preface

1. Plato, *Timaeus,* 27c, Benjamin Jowett, translator, in Edith Hamilton and Huntington Cairns, editors, *The Collected Dialogues of Plato.* Princeton: Princeton University Press, 1969, p. 1161.

2. Pindar, *Greek Lyrics,* Richmond Lattimore, translator. Chicago: University of Chicago Press, 1975, p. 63.

3. "This is not crazy enough!" was Bohr's "constant refrain," says Victor Weisskopf. (In Laurie M. Brown and Lillian Hoddeson, editors, *The Birth of Particle Physics.* Cambridge: Cambridge University Press, 1983, p. 265.) Abraham Pais recalls that Wolfgang Pauli gave a speculative talk on elementary particle physics at Columbia University on January 31, 1958, after which "Pauli turned to Bohr and said, 'You probably think these ideas are crazy.' Bohr replied, 'I do, but unfortunately they are not crazy enough.' " (Pais, *Niels Bohr's Times in Physics, Philosophy, and Polity.* Oxford: Oxford University Press, 1991, p. 30.)

4. Steven Weinberg, *Dreams of a Final Theory.* New York: Random House, 1992, p. 129. The italics are his.

5. Scientists *are* banished, of course, and sometimes for the wrong reasons. We shall see how this happened, for instance, to David Bohm, who was persecuted by politicians for his Marxist views, and how Hugh Everett, like Bohm the author of a new interpretation of quantum mechanics, exiled himself from physics when his work was ignored. But most often they are disregarded if they doggedly pursue research widely viewed as fruitless. Occasionally they turn out to be right and the wider community to be wrong: The textbook case is that of Alfred Wegener, whose (deeply flawed) account of continental drift was hooted down at a 1928 meeting of the American Association of Petroleum Geologists and not revived (albeit with a much better physical explanation) until the 1950s. Seldom, however, are scientists ostracized for simply making a mistake or challenging an accepted theory.

6. Martin Rees, Second Hitchcock Lecture, University of California, Berkeley, February 28, 1995.

7. John Locke, *Concerning Civil Government, Second Essay,* Chapter IV; first published in 1690.

8. So powerful was the influence of Locke's putatively scientific reasoning that Jefferson, who seldom bothered defending himself against criticism, felt it necessary to respond to a charge by one Richard H. Lee that he had copied the Declaration of Independence from Locke's *Concerning Civil Government.* Jefferson, who admittedly regarded Locke's essay as "perfect as far as it goes," wrote: "I know only that I turned to neither book nor pamphlet while writing" the declaration. (In Carl Becker, *The Declaration of Independence: A Study in the History of Political Ideas.* New York: Harcourt, Brace, 1922, p. 25.)

9. *The San Francisco Chronicle,* May 25, 1992, p. 35.

10. Stephen Hawking, *A Brief History of Time.* New York: Bantam, 1988, p. 175.

1: The Shores of Light

1. Quoted by Lance Morrow in *Time,* June 10, 1991, p. 48.

2. Kepler, Letter to Fabricius, quoted in Max Caspar, *Kepler.* New York: Dover, 1993, p. 170.

3. To be explicit, the third law reveals that if P is the time it takes any planet to orbit the sun and R is the semimajor axis (i.e., half the long axis) of the orbit, then:

$$\frac{P\,(\text{Any planet})^2}{P\,(\text{Earth})^2} = \frac{R\,(\text{That planet})^3}{R\,(\text{Earth})^3}$$

Jupiter, for example, orbits the sun once every 11.86 earth-years. Knowing this, we can derive the size of Jupiter's orbit, which for convenience we will express in multiples of the distance from Earth to the sun, which astronomers call one ***astronomical unit:***

$$\frac{11.86^2}{1^2} = \frac{R\,(\text{Jupiter})^3}{1^3}$$

$11.86^2 = 11.86 \times 11.86 = 141$, so:

$$\frac{141}{1} = \frac{R\,(\text{Jupiter})^3}{1}$$

Therefore the semimajor axis of Jupiter's orbit is the cube root of 141, or 5.2 astronomical units. One astronomical unit = 93 million miles, so we conclude that the distance from Jupiter to the sun, on average, is 5.2×93 million miles = 483.6 million miles.

4. Galileo, *Dialogues Concerning Two New Sciences,* Henry Crew and Alfonso deSalvio, translators. New York: Dover, 1954, p. 1.

5. Einstein was himself as instinctively antiauthoritarian as any sci-

entist who ever lived. When he was a small boy, he burst into tears at his first sight of a military brass band, and as an adult joked that "to punish me for my contempt for authority, Fate made me an authority myself." (In Banesh Hoffman, *Albert Einstein: Creator and Rebel.* New York: Viking, 1973, p. 24.)

6. Einstein, "Johannes Kepler," in his *Out of My Later Years.* Secaucus, N.J.: Citadel Press, 1979, p. 226.

7. Einstein, Herbert Spencer lecture, Oxford, June 10, 1933; in *Mein Weltbild.* Amsterdam: Querido Verlag, 1934.

8. Newton's achievement with regard to gravitation can be presented most concisely by summarizing his restatement of Kepler's three laws. The Newtonian version reads roughly this way:

The First Law: If two bodies interact gravitationally, each will describe an orbit that is a conic section about the common center of mass of the pair. If the bodies are permanently associated, their orbits will be ellipses. If not, their orbits will be hyperbolas.

The Second Law: If two bodies revolve about each other under the influence of a central force, a line joining them will sweep out equal areas in the orbital plane in equal intervals of time.

The Third Law: If two bodies revolve mutually about each other, the sum of their masses times the square of their period of mutual revolution is in proportion to the cube of the semimajor axis of the relative orbit of one about the other. Stated in standard units, the third law states that if R is the radius of an orbit in astronomical units, P is the period of the orbit in years, and m_1 and m_2 are the masses of the orbiting objects, expressed as a fraction of the mass of the sun, then

$$m_1 + m_2 = \frac{R^3}{P^2}$$

Armed with these Newtonian laws, astronomers knowing only the dimensions and periods of given orbits can derive the masses of planets orbited by their satellites, of stars orbited by other stars in binary star systems, of galaxies (from the orbits of their outermost stars), and of entire clusters of galaxies (from the motions of outer galaxies in the clusters).

9. Interestingly, general relativity breaks down if called upon to predict events in the extremely high densities that would have characterized the universe during the first fraction of a second of time—to be precise, during the first 10^{-43} second, when the newborn universe was less than 0.0000000000000000000000000000000000000000001 second old. The theory thus "predicts its own downfall," as one relativist puts it. One way of interpreting this aspect of the theory, a path explored by the British theorists Stephen Hawking and Roger Penrose, is to say that in an expanding universe all roads through cosmic history lead back to a point

of infinite gravitation—a *singularity,* a point at or near the beginning of time. So the very breakdown of relativity may be taken as evidence that the big bang actually occurred.

10. Leibniz paraphrased by J. A. Wheeler in Paul Buckley and F. David Peat, *A Question of Physics.* Buffalo: University of Toronto Press, 1979.

2: The Expansion of the Universe

1. Plato, *Timaeus,* Benjamin Jowett, translator, in Edith Hamilton and Huntington Cairns, editors, *The Collected Dialogues of Plato.* Princeton: Princeton University Press, 1969.

2. Robert P. Kirshner, "Exploding Stars and the Expanding Universe," *Quarterly Journal of the Royal Astronomical Society* 32, 233–244 (1991).

3. This is not to say that prophets don't make accurate prophecies. Michael Tanner reminds us that Nietzsche once wrote, "I have a terrible fear that one day someone will call me holy." Tanner adds, "Sure enough, at his funeral his closest friend, Peter Gast, did exactly that." (Review of Alexander Nehamas's *Nietzsche: Life as Literature, Times Literary Supplement,* May 16, 1986, p. 519.)

4. The history of this subject is complex. For a fuller discussion see Ferris, *The Red Limit* and *Coming of Age in the Milky Way;* Edward R. Harrison, *Cosmology: The Science of the Universe;* and B. Bertotti et al., editors, *Modern Cosmology in Retrospect.*

5. The Dutch astronomer Willem de Sitter in 1917 proposed a cosmology that incorporates redshifts and expansion. But his model, while valuable in many ways, contains no fewer than three different explanations of redshift, none of which invokes the expanding universe as it is now conceived. Nevertheless, because de Sitter's work was much better known among astronomers than was Friedmann's, some attributed the observed redshifts to de Sitter effects. Hubble himself was initially of this opinion.

6. Einstein, note deleted prior to publication of a 1923 paper, in B. Bertotti et al., editors, *Modern Cosmology in Retrospect.* Cambridge, U.K.: Cambridge University Press, 1990, p. 102.

7. Conversely, the spectra of approaching objects display a Doppler blueshift. For instance, the Andromeda galaxy, gravitationally bound to the Milky Way, is approaching us and consequently displays a small blueshift.

8. *Time,* February 9, 1948.

9. Hubble, letter to de Sitter, quoted in Norris Hetherington, "Hubble's Cosmology," *American Scientist* 78, no. 2, 149 (1990).

10. See, e.g., the appendix to the fifth edition of Weyl's *Raum-Zeit-Materie.* Berlin: Springer, 1923. Also Weyl, "Zur allgemeinen Relati-

vitätstheorie," *Physikalische Zeitschrift* 24, 230–232 (1923). For a discussion, see G. F. R. Ellis, "The Expanding Universe: A History of Cosmology from 1917 to 1960," in Don Howard and John Stachel, editors, *Einstein and the History of General Relativity*. Boston: Birkhäuser, 1989, pp. 367–431.

11. It is interesting to note that the effect of deceleration is to slowly decrease the value of the Hubble constant—which, therefore, is not, technically speaking, a constant but a variable. As the cosmologist Alan Guth writes, H_0 "was called a 'constant' by the astronomers, presumably because it remains approximately constant over the lifetime of an astronomer. The value of H_0 changes as the universe evolves, however, so from the point of view of a cosmologist it is not a constant at all." (Alan Guth, "The Birth of the Cosmos," to be published in Donald Osterbrock and Peter Raven, editors, *Origins and Extinctions*. New Haven: Yale University Press.)

12. The formula for omega is

$$\Omega = 2q_0 + 2/3\lambda \frac{c^2}{H_0^2}$$

in which q_0 is the deceleration parameter, λ is the cosmological constant, H_0 is the Hubble constant, and c is the velocity of light.

13. Imagine that you have been blowing up a big balloon for one hour, and that an observer—representing the human species—comes on the scene during the last minute of the hour. The observer measures your progress and finds, first, that during that one minute you increased the radius of the balloon by one centimeter and, second, that the radius of the balloon at the end of that minute is 100 centimeters. Naively, the observer deduces that you started inflating the balloon 100 minutes ago. But this would be an overestimate, because you are not blowing up the balloon as fast at the end of the hour as you were at the beginning. The observer needs to take into account that you may have been getting tired, that given a fixed lung volume it takes more breath to inflate a large balloon by the same fraction as a small balloon, and so on. Alternatively, you might be getting the hang of it and be blowing up the balloon more robustly now than in the past. This would be comparable to an expanding universe in which the cosmological constant was actually accelerating the expansion rate. The observer has two basic ways of improving his estimate. One way is to construct a theory in which he estimates how much the balloon's inflation would necessarily slow down owing to your fatigue, and other factors. The other would be to look for direct evidence, such as a videotape or set of snapshots, showing how large the balloon was at various times during the past hour. The first approach approximates what cosmologists do when they take their best estimates of the cosmic matter

density, apply the resulting braking force, determine how much faster the expansion rate was at past epochs, and arrive at an estimate of the age of the universe. The other approach is to measure the distance and recession velocity of galaxies at many different distances, including some billions of light-years away. The nearby ones tell us how fast the universe is expanding today, while the remote ones tell us how much faster it was expanding billions of years ago.

14. P. J. E. Peebles, address at the University of California, Berkeley, March 9, 1995.

15. This is known as the "Dicke-Peebles timing argument," after the Princeton cosmologists Robert Dicke and P. J. E. Peebles. See Michael Turner, "The Hot Big Bang and Beyond," in *Proceedings of CAM-94* (Cancun, Mexico, September 1994) and Menas Kafatos and Yoji Kondo, editors, *Examining the Big Bang and Diffuse Background Radiation*. Boston: Kluwer Academic, 1996, p. 301. For more, see also chapter 5 of this book.

16. For a fuller discussion, see Michael Rowan-Robinson, *The Cosmological Distance Ladder: Distance and Time in the Universe*. New York: Freeman, 1985.

17. As always, there are complications, starting with the fact that even the nearest Cepheids are too far from Earth for their distances to be measured by parallax. So this whole line of attack is based on indirect methods, resulting in some residual uncertainty as to the intrinsic brightness of Cepheids. But Cepheids have played a major role in charting cosmological distances ever since Henrietta Swan Leavitt discovered the Cepheid period-magnitude relation that Shapley and Hubble thereafter exploited.

18. It is perhaps unnecessary to point out that there are many exceptions to this generalization. Among the radicals could be found scientists of Sandage's generation, notably Halton Arp, Geoffrey Burbidge, and Sir Fred Hoyle, and there are many young astronomers who favor the old universe. For that matter, generations themselves are an artificial construct. But the generality is at least coarsely valid.

19. There was plenty of evidence that Sandage knew how to use a telescope. When, for instance, astronomers working with the giant telescope at the Cerro Tololo Observatory in Chile used modern digital technology to measure the apparent magnitudes of key stars, they found that Sandage's magnitude estimates, made over previous decades from photographic plates, an inherently less precise technique, came within one three-hundredths of a magnitude of theirs, a stunning affirmation of his skill at extracting accurate data from the sky.

20. See, e.g., a talk by Sandage in which he cites nine methods for deriving the distance to the Virgo cluster. J. Bagger et al., editors, *Particles, Strings and Cosmology: Proceedings of the Johns Hopkins Workshop on*

Current Problems in Particle Theory 19 and the PASCOS Interdisciplinary Symposium 5, Baltimore, 1995, March 22–25. River Edge, N.J.: World Scientific, 1996.

21. H. Jergen and G. A. Tammann, "The Local Group Motion Toward Virgo and the Microwave Background," *Astronomy and Astrophysics* 276, 1–8 (1993).

22. To put this in terms of general relativity, superclusters like Virgo occupy hollowed-out depressions in the curvature of spacetime. These depressions are enlarged somewhat by cosmic expansion, but were we to measure the expansion rate solely within the local dale we should conclude that the universe is expanding more slowly than is really the case. Some astronomers like to think of the universe in terms of "tiles"—regimes in which expansion is retarded (as with the Virgo Supercluster) or even nonexistent (as with the Local Group). They regard cosmic expansion as moving the tiles apart. Large tiles, like Virgo and other superclusters of galaxies, grow in size as they move apart in the general expansion, while smaller tiles, like the Local Group, don't grow at all. Another way to say this is that the value of the Hubble constant gets larger as one goes up in scale. That's the essence of the problem: To measure pure Hubble flow may require sampling a vast piece of the universe, one with a radius of several hundred million light-years.

23. Wendy L. Freedman et al., "Distance to the Virgo Cluster Galaxy M100 From Hubble Space Telescope Observations of Cepheids," *Nature* 371, 761 (1994).

24. Wendy Freedman, conversation with T.F., April 3, 1995.

25. Allan Sandage, in Malcolm W. Browne, "Age of Universe Is Now Settled, Astronomer Says," *The New York Times*, national edition, March 5, 1996, p. B7.

26. Sandage's recent estimates for H_0 have remained around 50 to 55 while the Young Turks' have been coming down—Freedman et al. by 1996 were reporting 70—so a purely sociological analysis would favor Sandage. But, then, the universe is not known to be a sociological system.

27. Tired light was first suggested by the Swiss-American astronomer Fritz Zwicky. See *Proceedings National Academy of Sciences* (U.S.) 15, 773 (1929).

28. For a recent defense of the steady state (or "C-field") theory by the able astrophysicist who has championed it for forty years, see Sir Fred Hoyle's autobiography, *Home Is Where the Wind Blows: Chapters from a Cosmologist's Life*. Mill Valley, Cal.: University Science Books, 1994, pp. 401ff.

29. For an explication of the plasma theory, see Eric J. Lerner, *The Big Bang Never Happened: A Startling Refutation of the Dominant Theory of the Origin of the Universe*. New York: Times Books, 1991.

3: *The Shape of Space*

1. Willem de Kooning, *Collected Writings.* Madras, India: Hanuman Books, 1988, p. 121.

2. John Archibald Wheeler, *A Journey into Gravity and Spacetime.* New York: Scientific American Library, 1990, p. 39.

3. While walking with Heisenberg, the physicist Felix Bloch, who had just read Weyl's *Space, Time and Matter,* felt moved to declare that space is simply the field of linear equations. Heisenberg replied, "Nonsense. Space is blue and birds fly through it." "What he meant," Bloch writes, "was that it was dangerous for a physicist to describe Nature in terms of idealized abstractions too far removed from the evidence of actual observation. In fact, it was just by avoiding this danger in the previous description of atomic phenomena that he was able to arrive at his great creation of quantum mechanics." Felix Bloch, *Physics Today* 29 (12), 27 (1976).

4. Were the spherical universe static, an observer who looked far enough and waited billions of years could see the back of his head—the light beam having circled the universe to arrive at the front of his telescope. But relativity does not permit static universes, and the universe we live in is expanding. So we cannot see ourselves through outward-pointing telescopes, and never will.

5. Jorge Luis Borges, "Of Exactitude in Science," in his *A Universal History of Infamy,* Norman Thomas di Giovanni, translator. New York: Dutton, 1972; [falsely] attributed by Borges to *Travels of Praiseworthy Men,* 1658, by "J. A. Suarez Miranda."

6. In Abraham Pais, *"Subtle Is the Lord . . ." The Science and the Life of Albert Einstein.* New York: Oxford University Press, 1982, p. 212.

7. Strictly speaking, I should use the verb "model," not "map," as the latter term properly is employed only when the surface to be mapped is already known. But "map" seems clearer when the point is to compare terrestrial to cosmological techniques of minimizing spatial distortion.

8. Those of us who are frustrated in the attempt to visualize curved space may take comfort in the fact that we have a lot of distinguished company. Bertrand Russell regarded this feat as "impossible." John Archibald Wheeler, the dean of American relativists, admits that he can do so only by suppressing one of the four dimensions. Immanuel Kant elevated Euclidean space to the status of *a priori* knowledge, meaning that it could be deduced from logical precepts without recourse to experience. This in turn led philosophers at Oxford University to reject relativity. The Oxford astronomer E. A. Milne went to the trouble of constructing a Euclidean cosmology to contradict relativity. But shocks like these are to be expected. It would be suspiciously comfortable if science, probing ten orders of magnitude out into the realm of the galaxies and down into

the diminutive realm of the atoms, did *not* learn things that upset the assumptions garnered in several million years of human evolution limited to the human scale of space and time.

9. The theory predicts that light coming from massive objects will lose energy as it climbs out of the local gravitational well, producing a "gravitational redshift." This effect has been observed in studies of white dwarf stars, which are compact objects with powerful gravitational fields, and by bouncing radar signals off Mercury, Venus, and Mars. Time delays in radio signals coming from the *Voyager* space probe as it passed the giant planet Saturn confirmed the relativistic bending of space by that planet's mass.

10. In Robert Osserman, *Poetry of the Universe*. New York: Anchor, 1995, p. 91.

11. As Howard L. Resnikoff puts it, in his *The Fusion of Reality* (New York: Springer-Verlag, 1989, p. 96): "Fermat's classical variational principle of 'least time' and Maupertuis and Hamilton's principle of 'least action' express [the] parsimony of nature in a mathematical form: the evolution of a physical system follows that path amongst all conceivable alternatives that extremizes, i.e., *maximizes* or *minimizes*, a suitable *cost function*, such as *time, action*, or *energy*. Thus, the path of a ray of light through an optically inhomogeneous medium *minimizes* the *time* required to pass from the initial position to its emergent point."

Relevant to our consideration of entropy later in this chapter is Resnikoff's association of entropy with the principle of least action: "The second law of thermodynamics is a dynamic version of this general physical principle. It can be interpreted as asserting that an isolated physical system will evolve toward an equilibrium configuration for which the entropy of the system is as large as possible. Once the system has reached a state of maximum entropy, it will remain in it, or in some state of equal entropy, thereafter. From the informational point of view, the system evolves to a state about which least is known *a priori*, and therefore about which a measurement will yield as much information as possible."

12. Michio Kaku, *Hyperspace*. New York: Anchor, 1994, p. 35.

13. Kip S. Thorne, *Black Holes and Time Warps*. New York: Norton, 1994, p. 477.

14. John Archibald Wheeler, *A Journey into Gravity and Spacetime*. New York: Scientific American Library, 1990, p. 210.

15. Richard H. Price and Kip S. Thorne, "The Membrane Paradigm for Black Holes," *Scientific American*, April 1988, p. 72.

16. As so often happens in astrophysics and other vital sciences the name *quasar* is a misnomer. It means "quasi-stellar objects." Blue-white pinpoints of light, quasars were at first mistaken for abnormal stars.

17. Mario Livio, PASCOS symposium, Johns Hopkins University, Baltimore, March 25, 1995.

18. Kip S. Thorne, *Black Holes and Time Warps*. New York: Norton, 1994, p. 524.

19. Charles W. Misner, Kip S. Thorne, and John Archibald Wheeler, *Gravitation*. San Francisco: Freeman, 1973, p. 863.

20. The question is sometimes asked why, if the law of increasing entropy is valid, there can be such a thing as life and learning here on Earth. The answer is that we use energy from the sun, from nuclear reactors, and from heat deep in the earth ("geothermal" energy). All these sources are nuclear. They are due, respectively, to nuclear fusion, nuclear fission, and nuclear beta decay (radioactivity is what keeps the earth hot at its core). Since atomic nuclei were assembled in the big bang, all energy sources can be traced back to the origin of the universe. The question of why there is any available energy in the first place is thus a subset of the question of how or why the universe came into being. Thanks to these energy sources, entropy can decrease locally even while it increases on the cosmic scale. One might go so far as to say that the excitement generated by life, art, science, and the spectacle of a bustling city with its libraries and theaters is at its root the excitement of seeing the law of entropy being defeated—in one place at least, for a while.

21: John Archibald Wheeler, *A Journey into Gravity and Spacetime*. New York: Scientific American Library, 1990, p. 221.

22. Stephen Hawking, interviewed by T.F., University of California, Santa Barbara, April 9, 1983.

23. Stephen Hawking, conversation with T.F., California Institute of Technology, March 28, 1983.

24. Kip S. Thorne, *Black Holes and Time Warps*. New York: Norton, 1994, p. 423.

25. Stephen Hawking, "Black Hole Explosions?" *Nature* 248, March 1, 1974, p. 30.

26. Stephen Hawking, conversation with T.F., California Institute of Technology, March 28, 1983.

27. Richard Feynman, talk given at the University of Southern California, December 6, 1983.

28. Stephen Hawking, interviewed by T.F., Santa Barbara, California, April 9, 1983.

29. Hawking continues to hold the position that information dumped into black holes is lost forever. "This loss of information . . . introduce[s] a new level of uncertainty into physics over and above the usual uncertainty associated with quantum theory," he argues. (Stephen W. Hawking and Roger Penrose, "The Nature of Space and Time," *Scientific American*, July 1996, p. 62.)

30. In this book, I use the terms *time machine* and *time travel* to refer solely to devices capable of transporting one into the past. Venturing into the future is unavoidable: You and I travel one second farther into

the future every second. Nor does travel into the future raise any paradoxes. Time travel into the past, however, is a very different and more problematical matter.

31. Hawking, in Michio Kaku, *Hyperspace*. New York: Anchor, 1994, p. 235. A young physicist, Neal Katz, proposed an elegant way of testing whether our descendants will ever travel back to the twentieth century. It was to publish a paper in an archival journal such as *Physical Review Letters*—so that, in the words of the paper, "copies of this article must be preserved for all time"—containing the request, written in boldface capital letters, "IF TIME TRAVEL IS POSSIBLE AT ANY FUTURE DATE PLEASE CONTACT THE AUTHORS ON [PUBLICATION DATE], 1992." These words were followed by the authors' campus addresses. "Simplicity itself," comments the cosmologist Eric Linder. Alas, the paper was neither submitted nor published.

32. Richard Gott, interviewed by T.F., Irvine, California, March 27, 1992.

4: Blast from the Past

1. William Blake "Auguries of Innocence," in Geoffrey Keynes, editor, *Poetry and Prose of William Blake*. London: Nonesuch, 1927, p. 118.

2. Werner Heisenberg, "The Nature of Elementary Particles," *Physics Today* 29, 32 (1976).

3. *The New York Times,* January 12, 1933.

4. *Los Angeles Times,* January 12, 1933.

5. In Murray Gell-Mann, "From Renormalizability to Calculability," Second Shelter Island Conference, May 1983, ms. p. 25.

6. George Gamow, *Nature,* October 30, 1948, p. 680.

7. Ibid.

8. Ralph A. Alpher and Robert C. Herman, "Evolution of the Universe," *Nature* 162, 774-775 (1948).

9. The COBE temperature measurement of the CMB is accurate to within 0.004 degrees Kelvin. This means that, interestingly enough, one of the most accurately determined quantities in science is the intensity of a wall of fire 10 billion light-years away.

10. The term "big bang" was coined with derisive intent by Fred Hoyle, and its endurance testifies to Sir Fred's creativity and wit. Indeed, the term survived an international competition in which three judges—the television science reporter Hugh Downs, the astronomer Carl Sagan, and myself—sifted through 13,099 entries from 41 countries and concluded that none was apt enough to replace it. No winner was declared, and like it or not, we are stuck with "big bang." (See Ferris, "Needed: A Better Name for the Big Bang," *Sky & Telescope,* August 1993.)

11. Fred Hoyle, *Home Is Where the Wind Blows: Chapters from a Cosmologist's Life*. Mill Valley, Cal.: University Science Books, 1994, p. 417.

12. E. Margaret Burbidge, Geoffrey R. Burbidge, William A. Fowler, and Fred Hoyle, "Synthesis of the Elements in Stars," *Reviews of Modern Physics* 29, 547–650 (1957). Similar work was done independently by Alastair Cameron.

13. Gary Steigman, "Big Bang Nucleosynthesis: Consistency or Crisis?" Ohio State University, preprint OSU-TA-22/94, 1994, p. 2.

14. It took 180 seconds for primordial helium to form, which is why Steven Weinberg titled his classic book on BBN *The First Three Minutes* (New York: Basic Books, 1977).

15. In John D. Barrow and Frank Tipler, *The Anthropic Cosmological Principle*. Oxford: Oxford University Press, 1986, p. 339.

16. Not all researchers agree that intergalactic clouds are necessarily pristine. According to one theory, the first stars formed very early in cosmic history, before the galaxies did. Such stars would have built heavy elements ("metals") and blown them into intergalactic space, contaminating the clouds. A rough-hewn but popular nomenclature classifies younger, metal-rich stars as Pop (for "population") I and older, metal-poor stars as Pop II. These hypothetical earliest of stars would be classed Pop III. Observational evidence for the existence of Pop III stars was published in the April 1995 issue of the *Astronomical Journal* by astronomers Len Cowie and Antoinette Songaila, of the University of Hawaii. Using the Keck telescope, they found traces—about one part in a million —of ionized carbon in intergalactic clouds about half the present age of the universe. Since carbon is made in stars, not in the big bang, this result suggests that stars did contaminate the intergalactic medium early in cosmic history. If, on the other hand, studies continue to show that distant intergalactic clouds contain the proportions of helium that the standard BBN theory attributes to the big bang, this would mean that contamination by Pop III stars was not sufficient to threaten the theory. Also it is difficult to reconcile Pop III models with COBE measurements of the cosmic microwave background radiation.

17. An atom emits a photon when one of its electrons drops down from one shell to an inner one. When an atom absorbs a photon, an electron jumps up a shell. An electron in the innermost shell is said to be in its *ground state*. Lyman alpha spectral lines are generated when an electron jumps between the second and the first (meaning the innermost) shell.

18. A. Songaila et al., *Nature* 368, 599 (1994). Similar results, obtained using the four-meter telescope at Kitt Peak National Observatory in Arizona, were reported by R. F. Carswell et al., *Monthly Notices Royal Astronomical Society*, 268 (1994).

19. In John Noble Wilford, "Primordial Helium, Created in Big Bang, Detected at Long Last," *The New York Times,* June 13, 1995, p. B5.

20. Gary Steigman, David Schramm, and James Gunn in 1977 constrained the number of neutrino families to no more than four (*Physics Letters: Part B,* 66, 202). By 1988–1990, studies by Steigman, Schramm, and others had reduced the number to fewer than four—meaning, in effect, the three known families.

21. Every baryon (indeed, every subatomic particle) has an *antimatter* partner, which has the same mass but opposite charge and also various opposite quantum numbers. When matter and antimatter particles meet, they annihilate. This makes antimatter handy in particle colliders: Experimental physicists manufacture antiprotons, send them through the collider ring in the opposite direction from the protons, then steer them together at detector sites along the ring where they collide, producing high-energy interactions that can replicate events that would have taken place in the big bang. But while antimatter may be found in storage rings at CERN and other such colliders, there is virtually no antimatter in nature. This can be demonstrated in many ways. Were, say, the planet Jupiter made of antimatter, its interaction with particles in the solar wind would produce powerful x-ray emissions, and these are not observed. Were there a significant amount of antimatter in the Milky Way galaxy, then many cosmic rays—subatomic particles that strike Earth after having traveled considerable distances through our galaxy—would be made of antimatter; at least 999 of every 1,000 cosmic rays are made of ordinary matter. Were other galaxies made of antimatter, we would see X rays and gamma rays coming from their interaction with intergalactic clouds, and these are not seen, either. So the universe is made of matter, not antimatter. This is known as the *baryon asymmetry* of the universe. In the standard big bang model it is explained by hypothesizing that the universe began with approximately equal amounts of matter and antimatter, most of which annihilated, but that owing to the chance nature of quantum mechanics, the amount of matter and antimatter could not have been exactly equal. So a residue of one kind of particle survived, while the other became extinct. We call the surviving particles matter, and search the cosmos in vain for any appreciable amount of antimatter. The initial inequity could have been quite small—only about one part in a billion—and still have produced this result.

5: The Black Taj

1. F. D. Reeve, "Coasting," in *The American Poetry Review,* July–August 1995, p. 38.

2. Robert Browning, "Epilogue," to *Asolando: Fancies and Facts.*

In *The Poetical Works of Robert Browning*. Boston: Houghton Mifflin, 1974, p. 1007.

3. The myth of the black Taj was reported as fact by the French travel writer Jean-Baptiste Tavernier, who claimed to have witnessed the construction of the Taj, and is routinely repeated to this day by guides at the site. But "Although Tavernier claims that a second mausoleum was begun, no other early sources mention it nor have any traces of such foundations been found among the ruins on the other side of the Yamuna." (Janice Leoshki, "Mausoleum for an Empress," in Pratapaditya Pal, et al., *Romance of the Taj Mahal*. London: Thames and Hudson, and Los Angeles, Los Angeles County Museum of Art, 1989, p. 77.)

4. David Weinberg, "The Dark Matter Rap: Cosmological History for the MTV Generation." Audiotape sent to T.F., October 1994.

5. In Ronald Florence, *The Perfect Machine: Building the Palomar Telescope*. New York: Harper, 1994, p. 151.

6. In Wallace and Karen Tucker, *The Dark Matter*. New York: Morrow, 1988, p. 83.

7. Vera Rubin, talk given at the National Academy of Sciences Colloquium on Physical Cosmology, Irvine, California, March 27–28, 1992.

8. Vera Rubin, "Dark Matter in Spiral Galaxies," *Scientific American*, June 1983, p. 96.

9. A. D. Chernin, A. V. Ivanov, A. V. Trofimov, and S. Mikkola, *Astronomy & Astrophysics* 281, 685–690 (1994).

10. Sandra Faber, talk given at the National Academy of Sciences Colloquium on Physical Cosmology, Irvine, California, March 27–28, 1992.

11. Critical density, assuming the standard big bang model employed in this book, amounts to only 10^{-29} gram of matter per cubic centimeter of space—much thinner than a high-quality laboratory vacuum. Even a critical density universe is mostly space.

12. Gerard Jungman, Marc Kamionkowski, and Kim Griest, "Supersymmetric Dark Matter." *Physics Reports* 267, March 1996, p. 206.

13. Radical feminists alert for examples of "phallocentrism" in the physical sciences will be exercised to learn that a computer program employed to convert data on large-scale bulk motions of galaxies into density maps is called POTENT.

14. Michael Turner, "Dark Matter: Theoretical Perspectives," *Proceedings National Academy of Sciences* (U.S.) 90, 4,828, June 1993.

15. In Dennis Overbye, "The Shadow Universe," *Astronomy*, May 1985, p. 24.

16. Gerard Jungman, Marc Kamionkowski, and Kim Griest, "Supersymmetric Dark Matter." *Physics Reports* 267, March 1996, p. 222.

17. Since electron neutrinos are routinely produced by radioactive

decay, and radioactive minerals are easy to come by, it might seem reasonable to look for neutrino mass by monitoring radioactive atoms and in effect weighing them before and after they emit a neutrino, thus estimating their mass loss. That was the approach employed by a Moscow team that studied the decay of tritium in valene molecules, startling the physics community with an announcement, in 1980, that they had found a rather hefty mass for the electron neutrino. But such experiments are actually very difficult to evaluate accurately, in part because the binding energies of the molecule approximate that suspected for the neutrino. Other laboratories have been unable to reproduce the Moscow result, and some have arrived at upper limits for electron neutrino mass below that claimed in Moscow.

18. A broad survey of cosmology, such as this book aspires to be, must necessarily skip over many of the details of experiments and theories alike. But perhaps a brief summary can suggest something of the sophistication of the Homestake mine neutrino detector. High-energy solar neutrino collisions occur in the tank at a rate of about one or two per day. They contaminate the perchloroethylene in the tank with argon-37, a noble gas with a half-life of 35 days. A helium purge is employed to remove the argon gas from the cleaning fluid. The gas is then put through a series of filters, culminating in a charcoal trap that is first cooled by liquid nitrogen, then heated and swept with helium. About 95 percent of the argon is recovered this way. (To check on the purging efficiency, a known quantity of a tracer gas, argon-36 or argon-38, is introduced into the tank at the beginning of each run.) The extracted gas is passed through a hot titanium filter to remove reactive gases. Other noble gases are then separated out through the use of a gas chromatograph. The purified argon is monitored in a radiation counter for ten half-lives, or about one year per sample. This is the standard operating procedure for an experimental apparatus routinely described, at the popularization level of resolution, as a tank of cleaning fluid.

19. Thomas J. Bowles, "Neutrino Mass," in Carl W. Akerlof and Mark A. Srednicki, editors, *Texas/PASCOS '92: Relativistic Astrophysics and Particle Cosmology*. New York: The New York Academy of Sciences, 1993, p. 80.

20. Many researchers have commented on the striking coincidence between the new physics envisioned by supersymmetry theory, which works from the small scale, and cosmology, which works on the large scale. For instance: "The remarkable fact is that for [a value of] Omega [of] approximately one, as required by the dark matter problem, the annihilation cross section . . . for any thermally created particle turns out to be just what would be predicted for particles with electroweak scale interactions," i.e., WIMPs. Kim Griest, "The Particle- and Astro-physics of Dark Matter," Plenary talk presented at Snowmass 94 (Particle and Nu-

clear Astrophysics and Cosmology in the Next Millennium, June 29–July 14, 1994, Snowmass, Colorado).

21. In Robert P. Crease, Jr., and Charles C. Mann, *The Second Creation*. New York: Macmillan, 1986, p. 417.

6: The Large-Scale Structure of the Universe

1. Dante, *Inferno,* final line, Allen Mandelbaum, translator. Berkeley: University of California Press, 1980, p. 300.

2. Dante, *Paradiso,* canto 33, Lawrence Grant White, translator. New York: Pantheon, 1948, p. 188.

3. Ya. B. Zeldovich, "The Structure of the Universe," *UNESCO Courier,* September 1984, p. 24.

4. Ya. B. Zeldovich and I. D. Novikov, *The Structure and Evolution of the Universe (Relativistic Astrophysics,* vol. 2). Chicago: University of Chicago Press, 1983, p. xvii.

5. In Marcia Bartusiak, "Mapping the Universe," *Discover,* August 1990, p. 62.

6. Harlow Shapley in the 1930s referred to "clouds" of galaxies, many of which today are classified as superclusters. For modern definitions see, e.g., in R. Brent Tully and J. Richard Fisher, *Nearby Galaxies Atlas.* New York: Cambridge University Press, 1987.

7. Alan Dressler, "Galaxies, Properties in Relation to Environment." In Stephen P. Maran, editor, *The Astronomy and Astrophysics Encyclopedia*. New York: Van Nostrand Reinhold, 1992, p. 263.

8. In Marcia Bartusiak, "The Universe, By and Large," *Mosaic,* vol. 15, no. 2, 1984, p. 3.

9. Ibid.

10. Margaret Geller, interviewed by T.F., September 14, 1995.

11. Ibid.

12. Philip Morrison and Phyllis Morrison, *Powers of Ten*. New York: Freeman, 1982.

13. Benoit B. Mandelbrot, *The Fractal Geometry of Nature*. New York: Freeman, 1983.

14. Referred to here is a pencil beam survey conducted by Alex Szalay, David Koo, and Richard Kron, who looked straight up through the north pole of our galaxy (to minimize interference by the Milky Way); subsequent probes by Szalay and Koo with T. J. Broadhurst and Richard Ellis, looking down through the south galactic pole; and two more, conducted with Jeff Munn, that aimed 40 and 60 degrees away from the north-south galactic axis. See, e.g., T. J. Broadhurst, R. S. Ellis, D. C. Koo, and A. S. Szalay, *Nature* 343, 726 (1990).

15. T. J. Broadhurst, R. S. Ellis, D. C. Koo, and A. S. Szalay, "Large-Scale Distribution of Galaxies at the Galactic Poles," *Nature* 343, February 22, 1990, p. 728.

16. In Ann K. Finkbeiner, "Mapmaking on the Cosmic Scale," *Mosaic,* vol. 21, no. 3, Fall 1990, p. 16.
17. R. C. Kraan-Korteweg and P. A. Woudt, "An Optical Galaxy Search in the Hydra/Antlia and the Great Attractor Region," University of Groningen, Kapteyn Institute Preprint no. 148, July 1994, p. 1.
18. Vera Rubin, interviewed by T.F., 1986.
19. Alan Dressler, *Voyage to the Great Attractor: Exploring Intergalactic Space.* New York: Knopf, 1994, p. 253.
20. Ibid., p. 211.
21. Ibid., p. 253.
22. P. J. E. Peebles, *Principles of Physical Cosmology.* Princeton: Princeton University Press, 1993, p. 528.
23. Ibid., p. 118.
24. E. R. Harrison, *Physical Review D* 1, 2726 (1970). Ya. B. Zeldovich, "A Hypothesis, Unifying the Structure and the Entropy of the Universe," *Monthly Notices Royal Astronomical Society* 160, 1P (1972).
25. Poe, "The Power of Words," *United States Magazine and Democratic Review,* June 1845, quoted in Edward Harrison, *Darkness at Night: A Riddle of the Universe.* Cambridge: Harvard University Press, 1987, p. 146.
26. The Proust is a paraphrase—or, more exactly, a translation from the French to the Russian to English—by A. A. Ruzmaikin and D. D. Sokoloff in their preface to Zeldovich, *The Almighty Chance.* Teaneck, N.J.: World Scientific, 1990, p. vi.
27. Andrei Sakharov, "A Man of Universal Interests," *Nature* 331, February 25, 1988, p. 672.
28. In Kim A. McDonald, "New Discoveries of Large Structures in Cosmos Challenge Leading Ideas of Universe's Origin," *Chronicle of Higher Education* 37, January 30, 1991, p. A11.
29. J. Richard Gott III, interviewed by T.F., 1992.
30. J. Richard Gott III, at a Princeton workshop, June 1992, called to discuss the COBE quadrupole results.
31. In *Maclean's,* May 4, 1992.
32. By 1996, an analysis of the accumulating COBE data had produced a result that further strengthened this conclusion. Where the Harrison-Zeldovich spectrum is set as $n = 1$, the COBE data indicate that $n = 1.2$, plus or minus 0.3.
33. In the *San Francisco Chronicle,* October 1, 1992.
34. Joel Primack, "Cosmology After COBE," *Beam Line,* Winter 1992, p. 2.
35. Alan Dressler, "The Great Attractor: Do Galaxies Trace the Large-Scale Mass Distribution?" *Nature* 350, April 4, 1991, p. 397.

7: Cosmic Evolution

1. Kant, *Universal Natural History and Theory of the Heavens,* W. Hastie, translator. Ann Arbor: University of Michigan Press, 1969, p. 145.

2. Browning, *Cleon,* 1855. In G. Robert Stange, editor, *The Poetical Works of Robert Browning.* Boston: Houghton Mifflin, 1974, p. 359.

3. For a fuller discussion, see section two, "Time," of my *Coming of Age in the Milky Way.* New York: Morrow, 1988. Also J. T. Fraser, editor, *The Voices of Time.* Amherst, University of Massachusetts Press, 1981.

4. In Daniel C. Matt, *The Essential Kabbalah: The Heart of Jewish Mysticism.* San Francisco: Harper, 1994, p. 31.

5. For more on the concept of evolution, see, e.g., R. C. Lewontin, "Evolution," in David L. Sills, editor, *International Encyclopedia of the Social Sciences.* New York: Macmillan, 1968, vol. 5, pp. 202ff.

6. Darwin, *On the Origin of Species.* New York: Gryphon, 1995 (facsimile of the John Murray edition of 1859), p. 490.

7. In Philip Appleman, editor, *Darwin.* New York: Norton, 1970, p. 1.

8. The comet zone is thought to have two major components, an inner disk and an outer shell. The disk is known as the Kuiper belt, after the astronomer Gerard Kuiper, who in 1951 hypothesized that there lies, beyond the planets, a ring of comets resembling a gigantic version of Saturn's rings. The shell is the Oort cloud, with its inner surface beginning toward the outer part of the Kuiper belt and its extremities tapering off roughly two light-years from the sun.

9. Pierre-Simon de Laplace, *The System of the World,* Henry Harte, translator, vol. 2, 1830. In Harlow Shapley and Helen E. Howarth, editors, *A Source Book in Astronomy.* New York: McGraw-Hill, 1929, p. 155.

10. Kant, *Universal Natural History and Theory of the Heavens,* vol. 1, p. 317, in Ernst Cassirer, *Kant's Life and Thought.* New Haven: Yale University Press, 1981, p. 48.

11. Charles Darwin, "Prefatory Notice: Studies in the Theory of Descent," in August Weismann, *Studies in the Theory of Descent.* London: Sampson Low, 1882, pp. v–vi; reprinted in Paul H. Barrett, editor, *The Collected Papers of Charles Darwin.* Chicago: University of Chicago Press, 1977, vol. 2, p. 281.

12. Archibald MacLeish, "Ars Poetica."

13. Jon Morse, in "Stellar Disks and Jets," *Hubble Space Telescope News,* NASA STSci-PR95-24, June 6, 1995, p. 5.

14. William Herschel, quoted in Heinz Pagels, *Perfect Symmetry: The Search for the Beginning of Time.* New York: Simon & Schuster, 1985, p. 28.

15. Butler and Marcy attached to their telescope a tube filled with gas, so that the spectrograph at the end of the tube recorded the spectrum of the gas, which of course was at rest relative to the spectrograph, along with that of the stars, which are in motion. Sifting out the small, planet-induced variations in each star's motion took years of computer programming, but once the programs had been debugged, the two astronomers looked forward to the possible discovery of further planets from data already collected and stored. Their optimism proved justified, when, in April 1996, they discovered a fourth planet, orbiting a star called HR3522, located 40 light-years from Earth.

16. In Charles Petit, "Bay Area Team Finds Two Planets," *San Francisco Chronicle,* January 18, 1996, p. A13.

17. John Noble Wilford, "Life in Space? Two New Planets Raise Thoughts," *The New York Times,* January 18, 1996, p. A10.

18. To date, about a dozen Mars meteorites and as many from the moon have been found. That there should be equal numbers of them is puzzling, in that the moon is much closer than Mars and has a lower escape velocity, so one would expect there to be far more moon rocks here. Results of a NASA-funded computer simulation conducted at Cornell University and released in 1996 indicate that 40 percent of the rocks knocked off the moon wind up on Earth, while for Mars the figure is a tenth that. The simulations also suggest that moon rocks get to Earth rapidly—typically within fifty thousand years—while material from Mars takes up to 15 million years to arrive here. (The fleetest Mars meteorite yet found spent seven hundred thousand years in transit.) Tantalizingly, the simulations present the possibility that a few pieces of the planet Mercury have hit Earth. A Mercury meteorite, if one could be found, would be very valuable: As it is, the Mars and moon meteorites, along with the less than half a ton of moon rocks returned by three automated Russian *Luna* missions and American astronauts participating in the six *Apollo* landings, constitute humanity's entire collection of stuff from other worlds.

19. The term originates with a pun in Milanese dialect, coined by the Italian astrophysicist Giovanni Bignami and arising from the difficulty astronomers had in locating the pulsar by charting high-energy photons called gamma rays and X rays using the low-resolution detectors with which it was first detected. *Gh'e' minga* is Milanese argot for "It's not there" or "It doesn't exist." The Geminga pulsar was eventually pinpointed, in 1979–1981, with the relatively high-resolution Einstein x-ray satellite, and it was detected optically with ground-based telescopes a decade later. It has since been observed by scientists using the ROSAT x-ray satellite and NASA's Compton Gamma Ray Observatory, and Energetic Gamma Ray Experiment Telescope satellites.

20. Richard Griffiths, in NASA/STScI paper GALAXY8-10JH.

The team's findings appear in the August 10, 1995, *Astrophysical Journal Letters*. The initial blue-galaxy image is STScI-PRC94-39B.

21. Heinz Pagels, *Perfect Symmetry: The Search for the Beginning of Time*. New York: Simon & Schuster, 1985, p. 28.

22. Alan Dressler, "Galaxies Far Away and Long Ago," *Sky & Telescope*, April 1993, p. 24.

23. A discovery supporting the dinosaurs-to-birds theory was announced just as this chapter was being completed, when scientists from the American Museum of Natural History in New York City and the Mongolian Academy of Sciences announced that they had found, in the Gobi desert, the remains of a female Oviraptor that died while sitting on a clutch of eggs in a tucked-leg posture identical with that exhibited by chickens and other modern birds. "Skeletal evidence clearly indicates that birds are a kind of living dinosaur," said Dr. Luis M. Chiappe, of the American Museum. "The unique anatomical characteristics uniting these two groups have been well documented and recognized for years. This discovery proves for the first time that birds and dinosaurs also share complex behaviors." (In John Noble Wilford, "Fossil of Nesting Dinosaur Strengthens Link to Modern Birds," *The New York Times*, national edition, December 21, 1995, p. A22.)

24. Bergson is generally remembered for having argued on behalf of *vitalism*, the belief that life cannot be attributed to mechanistic forces, as Darwinism would have it, but springs from a "life force" distinct from those known to physics. Since this *élan vital* cannot by definition be identified with any physical substance (such as blood), the doctrine of vitalism is irrefutable, and therefore unscientific. It enjoys virtually no support in the biological sciences today. I mention this in order to stress that my discussion of creative evolution has nothing to do with vitalism, and is meant, rather, to address the question of whether the concept of evolution is unscientific insofar as the products of evolution are unpredictable.

25. Pierre-Simon de Laplace, *Essai philosophique sur les probabilités*. Paris: Gauthier-Villars, 1821, p. 3. Translated by David Layzer, in his *Cosmogenesis*. Oxford: Oxford University Press, 1990, p. 5. Layzer notes that Einstein held a similar deterministic view: To the scientist, Einstein wrote, "the future . . . is every whit as necessary and determined as the past." (Einstein, "The Religious Spirit of Science," in his *Ideas and Opinions*. New York: Bonanza Books, 1954, p. 40.)

26. David Layzer, *Cosmogenesis*. Oxford: Oxford University Press, 1990, p. 302.

27. The aphorism "Ontogeny recapitulates phylogeny" has been widely criticized, for reasons irrelevant to our purposes here. My more limited point is the same one made by the evolutionary biologist Stephen J. Gould—"I am convinced that the vast majority of supposed recapitula-

tions represent nothing but the conservative nature of heredity"—in his authoritative study *Ontogeny and Phylogeny.* (Cambridge: Harvard University Press, 1977, p. 4).

28. Civilization itself embodies a more advanced form of conservancy: One need not recapitulate the airframe of a 707 before designing a new jetliner, or build a shack before building a skyscraper, precisely because records are kept of how to do these things without needlessly repeating errors. But that is another story.

29. The wording comes from the April 20, 1981, manuscript by David Layzer of what was to be published as his book *Constructing the Universe* (New York: Freeman, 1984). Layzer altered it in the finished work but I have taken the liberty of retaining his original language here, as I find that it gets to the point admirably, for my purposes if not for his.

30. Such a change may be global—as is the case with the cosmic catastrophes thought to have caused mass extinctions—or local. In many instances, it is simply a matter of some members of a species finding themselves in an extreme and isolated environment that puts a premium on innovation if they are to survive. As the biologist Ernst Mayr notes, "The most rapid evolutionary change does not occur in widespread, populous species, as claimed by most geneticists, but in small founder populations [which] have unique opportunities to enter new niches and to select novel adaptive pathways." (Ernst Mayr, "Speciational Evolution Through Punctuated Equilibria," in his *Toward a New Philosophy of Biology.* Cambridge: Harvard University Press, 1988, p. 461. See also Mayr, "Change of Genetic Environment and Evolution," in J. Huxley, editor, *Evolution as a Process.* London: Allen & Unwin, 1954, pp. 157–180.) Mass dieouts may be thought of as events that put virtually every species—especially those on dry land—in extreme, isolated situations, thus promoting spasms of rapid evolution.

31. J. A. Wheeler, conversations with T.F. See also "There is no law except the law that there is no law," in John D. Barrow and Frank J. Tipler, *The Anthropic Cosmological Principle.* New York: Oxford University Press, 1986, p. 224.

32. Murray Gell-Mann, *The Quark and the Jaguar.* New York: Freeman, 1994, p. 371.

33. Henry Adams, *The Education of Henry Adams,* in Ernest Samuels and Jayne N. Samuels, editors, *Adams.* New York: Viking/Library of America, 1983, p. 1132. Adams's exact words are, "Chaos was the law of nature; Order was the dream of man."

34. Ibid., p. 1084.

35. Jacques Monod, *Chance & Necessity.* New York: Vintage, 1971, p. xiii.

36. Ibid., p. xv.

37. In John D. Barrow and Frank R. Tipler, *The Anthropic Cosmological Principle*. New York: Oxford University Press, 1986, p. 84.

38. In Francis Darwin, *The Life and Letters of Charles Darwin*. New York: Basic Books, 1959, vol. 1, p. 105.

8: Symmetry and Imperfection

1. Anna Wickham, "Envoi," in *Selected Poems*. London: Chatto and Windus, 1971.

2. Steven Weinberg, interviewed by T.F., Austin, Texas, March 28, 1985.

3. Hermann von Helmholtz, "Autobiographical Sketch," 1891, and lectures at Berlin University, quoted in Christa Jungnickel and Russell McCormmach, *Intellectual Mastery of Nature*. Chicago: University of Chicago Press, 1986, vol. 1, p. xxiii. Helmholtz added that science bestows the "only gift of prophecy that is given to man."

4. Einstein's failure to arrive at a unified theory of gravity and electromagnetism via a top-down approach provides a notable negative example of just how right he was. His biographer Abraham Pais invokes Richard Feynman's classification of scientists as explorers and philosophers as tourists: "The tourists like to find everything tidy; the explorers take Nature as they find her," Feynman remarked, with forgivable exaggeration. In Pais's view, Einstein in later life turned from science to philosophy. Einstein's customary response to such criticisms was to smile pleasantly and suggest that perhaps he had earned the right to make his mistakes. (Abraham Pais, *Einstein Lived Here*. New York: Oxford University Press, 1994, p 130.)

5. In Michio Kaku, *Hyperspace*. New York: Anchor, 1994, p. 124.

6. W. M. Elsasser, *Memoirs of a Physicist in the Atomic Age*. Bristol, U.K.: Hilger, 1978, p. 51.

7. *Wisconsin State Journal*, April 31, 1929, quoted in Helge S. Dragh, *Dirac: A Scientific Biography*. Cambridge, U.K.: Cambridge University Press, 1990, p. 73. I have deleted some of the reporter's words—but none of Dirac's—from this excerpt.

8. C. P. Snow, "The Classical Mind," in Jagdish Mehra, editor, *The Physicist's Conception of Nature*. Boston: Kluwer, 1973, p. 810.

9. Paul Dirac interview, *Archives for the History of Quantum Physics*, American Institute of Physics, New York, May 7, 1963, p. 15; quoted in Robert P. Crease, Jr., and Charles C. Mann, *The Second Creation*. New York: Macmillan, 1986, p. 76.

10. In E. Salaman and M. Salaman, "Remembering Paul Dirac," *Endeavour*, May, 1986, pp. 66–70, quoted in Helge S. Dragh, *Dirac: A Scientific Biography*. Cambridge, U.K.: Cambridge University Press, 1990, p. 3. Dragh notes that the date of Dirac's trip to Russia was actually 1928, not 1927.

11. Rudolf Peierls, Address to Dirac Memorial Meeting, Cambridge, in J. G. Taylor, editor, *Tributes to Paul Dirac.* Bristol, U.K.: Hilger, 1987, p. 37.

12. Jagdish Mehra, in Jagdish Mehra, editor, *The Physicist's Conception of Nature.* Boston: Kluwer, 1973, p. 818.

13. Paul Dirac, "The Evolution of the Physicist's Picture of Nature," *Scientific American,* May 1963, p. 47. This remark has prompted discussion, with several thinkers objecting that there are many beautiful conceptions that are not true and many true ones that do not, at least initially, look all that beautiful. Bertrand Russell suggested that science, by filtering out those parts of nature not accessible to mathematical reasoning, bestows upon the universe a tincture of rational beauty that comes more from the mathematics than from some central quality of the universe. (For a discussion see, e.g., Judith Wechsler, editor, *Aesthetics in Science.* Cambridge: MIT Press, 1978; and Edward Rothstein, *Emblems of Mind: The Inner Life of Music and Mathematics.* New York: Times Books, 1995, p 151.) Perhaps illumination may be found in the "process" argument that it is the experience of *comprehension* that is beautiful, not necessarily nature itself, which in any event can through science be apprehended neither as a whole nor in essence.

14. Many years later, Murray Gell-Mann asked Dirac why he had not gone ahead and predicted the existence of the positron, since it was mandated by his equation for the electron. "Pure cowardice," Dirac replied. (Murray Gell-Mann, *The Quark and the Jaguar.* New York: Freeman, 1994, p. 179.)

15. In Robert P. Crease, Jr., and Charles C. Mann, *The Second Creation.* New York: Macmillan, 1986, p. 197.

16. Ibid., p. 282.

17. This marvelous property of gluons—that they get stickier as the quarks are pulled apart—is known as *infrared slavery.* It arises from a symmetry relationship with electrodynamics. Specifically, electrodynamics, a U(1) theory, is *Abelian,* meaning that it obeys the rules of a mathematical group in which transformations wind up with the same result regardless of the order in which they are performed, while chromodynamics is *non-Abelian,* meaning that the order of transformations does influence the outcome.

18. Steven Weinberg, interviewed by T.F., Austin, Texas, March 28, 1985.

19. Ibid.

20. Ibid.

21. Howard Georgi, postscript to "A Unified Theory of Elementary Particles and Forces," in Richard A. Carrigan Jr. and W. Peter Trower, editors, *Particle Physics in the Cosmos.* New York: Freeman, 1989, p. 77.

22. P. A. M. Dirac, *Directions in Physics,* H. Hora and J. R. Shepanski, editors. New York: Wiley, 1978, p. 20. In Eduard Prugovecki,

"Foundational Problems in Quantum Gravity and Quantum Cosmology," *Foundations of Physics,* vol. 22, no. 6, 1992, p. 758.

23. Richard Feynman, *QED: The Strange Theory of Light and Matter.* Princeton: Princeton University Press, 1985, p. 128.

24. In James Gleick, *Genius: The Life and Science of Richard Feynman.* New York: Pantheon, 1992, p. 378.

25. Richard Feynman, *QED: The Strange Theory of Light and Matter.* Princeton: Princeton University Press, 1985, p. 8. Feynman is here expressing what Einstein called the "opportunism" of scientific research: The physicist most often works not from some grand design but by tackling an immediate problem that looks interesting—and what *makes* it interesting is, often, that it makes no sense. As Niels Bohr once said, "How wonderful that we have met with paradox. Now we have some hope of making progress."

26. Steven Weinberg, *Dreams of a Final Theory.* New York: Pantheon, 1992, p. 214.

27. Ibid., p. 216.

28. In John Horgan, "Edward Witten: The Pied Piper of Superstrings," *Scientific American,* November 1991, p. 46.

29. In P. C. W. Davies and J. Brown, editors, *Superstrings: A Theory of Everything?* Cambridge, U.K.: Cambridge University Press, 1988, p. 102.

30. Nambu, who was born in Tokyo and is now at the University of Chicago, has often come up with ideas too far ahead of their time for their significance to be appreciated or properly attributed. He predicted the existence of the Goldstone boson before Jeffrey Goldstone did, and postulated the existence of the omega particle—in a talk that Feynman dismissed by shouting, "In a pig's eye!"—a year before the omega was discovered. Witten says of Nambu, "People don't understand him, because he is so farsighted." Bruno Zumino of Berkeley claims, "I had the idea that if I can find out what Nambu is thinking about now, I'll be ten years ahead in the game. So I talked to him for a long time. But by the time I figured out what he said, ten years had passed." (Madhusree Mukerjee, "Strings and Gluons—the Seer Saw Them All," *Scientific American,* February 1995, pp. 37–39.)

31. In Ivars Peterson, "Strings and Mirrors," *Science News,* February 27, 1993.

32. In Gary Taubs, "Everything's Now Tied to Strings," *Discover,* November 1986.

33. Ibid.

34. Ibid.

35. Shakespeare, *King Lear,* I, 4, 144: "Can you make no use of nothing, nuncle?" "Why, no, boy; nothing can be made out of nothing." The aphorism originally is found in Lucretius.

36. Michio Kaku, *Hyperspace.* New York: Anchor, 1994, p. 170. The italics are Kaku's.

37. Ibid., p. 329.

38. In John Horgan, "Particle Metaphysics," *Scientific American,* February 1994, p. 105.

39. Paul Ginsparg and Sheldon Glashow, "Desperately Seeking Superstrings?" *Physics Today,* May 1986, p. 7.

40. In John Horgan, "The Pied Piper of Superstrings," *Scientific American,* November 1991, p. 42.

41. Edward Witten, interviewed by T.F., Johns Hopkins University, March 23, 1995.

42. Edward Witten, "The Search for Higher Symmetry in String Theory," *Philosophical Transactions Royal Society London A* 31 (1989).

43. In Fred Golden, "Dangling Black Holes on a String," University of California, Santa Barbara, press release, June 19, 1995.

44. Andrew Strominger, interviewed by T.F., Padua, Italy, July 1983.

9: The Speed of Space

1. Alan H. Guth, "Inflation," *Proceedings National Academy of Sciences* (U.S.) 90, 4871 (1993).

2. Andrei Linde, interviewed by T.F., Stanford University, February 1, 1993.

3. The observant reader may by now have objected that for the universe to have expanded exponentially it must have exceeded the velocity of light, and have wondered whether this is forbidden by the special theory of relativity. The answer, curiously, is no. Although the special theory of relativity forbids two objects from being accelerated to velocities greater than light—or, to put it more succinctly, forbids information to be transferred faster than light speed—cosmic space can stretch at any velocity without violating the theory, since we are dealing not with velocities *in* space, but with the expansion of space itself.

4. Recently, inflationary models have been constructed in which the observed cosmic matter density is noncritical. (See, e.g., Andrei Linde, "Inflation with Variable Omega," preprint, plenary talk given at the Snowmass Workshop on Particle Astrophysics and Cosmology, 1995, to appear in *Proceedings,* edited by E. Kolb and R. Peccei. Also J. Richard Gott III, "Open, CDM Inflationary Universes," preprint.) Many theorists feel that omega = 1 inflationary models are the more natural and less forced possibility, but if the universe proves to have a noncritical mass density, that fact would not rule out inflation.

5. Andrei Linde, "Lectures on Inflationary Cosmology," based on lectures given in 1994 at the School on Particle Physics and Cosmology

at Lake Louise, Canada; at the Marcel Grossmann Conference, Stanford; at the Workshop on Birth of the Universe, Rome; at the Symposium on Elementary Particle Physics, Capri; and at the School of Astrophysics, Erice, Sicily. SISSA (Scuola Internazionale Superiore di Studi Avanzati) server preprint hep-th/9410082, October 11, 1994.

6. For a fuller and more technical discussion, see Andrei Linde, "Inflation and Quantum Cosmology," *Physica Scripta* T36, 34 (1991).

7. The putative scarcity of these exotic particles has not lessened theoretical interest in their properties. Linde, for instance, reports, "Recently we learned that the monopoles themselves may expand exponentially and become as large as a universe." (Andrei Linde, "The Self-Reproducing Inflationary Universe," *Scientific American,* November 1994, pp. 48–55.)

8. Democritus, Fragment 125.

9. Aristotle, *On Generation and Corruption,* 325a27.

10. The rigor of this and others of Aristotle's arguments against the atomists' void—e.g., in chapters 8 and 9 of his *Physics*—demonstrates something of his justly famous subtlety, and serves as a restorative to those moderns who lapse into the brash delusion that science has so eclipsed the ancients that their philosophies may now be ignored.

11. Hans Christian von Baeyer, *Taming the Atom: The Emergence of the Visible Microworld.* New York: Random House, 1992, p. 116.

12. A good idea has a thousand fathers, and various precursors to Guth have since been spotted in the floodlamps of hindsight. Relevant work had been done by Katsuhiko Sato in Japan, Martin Einhorn in the United States, and Demosthenes Kazanas of NASA. Andrei Linde, in his *Inflation and Quantum Cosmology* (New York: Academic Press, 1990, p. 7), ascribes the origin of the vacuum inflation idea to Erast Gliner of the Institute for Physics and Technology, Leningrad, in 1965. He also notes that Andrei Sakharov in that same year attempted to calculate the density perturbations that Gliner's model would produce and posited the radical notion of creation "from nothing," and credits Alexei Starobinsky with creating "the first semi-realistic version of inflationary cosmology," at about the same time that Guth arrived at his insights. Linde lists himself as among those who developed similar ideas before Guth did: "In the nineteen seventies I came to the realization that homogeneous classical scalar fields, which are present in all unified theories of elementary particles, can play the role of an unstable vacuum state, and that their decay can heat up the universe." (Linde, "Lectures on Inflationary Cosmology," SISSA preprint hep-th/9410082, October 11, 1994.) It should be added that Guth did his monopole research in collaboration with his Cornell colleague Henry Tye. But Guth developed the hypothesis in a relatively complete form, and did so independently, for which he certainly deserves ample credit.

13. Alan Guth and Paul Steinhardt, "The Inflationary Universe," in Paul Davies, editor, *The New Physics.* Cambridge, U.K.: Cambridge University Press, 1989, p. 48.

14. Andrei Linde, "Lectures on Inflationary Cosmology," SISSA preprint hep-th/9410082, October 11, 1994.

15. E. L. Turner, conference on COBE and cosmology, Princeton University, June 1992. Linde himself has had second thoughts in this regard, occasioned by his realization that ours may be one among many universes. "This changes our standard notions about what is natural and what is not," he writes. (Linde, "Lectures on Inflationary Cosmology.") Recently he remarked, "Sometimes we [theorists] are trying to take upon ourselves this role of creating, deciding for him [i.e., God] what is natural and what is unnatural. But the experimentalists tell us what is true. Sometimes we are ashamed of our presumptions." (Andrei Linde, interviewed by T.F., PASCOS 95 conference, Johns Hopkins University, Baltimore, March 25, 1995.)

16. Andrei Linde, "The Self-Reproducing Inflationary Universe," *Scientific American,* November 1994, pp. 48–55.

17. The same is true of weather maps that show isobars connecting points with identical air temperature or atmospheric pressure. When Thomas Jefferson urged that amateur scientists in various colonies follow his example and record the temperature and barometric pressure each day, to improve weather prediction in almanacs like the one published by his friend Benjamin Franklin, he was proposing that they map meteorological scalar fields.

18. Quoted in Bob Davis, "Inflation Theory Posits a Radical New View of Expanding Cosmos," *The Wall Street Journal,* January 2, 1991, p. 1.

19. There are some exotic inflationary models that make different predictions, but in this account I ignore them. Typically they involve more fine-tuning of parameters than is needed to produce inflation in the first place. Moreover, varying these parameters just a little tends to change the predictions a lot, which raises questions of how robust the predictions of the exotic models really are. In overlooking the exotic inflationary theories, I am guilty of a philosophical inconsistency, by here embracing a minimalist position that I sometimes eschew elsewhere in this book—e.g., in proposing that there is more than one kind of dark matter. Is this wise? Time will tell.

20. A. Linde, "Lectures on Inflationary Cosmology," SISSA preprint hep-th/9410082, October 11, 1994.

21. Lawrence Krauss and Michael Turner, "The Cosmological Constant Is Back," Fermilab preprint 95/063-A; SISSA preprint astro-ph/9504003, March 31, 1995.

22. Michael Turner and Frank Wilczek, "Is Our Vacuum Metasta-

ble?" *Nature* 298, August 12, 1982, p. 633. See also Piet Hut and Martin Rees, "How Stable Is Our Vacuum," *Nature* 302, April 7, 1983, p. 508. Also Sidney Coleman and Frank DeLuccia, "Gravitational Effects on and of Vacuum Decay," *Physical Review D* 21:3305 (1980).

10: The Origin of the Universe(s)

1. Leibniz, in John Archibald Wheeler, "Law Without Law," in Wheeler and Wojciech Hubert Zurek, editors, *Quantum Theory and Measurement*. Princeton: Princeton University Press, 1983, vol. 1, unpaginated manuscript.

2. Whitman, *Leaves of Grass*.

3. Stuart Bowyer, in Henry Margenau and Roy Abraham Varghese, editors, *Cosmos, Bios, Theos*. LaSalle, Ill.: Open Court, 1993, p. 32.

4. Charles H. Townes, in ibid., p. 123.

5. Kierkegaard, *Journals*, no. 206, entry for 1838. Wilde, Mr. Erskine, in *The Picture of Dorian Gray*, chapter 3. Leibniz, letter to *de l'Hospital*, September 30, 1695, quoted in Benoit B. Mandelbrot, *The Fractal Geometry of Nature*. New York: Freeman, 1983, p. 405.

6. Thomas Aquinas, *Summa Theologiae*, I, Q. 2, art. 3.

7. John William Miller, "The Paradox of Cause," in *The Paradox of Cause and Other Essays*. New York: Norton, 1978, p. 13.

8. Ibid., p. 15.

9. For a fuller account of Tryon's hypothesis, see my *Coming of Age in the Milky Way*. New York: Morrow, 1988, pp. 353ff.

10. Alexander Vilenkin, talk delivered at the Texas/PASCOS conference, University of California, Berkeley, December 13–19, 1992. For a paper based on the talk, see Vilenkin, "Quantum Cosmology," in Carl W. Akerlof and Mark A. Srednicki, editors, *Relativistic Astrophysics and Particle Cosmology*. New York: The New York Academy of Sciences, 1993, p. 271.

11. James B. Hartle, "Spacetime Quantum Mechanics and the Quantum Mechanics of Spacetime," lectures given at the 1992 Les Houches École, "Gravitation et Quantifications," July 9–17, 1992. SISSA preprint gr-qc/9304006, April 6, 1993.

12. John Archibald Wheeler, "Law Without Law," in Wheeler and Wojciech Hubert Zurek, editors, *Quantum Theory and Measurement*. Princeton: Princeton University Press, 1983, vol. 1, unpaginated manuscript.

13. Two years later, Hawking caught pneumonia and had to have a tracheotomy that eliminated his ability to speak. Since then he has communicated by means of a voice-synthesis computer that he manipulates with the two fingers of his right hand that have been spared his otherwise almost total paralysis. Remarkably, this has done little to dimin-

ish his sense of humor: He can get a laugh with a punch line that takes him five minutes of toil at the computer to deliver.

14. Stephen Hawking, address to the GR10 conference, Padua, July 1983.

15. The mathematical procedure that accomplishes this pleasant result comes from the Feynman sum-over-histories method. Hawking describes time in terms of *imaginary numbers.* At a first encounter this exotic term seems mystifying. (Presumably it marked the point, around page 134, that many readers set down Hawking's bestselling book *A Brief History of Time* and went off to make themselves a stiff drink.) But it's just a way of obtaining the square root of negative numbers. Students learn to square negative numbers by multiplying them to get a positive result: Thus the square of −3 and the square of +3 are both 9, and the square root of 9 is both 3 and −3. So if you want to obtain the square root of a negative number, you need a new set of integers, and that's what imaginary numbers are for. The square root of −9 is $3i$, meaning "imaginary three." In the Feynman method, the use of imaginary numbers removes all distinctions between time and space.

16. Stephen W. Hawking, *A Brief History of Time.* New York: Bantam, 1988, p. 136.

17. In John Gribbin, "The Birth and Death of the Universe," *UNESCO Courier,* May 1990, p. 38.

18. Research along these general lines is also being conducted by other investigators, among them Robert Griffiths, Erich Joos, Roland Omnès, Dieter Zeh, and Wojciech Zurek.

19. James B. Hartle, "The Quantum Mechanics of Cosmology," in S. Coleman, J. B. Hartle, T. Piran, and S. Weinberg, editors, *Quantum Cosmology and Baby Universes.* Singapore: World Scientific, 1991, pp. 67–68.

20. Ibid., p. 80. The humorist A. Whitney Brown makes much the same point: "History is a very tricky thing. To begin with, you can't get it mixed up with the past. The past actually happened, but history is only what someone wrote down." (A. Whitney Brown, *The Big Picture.* New York: Harper, 1991, p. 4.)

21. James B. Hartle, "Classical Physics and Hamiltonian Quantum Mechanics as Relics of the Big Bang," address at Nobel Symposium no. 79, "The Birth and Early Evolution of Our Universe," Gräftåvallen, Sweden, June 11–16, 1990. *Physica Scripta* T36, p. 232.

22. J. B. Hartle, "The Quantum Mechanics of Cosmology," in S. Coleman, J. B. Hartle, T. Piran, and S. Weinberg, editors, *Quantum Cosmology and Baby Universes.* Singapore: World Scientific, 1991, p. 101.

23. J. R. Lucas, "The Temporality of God," in Robert John Russell, Nancey Murphy, and C. J. Isham, editors, *Quantum Cosmology and the Laws of Nature.* Vatican City: Vatican Observatory Publications, and

Berkeley: The Center for Theology and the Natural Sciences, 1993, p. 243.

24. Andrei Linde, "The Self-Reproducing Inflationary Universe," *Scientific American*, November 1994, p. 35.

25. Andrei Linde, interviewed by T.F., Stanford University, February 1, 1993.

26. Andrei Linde, interviewed by T.F., Johns Hopkins University, March 24, 1995.

27. Andrei Linde, interviewed by T.F., Johns Hopkins University, March 25, 1995.

28. Andrei Linde, "Quantum Cosmology and Global Structure of the Universe," talk given at PASCOS/Hopkins 1995, symposium at Johns Hopkins University, March 25, 1995.

29. Andrei Linde, "The Self-Reproducing Inflationary Universe," *Scientific American*, November 1994, p. 38.

30. Ibid.

31. Andrei Linde, interviewed by T.F., Palo Alto, California, December 16, 1993.

32. Ibid. Linde, "Quantum Cosmology and Global Structure of the Universe," talk given at PASCOS/Hopkins 1995, symposium at Johns Hopkins University, March 25, 1995.

33. Is there a way to hold the wormhole open long enough to get the message safely through? That odd question, interesting as a problem in general relativity and quantum gravity, is the subject of a string of technical papers initiated by a letter from the astronomer Carl Sagan, who was writing a science fiction novel at the time, to the relativist Kip Thorne. For details see chapter 11 of Thorne's *Black Holes and Time Warps*. New York: Norton, 1994.

34. Andrei Linde, interviewed by T.F., Stanford University, February 1, 1993.

35. Ibid.

36. K. A. Bronnikov and V. N. Melnikov, "Vacuum Weyl Cosmologies in D Dimensions," SISSA preprint gr-qc/9410038, October 25, 1994.

37. Andrei Linde, "Quantum Cosmology and Global Structure of the Universe," talk given at PASCOS/Hopkins 1995, symposium at Johns Hopkins University, March 25, 1995.

38. Andrei Linde, interviewed by T.F., Palo Alto, California, December 16, 1993; Linde, "Quantum Cosmology and Global Structure of the Universe," talk given at PASCOS/Hopkins 1995, symposium at Johns Hopkins University, March 25, 1995.

39. Andrei Linde, *Inflation and Quantum Cosmology*. New York: Academic Press, 1990, p. 25.

40. Andrei Linde, interviewed by T.F., Johns Hopkins University, March 25, 1995.

11: Quantum Weirdness

1. Stein's last words were reported by her lifelong companion and secretary Alice B. Toklas, whose silence prompted them. (In Toklas, *What Is Remembered*. New York: Holt, Rinehart, 1963. Quoted in Dennis Flanagan, *Flanagan's Version*. New York: Vintage, 1988, p. 13.) Inasmuch as Ms. Stein is still routinely disparaged in newspapers and second-rate encyclopedias as a social butterfly famous solely for having known a lot of famous people, perhaps I may reassert the claim that she was both a subtle thinker and one of the most original writers of her time. Much of the essence of quantum philosophy is contained in her dying words and elsewhere, as in her remark that "The minute you or anybody else knows what you are you are not it, you are what you or anybody else knows you are and as everything in living is made up of finding out what you are it is extraordinarily difficult really not to know what you are and yet to be that thing." (Gertrude Stein, *Everybody's Autobiography*. New York: Random House, 1937, chapter 3.)

2. In T. A. Heppenheimer, "Bridging the Very Large and Very Small," *Mosaic,* Fall 1990, p. 33.

3. In Alan Burns, editor, *Gertrude Stein on Picasso*. New York: Liveright, 1970, p. 21.

4. Though that's pretty strange, especially in view of recent experiments confirming that "leaping" particles would have to move faster than light speed if they did cross the intervening gap, meaning that they really do disappear from one spot and reappear, instantaneously, at another.

5. In John Updike, "A Jeweler's Eye," *New York Review of Books,* October 29, 1995, p. 7.

6. In Murray Gell-Mann, *The Quark and the Jaguar*. New York: Freeman, 1994, p. 165. Bohr liked to joke about the difficulty of expressing quantum precepts in ordinary language by telling the following story: "A young rabbinical student went to hear three lectures by a famous rabbi. Afterwards he told his friends: 'The first talk was brilliant, clear and simple. I understood every word. The second was even better, deep and subtle. I didn't understand much, but the rabbi understood all of it. The third was by far the finest, a great and unforgettable experience. I understood nothing and the rabbi didn't understand much either.' " (In Abraham Pais, *Niels Bohr's Times in Physics, Philosophy, and Polity*. Oxford: Clarendon Press, 1991, p. 439.) Similar in spirit is Alice B. Toklas's remark, made as she and Gertrude Stein were leaving a dinner party in the company of Robert Hutchins, the president of the University of Chicago, that "Gertrude has said things tonight that will take her years to understand." (In Owen Gingerich, editor, *The Nature of Scientific Discovery: Symposium Commemorating the 500th Anniversary of the Birth of Nicolaus Copernicus*. Washington: Smithsonian Institution Press; New York: Braziller, 1975, p. 501.)

7. See, e.g., David Lindley, *Where Does the Weirdness Go?* New York: Basic Books, 1996; David Z. Albert, *Quantum Mechanics and Experience.* Cambridge: Harvard University Press, 1992; F. David Peat, *Einstein's Moon.* Chicago: Contemporary Books, 1990; Bas C. Van Fraassen, *Quantum Mechanics: An Empiricist View.* Oxford: Clarendon Press, 1991; Nick Herbert, *Quantum Reality.* New York: Anchor, 1985; and John Gribbin, *In Search of Schrödinger's Cat.* New York: Bantam, 1984.

8. Any particle will do. An entire atom was put in two quantum states in 1996, when scientists at the National Institute of Standards and Technology, using lasers, teased a beryllium atom into two spin states and separated them by the microscopically significant distance of 83 billionths of a meter.

9. Quantum mechanics is not inherently nonlocal. But it *exhibits* nonlocal behavior when examined on a classical level. Quantum weirdness is, remember, an interpretation issue. It's a matter of trying to understand, in classical terms, the strange side of nature revealed when quantum systems are amplified to a classical scale.

10. In Gerald Edelman, *Bright Air, Brilliant Fire: On the Matter of the Mind.* New York: Penguin, 1992, p. 216.

11. In Abraham Pais, *Niels Bohr's Times in Physics, Philosophy, and Polity.* Oxford: Clarendon Press, 1991, p. 349.

12. In Robert Scott Root-Bernstein, *Educating the Eye of the Mind.* Unpublished ms., 1985.

13. In P. N. Johnson-Laird, "The Ghost-Hunters," (London) *Times Literary Supplement,* December 14, 1984, p. 1441.

14. Albert Einstein and Leopold Infeld, *The Evolution of Physics.* New York: Simon & Schuster, 1938, p. 294.

15. In *The Cartoon Guide to Physics* CD-ROM. New York: HarperCollins Interactive, 1995.

16. Bohr, "The Unity of Human Knowledge," in his *Atomic Physics and Human Knowledge.* Bungay, U.K.: Richard Clay, 1963, pp. 9–10.

17. Leon Lederman, introducing Carlo Rubbia and congratulating him on having just won the Nobel Prize in physics, Santa Fe, New Mexico, November 3, 1984.

18. Einstein, statement on Schrödinger's "The Final Affine Laws" (1947), in Walter Moore, *Schrödinger: Life and Thought.* Cambridge: Cambridge University Press, 1989, p. 432.

19. The professional philosophers have not been entirely charitable toward Bohr, whose nonscientific writings many dismiss as vague and amateurish. Doubtless this is due in part to Bohr's having composed no formal (i.e., Cartesian) statement of his philosophy. But it may also represent the philosophers' wounded response to his blunt contempt for their profession. "It is hopeless to have any kind of understanding between scientists and philosophers directly," Bohr said. "All that philosophers

have ever written is pure drivel." (In Abraham Pais, *Niels Bohr's Times in Physics, Philosophy, and Polity.* Oxford: Clarendon Press, 1991, p. 421.)

20. Werner Heisenberg, "Quantum Theory and Its Interpretation," in S. Rozental, editor, *Niels Bohr: His Life and Work.* New York: North-Holland, 1967, p. 97.

21. Interview by Thomas Kuhn, in E. M. MacKinnon, *Scientific Explanation and Atomic Physics.* Chicago: University of Chicago Press, 1982, p. 375.

22. In Abraham Pais, *Niels Bohr's Times in Physics, Philosophy, and Polity.* Oxford: Clarendon Press, 1991, p. 170.

23. In A. P. French and P. J. Kennedy, editors, *Niels Bohr: A Centenary Volume.* Cambridge: Harvard University Press, 1985, p. 183.

24. In Abraham Pais, *Niels Bohr's Times in Physics, Philosophy, and Polity.* Oxford: Clarendon Press, 1991, p. 502.

25. Abraham Pais, "Niels Bohr and the Development of Physics," in M. Jacob, editor, *A Tribute to Niels Bohr on the Hundredth Anniversary of His Birth.* Geneva: CERN, 1985, preprint p.11.

26. In Abraham Pais, *Niels Bohr's Times in Physics, Philosophy, and Polity.* Oxford: Clarendon, 1991, pp. 426–427. Emphasis added by T.F.

27. The Nazi leader Hermann Goering may or may not have said, "Whenever I hear the word 'culture,' I reach for my revolver." The line appears in *Schlageter,* by the Nazi playwright Hanns Johst, where a storm trooper says, *Wenn ich Kultur höre . . . entsichere ich meinen Browning* ("I cock my Browning"). Hawking's joke was made in a conversation with T.F., in Pasadena, California, April 4, 1983. The complete exchange was as follows:

HAWKING: I regard [the many worlds interpretation] as self-evidently correct.

T.F.: Yet some don't find it evident to *them*selves.

HAWKING: Yeah, well, there are some people who spend an awful lot of time talking about the interpretation of quantum mechanics. My attitude—I would paraphrase Goering—is that when I hear of Schrödinger's cat, I reach for my gun.

T.F.: That would spoil the experiment. The cat would have been shot, all right, but not by a quantum effect.

HAWKING (laughing): Yes, it does, because *I myself am a quantum effect.* But, look: All that one does, really, is to calculate conditional probabilities—in other words, the probability of A happening, given B. I think that that's all the many worlds interpretation is. Some people overlay it with a lot of mysticism about the wave function splitting into different parts. But all that you're calculating is conditional probabilities.

28. This is known in the literature as the question of Wigner's friend, after the physicist Eugene Wigner, who asked whether the cat is dead or alive during the interval between when his friend opens the box

and observes the cat and when he tells Wigner about it. It is akin to quantum erasure, experiments which suggest that if Wigner's friend were to drop dead before he could announce the result, no observation would have been made. Is the cat then *still* neither dead nor alive, or did it *return* to a superposed, dead/alive state upon the death of Wigner's friend. Both alternatives are paradoxical—and this, again, is the point of the exercise, to deny the Copenhagen insistence that quantum mechanics provides a *complete* description of a system.

29. In Stephen Jay Gould, *New York Review of Books,* November 5, 1992.

30. A. Einstein, B. Podolsky, and N. Rosen, "Can Quantum-Mechanical Description of Physical Reality Be Considered Complete?" *Physical Review* 47, 777, 1935, p. 780.

31. David Z. Albert, "Bohm's Alternative to Quantum Mechanics," *Scientific American,* May 1994, p. 66.

32. To be sure, there have always been philosophers who maintained that the scientific enterprise has to do strictly with making accurate predictions or, even more strictly, with learning how to manipulate the forces of nature. Essentially this is a view of science as power. But many of the strongest minds in science and philosophy have objected to this position. They insist that science is primarily about knowledge, and that genuine knowledge is necessarily objective knowledge.

33. Disheartened by the chilly reception his ideas met with, Everett left physics after receiving his Ph.D. and found work with a defense contractor. Bryce DeWitt, a leading quantum-gravity theorist, promoted Everett's ideas from about 1968, and in the late 1970s Everett was invited to give some talks at the University of Texas, Austin, where Wheeler had joined the faculty. Everett arrived driving a Cadillac with longhorns adorning the hood and lectured with the intensity of an exiled intellectual. (He smoked so heavily that the auditorium's strict no-smoking rule had to be suspended so that he could speak.) Encouraged by an increasing willingness among scientists to take his ideas seriously, Everett was planning a return to physics when he died of a heart attack in 1982.

34. Einstein may have had in mind Proverbs 16:33: "The lot is cast into the lap; but the whole disposing thereof is of the Lord." To Max Born he wrote, "Quantum mechanics is certainly imposing. But an inner voice tells me that it is not yet the real thing. The theory says a lot, but does not really bring us any closer to the secret of the 'Old One.' I, at any rate, am convinced that *He* is not playing at dice." (Max Born, *The Born-Einstein Letters.* New York: Walker, 1971, p. 91.)

35. In Eduard Prugovečki, "Foundational Problems in Quantum Gravity and Quantum Cosmology." *Foundations of Physics,* vol. 22, no. 6, 1992, p. 766. As Prugovečki notes, the actual number of universes referred to as "many" in this interpretation is actually much

larger than 10^{100+}: In one formulation it is infinity raised to an infinite power.

36. In Max Jammer, *The Philosophy of Quantum Mechanics*. New York: Wiley, 1974, p. 278.

37. David Lindley, *Where Does the Weirdness Go?* New York: Basic Books, 1996, p. 109.

38. In Murray Gell-Mann, *The Quark and the Jaguar*. New York: Freeman, 1994, p. 170.

39. David Bohm, *Wholeness and the Implicate Order*. London, U.K.: Routledge, 1981, p. xiii.

40. David Z. Albert, *Quantum Mechanics and Experience*. Cambridge: Harvard University Press, 1992, p. 169.

41. Ibid., p. 161. "Instrumentalism" is the argument that physics deals not with physical reality but with patterns in observations that can be tested empirically—e.g., patterns in instrument readings. But even an instrumentalist must believe that there is *some* connection between his observations and physical reality; otherwise why bother doing physics? A gas-meter reader would soon lose interest if he thought that the readings he recorded had nothing to do with gas consumption.

42. David Bohm, *Wholeness and the Implicate Order*. London, U.K.: Routledge, 1981, p. 138.

43. In F. David Peat, *Einstein's Moon: Bell's Theorem and the Curious Quest for Quantum Reality*. Chicago: Contemporary Books, 1990, p. 113.

44. Actually, it bothered scientists even before relativity, since locality is essential to the mechanical, causal view of nature that predated Einstein. Isaac Newton, for one, was deeply troubled that his law of gravitation included no mechanical way of getting gravitational force across what he presumed to be empty space; this was *his* version of Einstein's worries about "spooky action at a distance."

45. F. David Peat, *Einstein's Moon*. Chicago: Contemporary Books, 1990, p. 124.

46. David Bohm, *Wholeness and the Implicate Order*. London, U.K.: Routledge, 1981, p. 148.

47. Richard Feynman, Nobel Lecture, December 11, 1965. This conversation with Wheeler took place in the 1940s, when Feynman was a graduate student at Princeton. As Feynman recounts it, Wheeler "explained on the telephone, 'Suppose that the world lines which we were ordinarily considering before in time and space, instead of only going up in time were a tremendous knot. Then, when we cut through the knot, by the plane corresponding to a fixed time, we would see many, many world lines and that would represent many electrons. Except for one thing. If in one section that is an ordinary electron world line, in the section in which it reversed itself and is coming back from the future, we

have the wrong sign to the proper time. . . . That's equivalent to changing the sign of the charge, and, therefore, that part of a path would act like a positron.' " (The editing of the published transcript has been altered slightly by T.F.)

48. Einstein, *The World as I See It.* New York: Philosophical Library, 1934.

49. Dante, *Purgatorio,* Dorothy Leigh Sayers, translator. New York: Penguin Books, 1969, 22:28.

50. John Archibald Wheeler, "Law Without Law," in Wheeler and Wojciech Hubert Zurek, editors, *Quantum Theory and Measurement.* Princeton: Princeton University Press, 1982, vol. 1, unpaginated manuscript edition.

12: A Place for Us

1. This fable has been repeated many times. The wording here is from Nevill Mott, "Science Will Never Give Us the Answers to All Our Questions," in Henry Margenau and Roy Abraham Varghese, editors, *Cosmos, Bios, Theos.* LaSalle, Ill.: Open Court, 1992, p. 69.

2. James Branch Cabell, *The Silver Stallion.* London: J. Lane, 1926, chapter 26.

3. Since modern science originated in Europe I am referring here mainly to Europeans, but the same could be said of many other prescientific cultures as well.

4. See Stillman Drake, *Galileo at Work.* Chicago: University of Chicago Press, 1978, p. 212.

5. Actually, the only attainable unambiguous answer is "Yes," since we shall never be able to prove that there is *no* life beyond Earth. No matter how many sterile worlds we reconnoitered, it would always be possible that the next one harbored life. But were many such searches to produce only negative results, and were new scientific findings to strongly indicate that the incidence of life is very improbable, we might well lose interest—if not in the question, at least in our ability to answer it.

6. Curiously, science still lacks an agreed-upon delineation of what life is. There are at least five viable definitions—physiological, metabolic, biochemical, genetic, and thermodynamic—and each is dogged by flaws and exceptions. This in itself may illustrate the difficulty of working with a phenomenon of which scientists have but one example.

7. One can imagine many intelligent life forms that could not or would not transmit radio signals. These would remain invisible to a SETI investigation. However, the objection that many alien civilizations would be "too advanced" to use radio is not terribly persuasive. Although the word "radio" has an obsolescent ring to us, invoking as it does memories of an entertainment medium largely supplanted by television, it really

means any communication that employs the electromagnetic spectrum. "Radio" in this sense is an efficient and far-reaching way to transmit just about any sort of information one can think of, including moving and still photographs, words and sounds, holograms and web pages, and so on. In any event, most SETI scientists assume that we are much more likely to detect a signal that was meant to be intercepted than to "eavesdrop" on the private conversations of alien worlds.

8. Paul Horowitz, Harvard SETI symposium, October 31, 1995.

9. Ernst Mayr, *Toward a New Philosophy of Biology.* Cambridge: Harvard University Press, 1988, p. 73. The one publicly funded American SETI project was canceled and the remaining ones are all privately funded; taxpayers' funds are no longer involved.

10. For a discussion, see William Calvin, *The Ascent of Mind: Ice Age Climates and the Evolution of Intelligence.* New York: Bantam, 1990.

11. Loren Eiseley, *The Star-Thrower.* New York: Times Books, 1978, pp. 120–121.

12. For a discussion, see Alan Cromer, *Uncommon Sense: The Heretical Nature of Science.* New York: Oxford University Press, 1993.

13. Even if communicative worlds are commonplace, it seems unlikely that they last *forever.* And if they typically are mortal, most must have flourished and subsided in the past. For a discussion of how some of their knowledge could have been preserved in the memory banks of galactic communications networks, see my *The Mind's Sky: Human Intelligence in a Cosmic Context.* New York: Bantam, 1992.

14. Some theorists have suggested that life might evolve more rapidly in a high-gravity universe with stars burning more furiously. But we don't have to turn up the strength of gravity very much before the universe would collapse too quickly to permit the formation of any stars at all, so this argument has only a narrow application.

15. Carter's immediate precursors include the English mathematician G. J. Whitrow, who in 1955 pointed out that space must be three-dimensional since life could not otherwise exist, and Robert Dicke, who in 1957 urged that the values of fundamental constants like gravity and the charge of the electron are "not random but conditioned by biological factors"—i.e., by the existence of life. (R. H. Dicke, *Reviews of Modern Physics,* 29:355, 363 [1957].)

16. In Tony Rothman, "A 'What You See Is What You Beget' Theory," *Discover,* May 1987, p. 91.

17. In Henry T. Simmons, "Redefining the Cosmos," *Mosaic,* March–April 1982, p. 19.

18. The mathematical physicist Frank Tipler, of Tulane University, has likened the universe to a computer and proposed that if it is closed and therefore destined to recollapse, all cognizant beings could be "resurrected" in a gigantic computer simulation made possible by the recon-

vergence of world lines near the end of time. Furthermore, he argues (on technical grounds that I'll not go into here) that such simulations could permit the resurrected intelligences to enjoy immortal life—even though, to an "outside" observer, it would all be over in the wink of an eye. Tipler's book making this extraordinary case, *The Physics of Immortality: Modern Cosmology, God, and the Resurrection of the Dead* (New York: Doubleday, 1994), is a striking mix of wistfulness and hardheaded science. He claims that his model generates "the notions of 'Holy Spirit,' 'grace,' 'heaven,' 'hell,' and 'purgatory,' " and that "it may also be possible to develop a Christology in the model," while maintaining that "either theology is pure nonsense, a subject with no content, or else theology must ultimately become a branch of physics." Tipler called his theory FAP, the final anthropic principle—that "laws of physics allow life to exist *forever.*" The science writer Martin Gardner memorably, if unkindly, dubbed it "CRAP—the Completely Ridiculous Anthropic Principle."

19. For a discussion, see Gerald Feinberg and Robert Shapiro, *Life Beyond Earth*. New York: Morrow, 1980.

20.
Full fathom five thy father lies;
Of his bones are coral made;
These are the pearls that were his eyes;
Nothing of him that doth fade
But doth suffer a sea-change
Into something rich and strange . . .
(SHAKESPEARE, "Ariel's Song," *The Tempest,* I, ii, 396ff.)

21. T. S. Eliot, "Little Gidding," part 5, in his *Four Quartets.*

Contrarian Theological Afterword

1. Gillian Conoley, "Beckon," *American Poetry Review,* March–April 1996, p. 9.

2. Victor Weisskopf, "There Is a Bohr Complementarity Between Science and Religion," in Henry Margenau and Roy Abraham Varghese, editors, *Cosmos, Bios, Theos.* La Salle, Ill.: Open Court, 1993, p. 127.

3. Fred Hoyle, "The Universe: Past and Present Reflections," *Engineering & Science,* November 1981, p. 12.

4. Bernard de Fontenelle, *A Plurality of Worlds,* John Glanville, translator. New York: Nonesuch Press, 1929, p. 20.

5. In John D. Barrow and Frank Tipler, *The Anthropic Cosmological Principle.* Oxford: Oxford University Press, 1986, galley proof p. 65.

6. The inescapability of this finding has recently been recapitulated in Richard Dawkins's aptly titled book *The Blind Watchmaker: Why the Evidence of Evolution Reveals a Universe Without Design.* New York: Norton, 1987.

7. Letter to J. Hooker, 1870, in F. Darwin and A. C. Seward, *More Letters of Charles Darwin*. New York: Appleton, 1903, 1:321. In Stanley Jaki, *The Road of Science and the Ways to God*. Chicago: University of Chicago Press, 1978, p. 293.

8. When giving public lectures on cosmology I am often asked about my own religious convictions. On such occasions I sometimes tell the story about a theologian who is asked by an old friend, "Do you believe in God?"

The theologian replies, "I can answer you, but I promise that you won't *understand* my answer. Do you want me to go ahead?"

"Sure."

"OK, the answer is 'Yes.' "

The point of the story has to do, of course, with the preposterous ambiguity of such terms as "believe" and "God." And that is one reason I try to avoid answering such questions. Nor do I see that a statement of my personal beliefs would do much to illuminate the issues under discussion in this book.

But, if only to avoid being coy about it, let me say that I'm an agnostic. The term is derived from the Greek *agnostos*, "unknowable." It was coined in 1869 by Thomas Henry Huxley, who sought to define his stance on religion in contributions to the Metaphysical Society, an organization of eminent English thinkers that met nine times a year to discuss philosophical and theological matters. Huxley meant it to oppose gnosticism, which asserts the primacy of mystical and esoteric faith over logic and reason. His position was comparable to that of Hume, who in the essay "Of Miracles" in his *Enquiry Concerning Human Understanding* writes that "a wise man . . . proportions his belief to the evidence."

There are two varieties of agnosticism.

"Weak" agnosticism consists of suspending one's opinion as to the existence of God—pending, I suppose, the introduction of further evidence. This stance seems wishy-washy and probably deserves its dismissal, by a contemptuous Friedrich Engels, as "shame-faced" atheism.

My position is "strong" agnosticism. It denies that God's existence can ever be disproved. There are many definitions of God, some of which seem to say nothing at all about God *except* that he exists. I hold that it is impossible to disprove all these definitions. If, to take an extreme example, science were one day to establish beyond reasonable doubt that the universe was created by a mad scientist in a basement laboratory, it would still be possible to posit that the prior universe in which that scientist lived was created by God. Moreover, that scientist, regardless of his inhuman brilliance, would be unable, in principle, to prove that God does not exist —or so I maintain. I might add that this view is not just a matter of logic-chopping, or a sly way of skirting theological issues. It is offered in good faith, with an honest appreciation of the merits of religion, science, and reason. It's not just that I don't know; I assert that we *cannot* know.

Strong agnosticism is open to disproof. Were God to appear on Earth tomorrow and work convincing miracles—an act that would imply deplorable taste on his part, but which is possible—then agnosticism, like atheism, would have been disproved.

As to the usefulness of these opinions of mine I remain a skeptic.

9. Keith Ward, *Rational Theology and the Creativity of God.* New York: Pilgrim Press, 1982, p. 73, in Paul Davies, *The Mind of God.* New York: Simon & Schuster, 1992, p. 180.

10. In Kantian language these are the realms of *analytic* and *synthetic* propositions, respectively.

11. Kant, *The Critique of Pure Reason,* J. M. D. Mieklejohn, translator. Chicago: University of Chicago Great Books, 1952, p. 180. Harrison Ford's character in the movie *Star Wars* makes much the same point. Told that he may, if he is of service, be rewarded with more money than he can imagine, he replies, "I can *imagine* a great deal." But the fact that he can imagine it doesn't mean it exists.

12. This version of the verse is from Amit Goswami, *The Self-Aware Universe,* in manuscript, pp. 129–130.

13. In Blanche Patch, *Thirty Years with G.B.S.* London: Gollancz, 1951. Quoted in Edward Harrison, *Cosmology: The Science of the Universe.* Cambridge, U.K.: Cambridge University Press, 1981, p. 179.

14. Thomas Chalmers, *On Natural Theology,* 18th edition, p. 35, in Paul Edwards, editor, *The Encyclopedia of Philosophy.* New York: Macmillan, 1967, vol. 1, p. 186.

15. Charles Bradlaugh, in Paul Edwards, "Atheism." Paul Edwards, editor, *The Encyclopedia of Philosophy.* New York: Macmillan, 1967, vol. 1, p. 177.

16. F. W. J. Von Schelling, *System of Transcendental Idealism,* in John D. Barrow and Frank J. Tipler, *The Anthropic Cosmological Principle.* New York: Oxford University Press, 1986, p. 138. Schelling's answer to his own rhetorical question is "Because God is a *life,* not a mere being."

17. Adlai Stevenson, speech delivered at Northwestern University, Evanston, Illinois, 1962. Francis Bacon notes that from ancient times believers have placed "the angels of love, which are termed seraphim" at the top of the heavenly hierarchy, even above "the angels of light, which are termed cherubim." (Bacon, *Advancement of Learning,* I, VI, 3.)

18. "You, speak with us, and we will listen; but let God not speak with us, lest we die." (Exodus 20:19.) "If Yahweh is with us, then . . . where are all His miracles that our fathers told us about?" (Judges 6:13 Authorized [King James] Version.)

Glossary

Absolute magnitude. See *magnitude.*
Absolute space. Newtonian space. Defined without reference to its matter/energy content.
Absolute zero. See *Kelvin.*
Absorption lines. Dark lines in *spectra.* Produced when light from a distant source passes through a gas cloud closer to the observer. Displays chemical composition, etc., of the cloud.
Acceleration. Increase in velocity over time.
Accelerator. Machine for speeding subatomic particles to high velocity, then colliding them with a stationary target. A machine that collides two beams of particles is a collider.
Aether. (1) Aristotle: The fifth element, of which the stars and planets are made. (2) Nineteenth-century physics: Hypothetical medium pervading space.
Agnosticism. Belief that (1) existence of God *has* not been established ("weak" version), or (2) it *cannot* be ("strong" version).
Andromeda galaxy. Spiral galaxy in the *Local Group,* sister to the *Milky Way.*
Angular momentum. Product of mass and angular velocity for a rotating object.
Anisotropy. Quality of being dependent upon direction. See *isotropy.*
Anthropic principle. Approach to cosmology that constrains fundamental constants of nature and other cosmic circumstances by demonstrating that were they otherwise, the universe could not support life and therefore would not be observed.
Antimatter. Matter made of *particles* with identical *mass* and *spin* as those of ordinary matter, but with opposite charge.
Asteroid. One of the small, rocky objects that orbit the sun. Unlike *comets,* asteroids have little or no ice.
Astronomical unit. Mean distance from Earth to the sun. One A.U. = 92.81 million miles = 499.012 *light-seconds.*
Astrophysics. Science of the physics of extraterrestrial objects.
Asymmetry. A violation of *symmetry.*
Atheism. Belief that God does not exist, or can be shown not to exist as defined in a particular way.

Atom. Fundamental unit of a chemical element. Consists of a *nucleus,* surrounded by *electrons.*

Background radiation. See *cosmic microwave background.*

Baryon(s). Class of particles that includes protons and neutrons—that is, "ordinary matter." Baryons are massive particles with half-integral spin. They respond to the *strong nuclear force.*

Baryon number. Total number of *baryons* in the universe, minus the total number of antibaryons. Indexes cosmic matter-antimatter *asymmetry.*

Bell's theorem. Way of testing whether quantum systems, interpreted classically, obey *locality.* Experiments indicate that they do not. See *classical physics.*

Big bang. (1) Theory that the universe originated as a *singularity.* (2) The singularity itself. (3) Loosely, the high-energy physics of the early universe.

Big bang nucleosynthesis (BBN). See *nucleosynthesis.*

Billion. Here, the American billion, equal to 1,000 million, or 10^9.

Binary star. (1) Double star system in which the two stars are bound together gravitationally. (2) A star in such a system.

Bipolar jets. Spurts of *plasma* emitted from the rotational poles of, for example, *protostars.*

Black-body radiation. Energy reemitted by an object capable of absorbing all energy that strikes it. When plotted against wavelength, produces a characteristic black-body curve.

Black hole. Object with a gravitational field so intense that its *escape velocity* exceeds that of light. A *singularity.*

Bohmian interpretation. See *quantum observership.*

Boson(s), gauge boson(s). Force-conveying subatomic particle with integer spin that does not obey the *Pauli exclusion principle.* See *supersymmetric.*

Boundary condition. Defines limit of applicability of a physics equation.

Broken symmetry. Asymmetrical state in which evidence of an earlier *symmetry* is discerned.

Carbon cycle. Nuclear fusion process in stars that both begins with and generates carbon-12. See *fusion, nuclear.*

Causation, causality. Doctrine that every new situation must have resulted from a previous state. See *determinism.*

CCD. Charge-coupled device, a digital imaging system.

Cepheid variable. Pulsating *variable star* whose period of brightness variation is directly related to its absolute *magnitude.* Used to measure distances of galaxies.

Charm. Fourth *flavor* of *quarks.*

Chromodynamics. Quantum theory of the color force that binds quarks together to form, for example, protons.

Classical physics. (l) Physics prior to the introduction of the quantum principle. Incorporates Newtonian mechanics and relativity, views energy as a continuum, is deterministic. See *determinism.* (2) Relativity theory, when compared with *quantum theory.*

Closed universe. Cosmological model in which the universe is geometrically spherical. Closed universes eventually stop expanding and thereafter collapse.

Cluster of galaxies. Association of roughly one hundred galaxies. ("Rich" clusters may hold a thousand or more galaxies.) See *Local Group; supercluster.*

CMB. See *cosmic microwave background.*

Collider. See *accelerator.*

Comet(s). Dirty iceberg in space, left over from the formation of the solar system. See *Oort cloud.*

Conservation laws. Laws that identify a quantity that remains unchanged after a transformation. Hence, an expression of *symmetry.*

Copenhagen interpretation. See *quantum observership.*

Cosmic background radiation. (1) Broadly, any isotropic distribution of particles or gravitational waves released when the universe was young. See *isotropy.* (2) Of photons, the only cosmic background radiation observed to date, the *cosmic microwave background (CMB).*

Cosmic matter density. Average number of *fermions* per unit volume of space throughout the universe. Thought to index global curvature of cosmic space.

Cosmic microwave background (CMB). Also known as cosmic background radiation. Microwave radio emission composed of *photons* released from primordial cosmic material as it thinned out owing to expansion of the universe, at the epoch of *photon decoupling.*

Cosmic ray. Subatomic particle, usually a *proton,* moving through space at a high velocity.

Cosmogony. Study of the origin of the universe.

Cosmological constant. Term in cosmological equations, symbolized by Greek letter lambda (λ), employed to denote theoretical antigravitylike force.

Cosmological distance ladder. Set of overlapping distance-measurement techniques used to find distances of galaxies and, hence, when combined with their *redshifts,* the expansion rate of the universe.

Cosmology. (1) The science concerned with discerning the structure and composition of the universe as a whole. Combines astronomy, astrophysics, particle physics, and a variety of mathematical approaches including geometry and topology. (2) A particular cosmological theory.

Coulomb barrier. Electromagnetic zone of resistance surrounding protons or other electrically charged *particle;* repels particles of like charge.

Critical density. Cosmic matter density at which value the universe would be geometrically midway between open and closed. See *closed universe; open universe. Omega* for a critical density universe equals one.
Dark matter. Matter whose existence is inferred on the basis of dynamical studies—for instance, orbits of stars in galaxies—but which does not show up as bright objects such as stars and nebulae.
Deceleration parameter. Quantity designating the rate at which the *expansion of the universe* is slowing down. A function of the *cosmic matter density.*
Decoupling. Separation of classes of particles from regular interaction with one another, as in the decoupling of photons from particles of matter that produced the *cosmic microwave background.*
Degree. Angle subtended in the sky. From the zenith to the horizon is 90 degrees.
Deism. Belief in God as creator of the universe, and not as having worked miracles since.
Detector. Device for recording the presence of subatomic particles in a collider or *accelerator* experiment.
Determinism. Doctrine that all events are the precisely predictable effects of prior causes. See *causation.*
Deuterium (heavy hydrogen). An isotope of hydrogen; nucleus has one neutron and one proton.
Dimension. A geometrical axis in space or time.
Dirac equation. Mathematical description of the electron, derived by Paul Dirac, that incorporates both quantum mechanics and special relativity.
DNA. Deoxyribonucleic acid, the macromolecule that carries and copies the genetic information requisite to life on Earth.
Doppler shift. Change in the apparent wavelength of radiation (e.g., light or sound) emitted by a moving body and caused by its motion or by the expansion of the intervening space.
Double star. See *binary star.*
Drake equation. Method of calculating the prevalence of intelligent extraterrestrial life. The formula is N = R*fgfpneflfifcL, in which R* = rate of star formation in our galaxy, fg = fraction of these stars that are "good suns" (i.e., stars like the sun), fp = fraction of these stars likely to have planets, ne = average number of "good earths" (Earthlike planets) per planetary system, fl = fraction of good earths on which life begins, fi = fraction of life-bearing planets on which intelligence arises, fc = fraction of intelligent societies that develop interstellar communication, and L = longevity of the communicative phase of these civilizations.
Dwarf star. Main-sequence star with a mass equal to or less than that of the sun. More generally, any star on or below the *main sequence* in the *Hertzsprung-Russell diagram.*
Dynamics. Study, in physics, of the motion and equilibrium of systems under the influence of force.

Dynamo. An electric generator that employs a spinning magnetic field to produce electricity.
Eclipse. Obscuration of light striking an astronomical object (such as the moon) by another such object (such as the earth).
EGG. "Evaporating gaseous globule." Dense knot in a *giant molecular cloud.*
Electrodynamics. Quantum theory of the behavior of the *electromagnetic force.*
Electromagnetic force (or interaction). Fundamental force of nature; involves electrically charged *particles.*
Electron(s). Light elementary particle with a negative electrical charge.
Electron shells. Zones in which the electrons in atoms reside. Their radius is determined by the quantum principle, their population by the *Pauli exclusion principle.*
Electronuclear force. Fundamental force thought to have functioned in the very early universe and to have combined the attributes thereafter parceled out to the *electromagnetic force* and the strong and weak nuclear forces. See *grand unified theories.*
Electron volt. Measure of energy, equal to 1.6×10^{-12} erg.
Electroweak theory. Theory demonstrating links between the *electromagnetic* and the weak nuclear forces.
Ellipse. A plane curve in which the sum of the distances of each point along its periphery from two points—its "foci"—are equal.
Emission lines. Bright lines produced in a *spectrum* by a luminous source, such as a star or a bright nebula. Compare *absorption lines.*
Emission nebula. See *nebula.*
Empiricism. An emphasis on sense data as a source of knowledge, in opposition to the rationalist belief that reasoning is superior to experience.
Escape velocity. The speed at which an object can leave another object behind, without being recalled by its gravitational force.
Euclidean geometry. See *geometry.*
Evolution. Here defined as an increase in the complexity and variety of a system over time.
Exclusion principle. See *Pauli exclusion principle.*
Expansion of the universe. Increase, with time, in the amount of space separating galaxies from one another, at a rate proportional to their distances.
Extreme black hole. Theoretical black hole of *particle* mass, its electrical charge equal to its mass.
Fermion(s). Particle with half-integral spin. Fermions obey the *Pauli exclusion principle,* which says that no two fermions can exist in an atom in the same quantum state. This restricts the number of electrons, which are fermions, permitted in each *electron shell.*
Field. Domain or environment in which the real or potential action of a force can be described mathematically at each point in space.

Fission, nuclear. Interaction in which nucleons previously united in an atomic nucleus are disjoined, releasing energy.
Flatness problem. The riddle of why the universe is neither dramatically open nor closed, but appears to be almost perfectly balanced between these two geometrical shapes.
Flavor. Designation of quark types—up, down, strange, charmed, top, and bottom. Flavor determines how the weak nuclear force influences quarks.
Force(s). Agency that causes a change in a system. In quantum mechanics, called interaction.
Fusion, nuclear. Interaction in which nucleons are forged together, creating new atomic nuclei and releasing energy.
Galactic disk. The plate-shaped region of a spiral galaxy, in which the spiral arms are found.
Galactic halo. A spherical aggregation of stars, globular star clusters, and thin gas clouds, centered on the nucleus of the galaxy and extending beyond the known extremities of the galactic disk.
Galaxy(ies). A large aggregation of stars, bound together gravitationally, of mass roughly 100 billion times that of the sun. There are three major classifications of galaxies—spiral, elliptical, and irregular—plus 50 galaxies that resemble spirals but display no spiral arms.
Galaxy cluster. See *cluster of galaxies.*
Gamma ray(s). Extremely short-wavelength *photon.*
Gauge theory. Account of forces that views them as arising from broken *symmetries.*
Geocentric cosmology. School of ancient theories that depicted the earth as situated at the center of the universe.
Geometry. Mathematics of lines drawn through space. Euclidean space is flat, the three-dimensional analogue of a plane. Non-Euclidean space is curved, the four-dimensional analogue of a sphere or a hyperbola.
Giant molecular cloud (GMC). Gas cloud where stars may form.
Giant star. High-luminosity star that lies above the *main sequence* on the *Hertz-sprung-Russell diagram.*
Globular cluster. See *star cluster, globular.*
Gluons. *Bosons* that carry the *strong nuclear force.*
Grand unified theories. Theories that purport to reveal identities linking the strong and electroweak forces.
Gravitational lensing. Multiple images, arcs, and other distortions in the light from quasars and other distant objects produced by the warping of space in the gravitational fields of foreground objects (usually galaxy clusters).
Gravity, gravitational force. Interaction experienced by particles that possess mass.
Group of galaxies. Gravitationally bound association of a few dozen galaxies.

GUT. See *Grand unified theories.*
Hadrons. Class of elementary particles, divided into *baryons* and mesons. Hadrons respond to the strong nuclear force; *leptons* do not.
Half-life. Time it takes for half of a given quantity of radioactive material to decay.
Halo, galactic. See *galactic halo.*
Heisenberg uncertainty. See *indeterminacy principle.*
Heliocentric cosmology. Class of models in which the sun was portrayed as standing at the center of the universe.
Hertz. A unit of frequency equal to one cycle (or wave) per second.
Hertzsprung-Russell diagram. One of a series of plots that reveal a relationship between the colors and absolute magnitudes of stars.
Higgs boson. See *Higgs field.*
Higgs field. Mechanism theorized to have imparted mass to particles. Mediated by the Higgs boson.
High-energy physics. See *particle physics.*
HI region. Galactic cloud of nonionized gas.
HII region. Galactic cloud of *ionized* gas.
Horizon problem. A quandary in standard big bang theory, which indicates that few of the particles of the early universe would have had time to be in causal contact with one another at the outset of cosmic expansion. Resolvable via *inflationary theory.*
Hubble constant. The rate at which the universe expands, equal to approximately 50 kilometers of velocity per *megaparsec* of distance.
Hubble diagram. Plot of galaxy *redshifts* against their distances; evidence of the *expansion of the universe.*
Hubble law. That distant galaxies are found to be receding from one another at velocities directly correlated to their distances apart.
Hubble Space Telescope. Large optical telescope in Earth orbit.
Hyperdimensional. Involving more than the customary four dimensions (three of space plus one of time) of relativistic spacetime.
Hypothesis. A scientific proposition that purports to explain a given set of phenomena. Less comprehensive and less well established than a theory.
Indeterminacy principle. Quantum precept indicating that the position and trajectory of a particle cannot both be known with perfect exactitude. Also known as Heisenberg indeterminacy, uncertainty principle.
Inertia. Quality of mass, such that any massive particle tends to remain at rest relative to a given reference frame, and to remain in constant motion once in motion, unless acted upon by a force.
Inflationary theory(ies). Proposes that the expansion of the very early universe proceeded much more rapidly than it does today—at an exponential rather than a linear rate.
Infrared Astronomical Satellite (IRAS). Studies the sky in *infrared* wavelengths.

Infrared light. Electromagnetic radiation slightly longer in wavelength than visible light.
Initial condition. (1) In physics, the state of a system at the time at which a given interaction begins. (2) In cosmology, a quantity inserted as a given in cosmogonic equations describing the early universe.
Intelligence. Defined in *SETI* as the ability and willingness to transmit electromagnetic signals across interstellar space. More generally, capable of science and technology.
Interaction. See *force.*
Interferometer. A device for observing the interference of waves of light or similar emanations caused by a shift in the phase or wavelength of some of the waves.
Inverse square law. In Newtonian mechanics, the rule that the measured intensity of light diminishes by the square of the distance of its source.
Ion. An atom with more or fewer electrons than normal.
Ionized. State of an atom's having fewer or more electrons than normal, leaving it with an electrical charge.
Isotope. Atom having the same number of protons in its nuclei as do other atoms of a given element, but different numbers of neutrons.
Isotropy. Quality of being the same in all directions. Compare *anisotropy.*
Jet. *Plasma* emitted by *quasars* and *protostars.*
K. Degrees *Kelvin.*
Kelvin. Measurement of temperature employed in astronomy. Zero degrees Kelvin equals absolute zero, the (unattainable) temperature at which thermal motion of atoms ceases. To convert Kelvin to Celsius, subtract 273.
Lambda. See *cosmological constant.*
Law. A scientific theory of sufficiently wide application that its violation is thought to be impossible.
Leptons. Elementary particles that have no measurable size and are not influenced by the strong nuclear force. Electrons, muons, and neutrinos are leptons.
Light. Electromagnetic radiation with wavelengths detectable, or close to detectable, by the eye.
Light-second(s). Distance light travels through a vacuum in one second. Equals 186,000 miles.
Light-year(s). The distance light travels in one year in a vacuum, equal to 5.8×10^{12} (some 6 trillion) miles.
Local Group. The association of galaxies to which the Milky Way galaxy belongs.
Local Supercluster. See *Virgo Supercluster.*
Locality. Classical assumption that change in physical systems requires presence of (mechanical) links between the cause and the effect. Nonlocality is exhibited by systems in which change evidently occurs without such links.

Lookback time. Phenomenon that, owing to the finite velocity of light, the more distant an object being observed, the older is the information received from it.
Lorentz contraction. Diminution in the length of an observed object along the axis of its motion, as perceived by an external observer who does not share its velocity.
Luminosity. Intrinsic brightness of a star or galaxy. See *magnitude.*
Machian. Of or pertaining to the outlook of the physicist Ernst Mach, who saw inertia as resulting from the cosmic matter distribution. The Machian rest frame of the universe is defined as the state of zero motion, against which all peculiar velocities, such as those introduced by superclusters pulling on neighboring clusters of galaxies, can be measured.
Magellanic clouds. Two galaxies that lie close to the Milky Way galaxy, visible in the southern skies of Earth.
Magnetic monopole. A massive particle with but one magnetic pole.
Magnitude. Brightness of a star or other astronomical object, expressed on a scale in which lower numbers mean greater brightness. Apparent magnitude indicates brightness of objects as seen from Earth. Absolute magnitude is defined as the apparent magnitude a star would have if viewed from a distance of 10 parsecs, or 32.6 light-years.
Main sequence. Zone on the *Hertzsprung-Russell diagram* where normal stars reside for most of their visible careers.
Many histories approach. *See* quantum observership.
Many worlds interpretation. *See* quantum observership.
Mass. The amount of matter in an object.
Mass density, cosmic. Average amount of matter in the universe per unit volume.
Mechanics. The study, in physics, of *forces.*
Megaparsec. One million (10^6) *parsecs.* Equals 3.26 million *light-years.*
Metals. In astrophysics, all elements heavier than helium.
Meteor. Chunk of rock, typically from size of a pea to that of a fist, that glows as it falls through Earth's atmosphere.
Meteorite. A *meteor* that has struck Earth.
MeV. One million (10^6) electron volts.
Microwave background. See *cosmic microwave background.*
Microwaves. Radio radiation with wavelengths of about 10^{-4} to one meter, equal to 10^9 to 10^{13} *Hertz.*
Milky Way. (1) The spiral galaxy in which the sun resides. (2) Glowing band of light bisecting the skies of Earth and produced by light from *stars* and *nebulae* in the galaxy's disk.
Million. A thousand thousand (10^6).
Missing matter. Alternate term for *dark matter.*
Molecular cloud. See *giant molecular cloud.*
Molecule. The smallest unit of a chemical compound.

Monopole. See *magnetic monopole.*
M31. See *Andromeda galaxy.*
Multiverse. Set of all universes.
Muon. Short-lived elementary particle with negative electrical charge. Muons are *leptons.* They are 207 times more massive than *electrons.*
Natural selection. Tendency of individuals better suited to their environment to survive and perpetuate their species, leading to changes in the genetic makeup of the species and, eventually, to the origin of new species.
Nebula(e). An indistinct, nonterrestrial object beyond the solar system. "Spiral nebulae" are *galaxies.*
Neutrino. Electrically neutral, massless particle that responds to the weak nuclear force but not the strong nuclear and electromagnetic forces.
Neutron(s). Electrically neutral, massive particle found in the nuclei of atoms. Composed of one up quark and two down quarks; mass 939.6 MeV, slightly more than that of the proton.
Neutron star. Star with a gravitational field so intense that most of its matter has been compressed into *neutrons.* Known as *pulsars* when rapidly rotating and emitting radio pulses.
Nonbaryonic. Pertaining to particles other than the *baryons* (protons, neutrons) that comprise ordinary matter. Candidate for some (maybe most) of the *dark matter.*
Non-Euclidean geometry. See *geometry.*
Nonlocality. See *locality.*
Nova. Star that brightens suddenly and to an unprecedented degree, creating the impression that a new star has appeared where none was before. Hence the name, from *nova,* the Latin word for *"new."* See *supernova.*
Nucleon(s). Proton or neutron; the constituents of atomic nuclei.
Nucleogenesis, nucleosynthesis. The fusion of nucleons to create the nuclei of new atoms. Takes place in the cores of stars and, at an accelerated rate, in supernovae. Big bang nucleosynthesis (BBN) is the synthesis of hydrogen, helium, and traces of other light atoms in the early universe.
Nucleus(i). (1) The central part of an atom, composed of protons and neutrons (which are made of quarks) and containing nearly all of each atom's mass. (2) The central region of a *galaxy.*
Observational cosmology. The application of observational data to the study of the universe as a whole.
Omega. Index of cosmic matter density and, therefore, of global shape of space. Defined as the ratio between actual density and the *critical density* required to close the universe.
Oort cloud. Spherical cloud of *comets,* centered on the sun.
Open cluster. See *star cluster, open.*
Open universe. State of the universe if its geometry is hyperbolic.

Oscillating universe. Cosmological model in which the universe is closed. Cosmic expansion eventually stops and the universe collapses, then rebounds into a new expansion phase. See *closed universe.*
Parallax. (1) Apparent displacement in the position of a star when it is viewed from two different locations. (2) Method exploiting this phenomenon to measure distances of stars; interstellar triangulation.
Parsec(s). Astronomical unit of distance, equal to 3.26 light-years.
Particle(s). Fundamental unit of matter or energy. The term is metaphoric, in that all subatomic particles also evince aspects of wavelike behavior.
Particle accelerator. See *accelerator.*
Particle physics. Branch of science dealing with the smallest-known structures of matter and energy. Also called high-energy physics.
Pauli exclusion principle. Quantum rule that no two *fermions* can occupy the same quantum state.
Period-luminosity function. Relationship between absolute magnitude and period of variability of *Cepheid variable* stars. See *magnitude.*
Phase transition. Abrupt change in the equilibrium state of a system, as evoked by the cooling of the early universe as it expanded. Breaks *symmetries.*
Photino. *Supersymmetric* partner of the *photon.*
Photon(s). Quantum of the electromagnetic force. Photons have zero rest mass and can therefore travel infinitely far.
Photon decoupling. The release of *photons* from constant collisions with massive particles that took place as the universe expanded and its matter density diminished. See *decoupling.*
Physics. The scientific study of the basic interactions of matter and energy.
Planck epoch, Planck time. The first instant following the beginning of the expansion of the universe, when the cosmic matter density was still so high that gravitational force acted as strongly as the other fundamental forces.
Planck's constant. The fundamental quantity of action in quantum mechanics.
Planet. An astronomical object more massive than an asteroid but less so than a star.
Plasma. A state in which matter consists of electrons and other subatomic particles without any structure of an order higher than that of atomic nuclei.
Positron. *Antimatter* twin of the *electron.*
Proper motion. Individual drifting of stars through space.
Protogalaxy. Galaxy in the process of formation.
Proton(s). Massive particle with positive electrical charge found in the nuclei of atoms. Composed of two up quarks and one down quark. Mass 938.3 MeV, slightly less than that of the neutron.

Proton decay. Spontaneous disintegration of the proton, predicted by *grand unified theories* but never observed experimentally.
Proton-proton chain. An important nuclear fusion reaction that occurs in stars. It begins with the fusion of two hydrogen nuclei, each of which consists of a single proton. See *fusion, nuclear.*
Protostar(s). Star in the process of forming.
Pulsar. See *neutron star.*
Quantum(a). Indivisible unit of energy, matter, or knowledge. The minimum amount of a quantity that can exist.
Quantum chromodynamics. Quantum theory of the strong nuclear force, which it envisions as being conveyed by quanta called *gluons.*
Quantum electrodynamics. Quantum theory of the electromagnetic force, which it envisions as being carried by quanta called *photons.*
Quantum genesis. Hypothesis that the origin of the universe may be understood in terms of a quantum theory.
Quantum leap. The disappearance of a subatomic particle—for example, an electron—at one location and its simultaneous reappearance at another.
Quantum mechanics. Set of all *quantum* theories accounting for dynamics of, especially, subatomic systems. Incorporates the *wave-particle duality* and the *indeterminacy principle.*
Quantum observership. Issue of the role played by observing (or "measuring") a quantum theory. Extracting one sort of information from a quantum system means that the behavior of other parts of the system is suddenly decided, even though the two parts may be separated by distances too great for a signal, traveling at light speed from one part, to have conveyed news of the change to the other part in time for the measurement to be completed. The Copenhagen interpretation accounts for the change by stating that the system has no definite state until it is measured. The Bohmian (after David Bohm) interpretation states that the system is deterministic and the news is conveyed in some nonlocal manner. The many worlds interpretation holds that both states exist and that the universe splits in two each time an observation is made. A variation is the "many histories" approach, in which the bifurcation need not be an actual division of the universe but is seen, rather, as a situation in which history took one of two statistically available states.
Quantum physics. Physics including the *quantum* principle. In this book, used interchangeably with the term *quantum mechanics.*
Quantum space. Vacuum with the potential to produce *virtual particles.*
Quantum theory. Any theory incorporating the *quantum* principle.
Quantum tunneling. A *quantum leap* over an otherwise insurmountable barrier.
Quark(s). Fundamental particle from which all *hadrons* are made.
Quasar(s). Pointlike source of light whose *redshift* indicates that it lies at distances of billions of light-years. Quasars are thought to be the *nuclei* of young galaxies.

Radio. Electromagnetic radiation with wavelengths of approximately 0.1 meter to 10^5 meters.
Radioactivity. Emission of particles by unstable elements as they decay.
Radio astronomy. Study of the universe at the radio wavelengths of electromagnetic energy.
Radiometric dating. Determination of the age of a substance containing radioactive elements by means of its radioactive *half-life.*
Radio telescopes. Sensitive radio antennae employed to detect the radio energy emitted by astronomical objects such as nebulae, galaxies, and pulsars.
Recombination. Capture of an electron by a proton. Predominant nuclear physics of the early universe at time of *photon decoupling.*
Red giant. Large star with an atmosphere that is relatively cool; therefore looks redder in color than does a main-sequence star.
Redshift(s). Displacement of the *spectral lines* in light coming from the stars of distant galaxies, thought to be produced by the expansion of cosmic space.
Relativistic. Approaching the velocity of light.
Relativity, general theory of. Einstein's theory of gravitation.
Relativity, special theory of. Einstein's theory of the electrodynamics of moving systems.
Renormalization. Mathematical procedure for the removal of nonsensical infinities from quantum mechanics equations.
Resonance. Excited state of a quantum system.
Satellite. Object orbiting another, more massive object.
Scalar fields. Class of fields, present in all *unified theories* of particle interactions, that describes fluctuations of matter and energy in a vacuum. Invoked by certain *inflationary theories* of the origin and expansion of the universe.
Science. Systematic study of nature, based on assumption that the universe is based on reason, that tests theories by experiment.
SETI. The Search for Extraterrestrial Intelligence, by the use of *radio telescopes* to listen for artificial radio signals from space. See *Drake equation.*
Singularity. A point of infinite curvature of space, where the equations of general relativity break down.
Sneutrino. *Supersymmetric* partner of the *neutrino.*
Solar system. The sun, its planets, and the asteroids and comets that, like the planets, orbit the sun.
Space. (1) Colloquially, the three-dimensional theater of ordinary experience. (2) Broadly, such a theater in any number of dimensions. (3) Abstractly, any imaginary geometrical construct with spacelike characteristics.
Spacetime. Four-dimensional arena in which events are depicted in general relativity. See *relativity, general theory of.*
Spectral lines. Bright and dark lines seen in *spectra* of stars and other luminous objects.

Spectrograph. A device for recording distribution of input by frequency —a *spectrum.*
Spectrum(a). Record of distribution of energy (e.g., light) by wavelength.
Spherical space. See *geometry.*
Spin. Intrinsic angular momentum of an elementary *particle.* Particles with integral spin (0, 1) are *bosons;* those with half spin, *fermions.*
Squark. *Supersymmetric* partner of the *quark.*
SSC. See *superconducting supercollider.*
Standard model. (1) In cosmology, basic big bang theory. (2) In quantum mechanics, the theories of the four forces.
Star(s). Astronomical object that sustains thermonuclear fusion reactions at its core, or did so at some time in the past.
Star, binary. See *binary star.*
Star, variable. See *variable star.*
Star cluster. Gravitationally bound aggregation of stars, much smaller than galaxies.
Star cluster, globular. Large, old aggregation of stars.
Star cluster, open. Relatively young, low-mass aggregation of stars.
Steady state. Theory that the expanding universe was never in a state of appreciably higher density—that is, that there was no "big bang." Hypothesizes that matter is constantly being created out of empty space.
Stellar evolution. (1) Building of heavy atomic nuclei from light ones in stars. (2) Changes undergone by stars due to this ongoing process.
String. See *string theory, superstring.*
String theory. Theory that subatomic particles actually have extension along one axis, and that their properties are determined by the arrangement and vibration of the strings.
Strong nuclear force (or interaction). Fundamental force of nature that binds quarks together, and holds *nucleons* (which are composed of quarks) together as the nuclei of atoms. Portrayed in *quantum chromodynamics* as conveyed by quanta called *gluons.*
Subatomic. Smaller in size than an atom.
Subatomic particles. See *particle.*
Sum over histories. Probabilistic interpretation of a system's past, in which quantum indeterminacy is taken into account and the history is reconstructed in terms of each possible path and its relative likelihood. See *indeterminacy principle.*
Supercluster. Cluster of clusters of galaxies. Typical diameter 100 million (10^8) light-years; mass, tens of thousands of galaxies.
Superconducting supercollider (SSC). A collider of unprecedented power, proposed but canceled while under construction in Texas. See *accelerator.*
Supergiant. Massive, bright star.

Supernova. Explosion of a star.
Superstring. (1) Theory that all *particles* are made of hyperdimensional space; also called *string theory.* (2) Particles, so viewed.
Supersymmetric. Pertaining to theories of supersymmetry (SUSY). Supersymmetry theories relate fermions (particles of half-integral spin, e.g., electrons, protons, and neutrinos) and bosons (particles with integral spin, e.g., photons and gluons).
Supersymmetry. See *supersymmetric.*
SUSY. See *supersymmetric.*
Symmetry(ies). State of having a quantity that remains invariant after a transformation.
Symmetry breaking. Loss of *symmetry* in a transformation.
Symmetry group. Mathematical group with a common property that unites its members and evinces a *symmetry.*
Telescope. Device for gathering and amplifying light or other energy emitted by astronomical objects.
Theism. Belief in God. In this book, *theist* means a believer in God, without reference to whether there are many gods (pantheism) or whether one's belief excludes revelation (deism).
Theory. Rationally coherent account of a wider range of phenomena than are customarily accounted for by a *hypothesis.*
Theory of everything. See *unified theory.*
Thermodynamics. The study of the behavior of heat (and, by implication, other forms of energy) in changing systems.
Thought experiment. Experiment conducted imaginarily but not in actuality.
Time. Dimension that distinguishes past, present, and future.
Triangulation. Measurement of the distance of a planet or nearby star by sighting its apparent position against background objects from two or more separate locations. See *parallax.*
Trillion. A thousand billion (10^{12}).
Tully-Fisher method. Exploits a relationship between the luminosity and spectral line widths of galaxies to estimate their distances.
Turnoff point. Upper termination of the *main sequence* in a *Hertzsprung-Russell diagram.* Indexes age of a star cluster: Massive stars, found high on the diagram, burn up fastest and exit the main sequence, creating the turnoff point. See *ZAMS.*
Ultraviolet light. Electromagnetic radiation of a wavelength slightly shorter than that of visible light.
Uncertainty principle. See *indeterminacy principle.*
Unified theory(ies). Single account, as yet unattained, of all fundamental interactions. Would eliminate disparities between *relativity* and *quantum theory.* See *force.*
Universe. Set of all observable and potentially observable phenomena.

Vacuum. (1) Classically, empty space between particles. (2) In quantum physics, arena populated by *virtual particles.*
Vacuum genesis. Hypothesis that the universe began as all *vacuum.*
Variable star. A star that changes in brightness periodically.
Virgo Cluster. Cluster of galaxies in constellation Virgo, located near center of the *Virgo Supercluster.*
Virgo Supercluster. *Supercluster* of galaxies in constellation Virgo. Contains *Virgo Cluster* and *Local Group.* Also known as the *Local Supercluster.*
Virtual particles. Short-lived particles that arise through quantum flux from a vacuum.
W particle. Massive *boson,* reproduced experimentally and thought to have been abundant in the early universe.
Wave function. Quantum mechanical equation that describes all relevant properties of a quantum system.
Wave-particle duality. Quantum realization that *quanta* and other "particles" exhibit characteristics of both particles and waves.
Weak nuclear force (or interaction). Fundamental force of nature that governs the process of radioactivity. It is currently accounted for by the *electroweak theory.*
World line. In *relativity,* the path traced out by an object in four-dimensional *spacetime.*
Wormhole. A tube made of space connecting two points separated in three-dimensional space.
X ray. Short-wavelength electromagnetic energy. The x-ray portion of the electromagnetic spectrum lies between the realms of *gamma rays* and of *ultraviolet light.*
Z particle. Massive *boson* thought to have been abundant in the early universe, when the unified electroweak force was manifest. See *electroweak theory.*
ZAMS, Zero-Age Main Sequence. The *main sequence* on the *Hertzsprung-Russell diagram* of a star cluster or other such association at the beginning of its evolution.

Acknowledgments

The Whole Shebang was written in San Francisco, Berkeley, and Sonoma County, from 1992 to 1996, and incorporates decades of prior research. Anyone undertaking such a work will, unless he is a hermit or a solipsist, incur debts of gratitude to more persons than he can succinctly name. So this list is woefully abridged.

Such errors and infelicities as this book may contain are, of course, my own responsibility, but that they are not more numerous or more egregious is due in part to the help of those who took the trouble to read the manuscript in various stages, notably Eric Linder, Sara Lippincott, Owen Laster, William Alexander, and its editor, Alice Mayhew.

For stimulating discussions on science, thanks especially to J. Richard Gott III, James B. Hartle, Stephen Hawking, Andrei Linde, Allan Sandage, Kip S. Thorne, Steven Weinberg, John Archibald Wheeler, and Edward Witten. For conversation, guidance, and moral support I am indebted to Jean Baird Ferris, Patrick Ferris, Lynda Obst, Hunter S. Thompson, and—as always—Cal Zecca.

I am grateful to Jeff Kao for drawing the illustrations, and to Meir Rinde for his cheerful, industrious aid in keeping mundane affairs in order while my head was in the stars.

Like any science writer, I have benefited from the remarkable openness of the international scientific community, which constitutes a kind of white hole of information and ideas, crowded with researchers happy to oblige a nosy reporter by patiently explaining and elaborating their ideas in the clearest terms they can muster. The world would be a happier place if more disciplines took their example to heart.

T.F.
Rocky Hill Observatory

Index

Page numbers in *italics* refer to illustrations.